Process Development Division 2014

Core Programming Area at the 2014 AIChE Spring Meeting and 10th Global Congress on Process Safety

New Orleans, Louisiana, USA
30 March - 3 April 2014

ISBN: 978-1-63439-077-4

Printed from e-media with permission by:

Curran Associates, Inc.
57 Morehouse Lane
Red Hook, NY 12571

CURRAN ASSOCIATES INC.
proceedings
.com

Some format issues inherent in the e-media version may also appear in this print version.

Printed by Curran Associates, Inc. (2014)

For permission requests, please contact AIChE
at the address below.

AIChE
3 Park Avenue
New York, NY 10016-5991

Phone: (203) 702-7660
Fax: (203) 775-5177

www.aiche.org

Additional copies of this publication are available from:

Curran Associates, Inc.
57 Morehouse Lane
Red Hook, NY 12571 USA
Phone: 845-758-0400
Fax: 845-758-2634
Email: curran@proceedings.com
Web: www.proceedings.com

TABLE OF CONTENTS

354012 Possible vs. Practical: Engineers Must Lead the Development of Practical Technologies

Abstract:

Chemical Engineering is need today more than ever. There is a strong societal desire for sources of energy and feedstocks that can sustainably support a high standard of living for all. Thermodynamic, material and economic challenges stand in the way. Most of the general public does not have the basic understanding of the energy flows, chemical transformations, and the scale of the systems that support our current way of life. These concepts are fundamental to chemical engineering. Chemical engineers should be on the vanguard of separating hype from truly practical options yet occationally we let society down. This talk will present several examples where the fundamentals of chemical engineering can be used to access the practicality of proposed solutions to energy and water issues, as well as explain why the venture capital model struggles in the materials market.

351063 Improving the Speed and Representation of Physical Properties Regression

Abstract:

Abstract: Improving the Speed and Representation of Physical Properties Regression

One of the more difficult challenges facing organizations that provide separations solutions is the need to improve our productivity in the area of regressing and continuously improving our physical property models. We are often faced with the broad range of tasks ranging from developing a thermodynamic model from inception to the prospect of including and incorporating new data into existing models. Rarely, if ever, is the task so simple as to require the regression of one or a few sets of data into an acceptable package. More often, the regression task requires the regression of many sets of data, with an ill defined picture of which interaction parameters should be used, and which set of parameters gives the optimum fit of the collection of data. In point of fact, the statement about "returning the optimum set of values" is not at all clear. Using many parameters may minimize the sum of the squares error, but may return a set of parameters that is so unstable as to be useless in practice.

There are a number of approaches to potentially solving this problem. Modern computing power and optimization techniques have made assessing all or a very large subset of all of the possible permutations and combinations for a particular property representation possible. We will present some of our efforts to improve the speed, accuracy, and physical representation of some of our properties work in this session.

John Pendergast [(1)] John Dowdle[(1)] Pradeep Suresh-Babu[(1)] Abu Hassan[(1)] Diego Cristancho[(1)]

[(1)] The Dow Chemical Company

352692 Shale Gas and the History of C1 Chemistry

Abstract:

The shale gas revolution has resulted in historically low prices for gas relative to oil on an energy content basis. This drives new incentives for converting C1 (methane) into commodity chemicals and liquid fuels. But enhanced opportunities in C1 chemistry have occurred before due to war, politics, and transport (stranded gas). This talk will focus on a history of key developments in C1 chemistry including methanation and the Fischer-Tropsch synthesis, hydroformylation, methanol synthesis and carbonylation chemistry, reductive and oxidative coupling of methane. It will also present current challenges and opportunities in producing liquid fuels and chemicals from methane.

Using Patterns of Invention in the Patent Literature to Predict Disruption

AIChE Spring Meeting, 2014

Jack Hipple
TRIZ and Engineering Training Services LLC
Tampa, FL
www.innovation-triz.com

Our typical view of disruption..

- A "black swan" event that is unpredictable
- Tough to forecast
- No point in planning
- Keep the antenna out
- Be nimble in reacting

We can actually study this!

- Patents are the documentation (**DATA**) of breakthroughs
  - The top few %
- We frequently only study the literature relevant to our field of science or technology

If we broaden our perspective...

- We find
  - Repeated patterns of invention that represent disruption
  - These patterns repeat themselves across virtually all technology areas
  - Challenge is to get rid of our ego, generalize our problems, and use what is already known
    - Our technology and challenges are seldom unique

Patterns of Evolution

1. Products and systems become more ideal through "trimming" and the resolution of contradictions

2. Non-uniform development of systems elements

3. *Products and systems become more dynamic, responsive, and controllable*

4. Products and systems oscillate between adding useful complexity and simplification

5. *Products and systems integrate into their super-systems*

6. *Products and systems evolve through the use of higher level fields (mechanical, thermal/acoustic, chemical, electronic, electromagnetic/optical)*

7. Products and systems evolve by mis-matching and dynamic mis-matching

8. Products and systems evolve toward less human involvement

A. Dynamism

- Products and systems become more dynamic (*over time, with or without you*)
  - Dynamism in terms of
    - Time
    - Situation
    - Condition
    - Need

Everyday examples

- Mortgages
  - 30 year, fixed rate
  - Rate vs. credit worthiness
  - 15,20,30 years
  - Variable rate (vs......)
  - Reverse
  - "Yourgage"
- Airline fares
  - Fixed, distance, vs. capacity, competition, competition fill rate
- Toll roads in CA
  - Time of day, capacity
- Conference fees vs. time
- Luggage wheels

WHAT SEPARATION PRINCIPLE DOES THIS ALSO REPRESENT?

1990, company A

United States Patent [19] **Sistare**

[11] **Patent Number:** 4,999,916
[45] **Date of Patent:** Mar. 19, 1991

[54] CIRCULAR SAW BASE ANGLE ADJUSTMENT MECHANISM
[75] Inventor: James R. Sistare, Pickens, S.C.
[73] Assignee: Ryobi Motor Products Corp., Pickens, S.C.
[21] Appl. No.: 568,840
[22] Filed: Aug. 17, 1990
[56] References Cited
U.S. PATENT DOCUMENTS

Primary Examiner—Douglas D. Watts

[57] ABSTRACT

A mechanism for adjusting the angle of the base of a circular saw relative the saw blade includes an adjustment bracket having an elongated adjustment aperture which extends over both ranges of angular adjustment. For the first range, the aperture follows a first circular arc at a first radial distance from the pivot axis of the base relative the housing and for the second range the aperture follows a second circular arc at a second radial distance from the pivot axis.

6 Claims, 2 Drawing Sheets

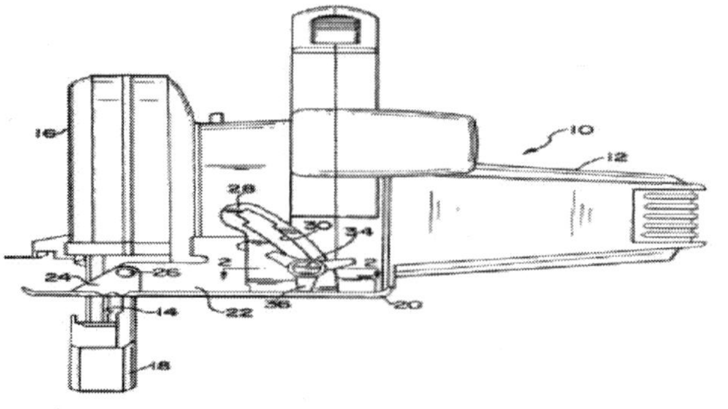

1992, company B

United States Patent [19]
Fuchs et al.

[11] **Patent Number: 5,271,155**
[45] **Date of Patent: Dec. 21, 1993**

[54] HAND CIRCULAR SAW

[75] Inventors: Wolfgang Fuchs, Filderstadt; Rainer Schilling; Heribert Schramm, both of Stuttgart; Claus Kemmer, Aichtal-Neuenhaus, all of Fed. Rep. of Germany

[73] Assignee: Robert Bosch GmbH, Stuttgart, Fed. Rep. of Germany

[21] Appl. No.: 918,270

[22] Filed: Jul. 22, 1992

[30] Foreign Application Priority Data

[56] References Cited

U.S. PATENT DOCUMENTS

FOREIGN PATENT DOCUMENTS

[57] ABSTRACT

A motor driven hand circular saw has a base plate arranged to be positioned over a workpiece, and a saw blade which is displaceable parallel and tiltably about a turning axle relative to the base plate for performing inclined or bevel cuts so that by tilting of the saw blade relative to the base plate about the axle, the saw blade and the axle are displaced parallel to the base plate with a stroke which is proportional to a respective tilting angle transversely to a direction downstream of a cutting point and counteracts a natural displacement of a saw engaging line of the saw blade on the workpiece.

21 Claims, 2 Drawing Sheets

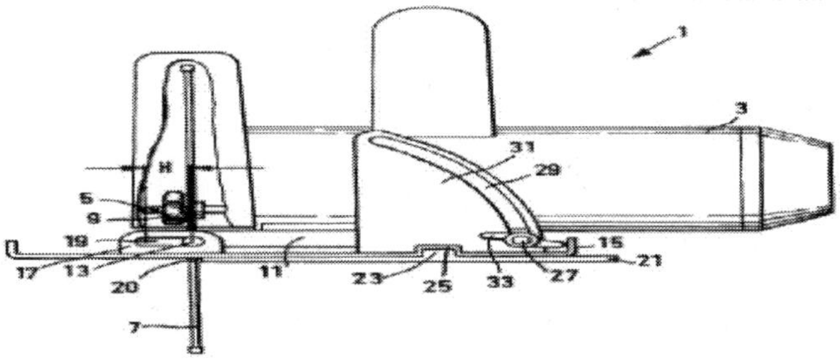

1993, company C (3 ½ years)

United States Patent [19]
Schilling

[11] **Patent Number: 5,452,515**
[45] **Date of Patent: Sep. 26, 1995**

[54] HAND CIRCULAR SAW WITH MITRE ADJUSTING DEVICE

[75] Inventor: Rainer Schilling, Stuttgart, Germany

[73] Assignee: Robert Bosch GmbH, Stuttgart, Germany

[21] Appl. No.: 170,424

[22] Filed: Dec. 20, 1993

[30] Foreign Application Priority Data

[56] References Cited

U.S. PATENT DOCUMENTS

FOREIGN PATENT DOCUMENTS

[57] ABSTRACT

A hand circular saw comprises a base plate, a saw blade arranged so that at least one of an angle position of the saw blade and a mitre angle is adjustable relative to the base plate, a clamping block supported by the base plate and provided with a turning guide slot, a clamping screw extending through the turning guide slot, and a supporting arm. The clamping block has a plurality of arresting pockets arranged along the turning guide slot, and the clamping screw carries at least one axially spring-biased displaceable arresting sleeve which is arrestingly positionable in a respective one of the arresting pockets.

12 Claims, 3 Drawing Sheets

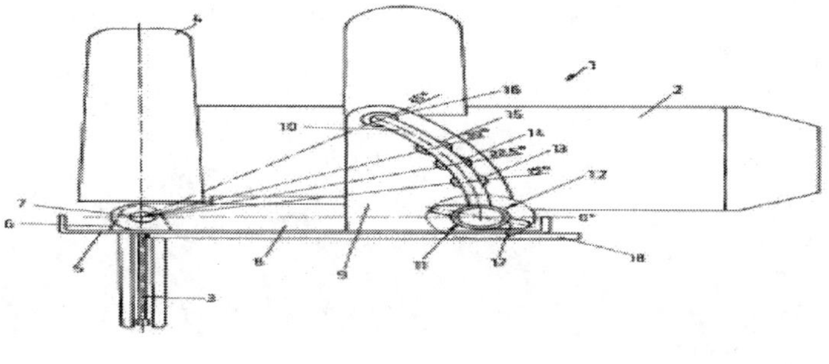

Ask yourself...

- Why should the GM 4-6-8 engine and the Intel I4 chip have separate patents?
  - Both use the concept of using only the power needed when it is needed

Consumer Products

Wal-Mart and worker flexibility

Wal-Mart Seeks New Flexibility In Worker Shifts

By KRIS MAHER (Wall Street Journal)
January 3, 2007; Page A1

...Early this year, Wal-Mart Stores Inc., using a new computerized scheduling system, will start moving many of its 1.3 million workers from predictable shifts to a system based on the number of customers in stores at any given time. The move promises greater productivity and customer satisfaction for the huge retailer but could be a major headache for employees.

The change is made possible by a software system that can crunch an array of data, part of a shift toward computerized management tools that can help pare costs and boost companies' bottom lines. But it also could demand greater flexibility and availability from workers in place of reliable work shifts -- and predictable paychecks.

Dynamic Coke™ Machine: 125 Flavors

In what other way could a pop machine be "dynamic"?

In chemical processing...

- Reflux ratio
- Response to raw materials
  - Water pressure and temperature, steam pressure
- Refinery operation
  - Crude content, product demand
  - Volatility of refined products
- Wastewater treatment
- Flare operation and control
- Pricing vs. raw material costs
- Drug delivery; controlled release

B. Upward system integration

- Functions gradually become integrated into their super-systems
  - Headlights, warning, visualization systems
  - Food products; nutrition
  - Office machines
    - Fax, copy, Email
  - Smart phones
  - Bifocals, sunglasses
  - Receipts and packages plus ads
  - Integrated gift wrapping
  - Terminix™ installing insulation

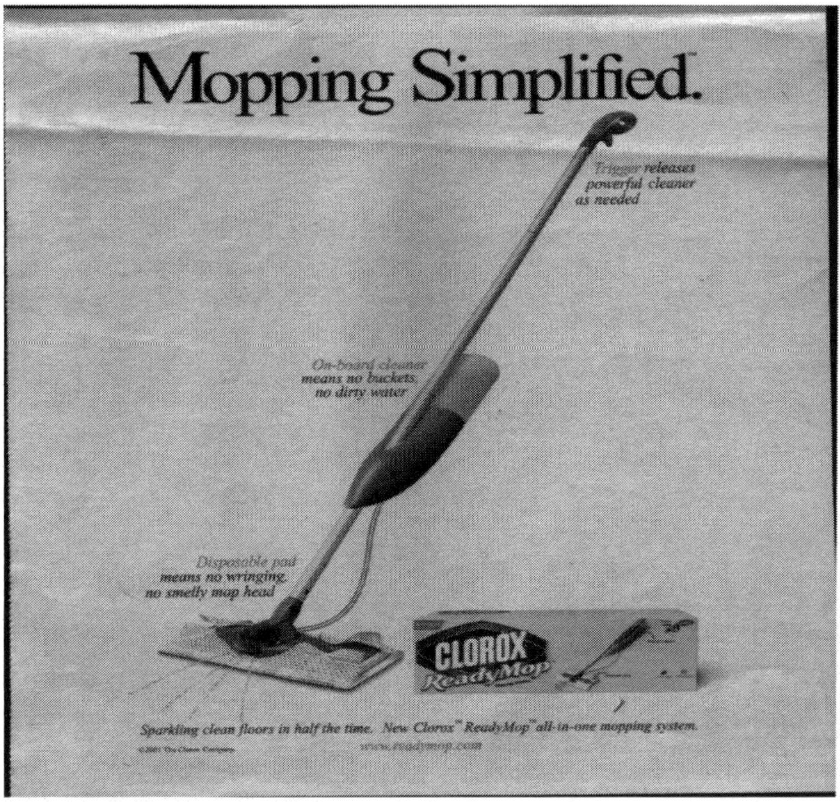

The "Tweel"—Michelin: The Tire Performs its Function and Doesn't Exist

TiroGage: The Stem is the gage

Sweetener Plus Flavor

Radar pitch baseball

Cell Phones Detect Hazardous Chemicals

C. Field progression

- Products and systems evolve with higher level fields
  - Mechanical, thermal/acoustic, chemical, electronic, electromagnetic/optical
  - Communication
    - Drums, smoke signals, letter writing, phones, wireless phone
  - Cooking
    - Heating, chemical preservatives, microwaves
  - "Indicating"
    - Pointers, laser pointers
  - "Photography" (Image capturing?)
    - Etching, thermal etching, ink, wet chemical photography, electronic photography

In chemical processing...

- Polymer processes
  - Thermal, catalytic, microwaves
- Process monitoring and control
- Liquid/solid separations
  - Surfactants, magnetic

"Nine Box" Analysis

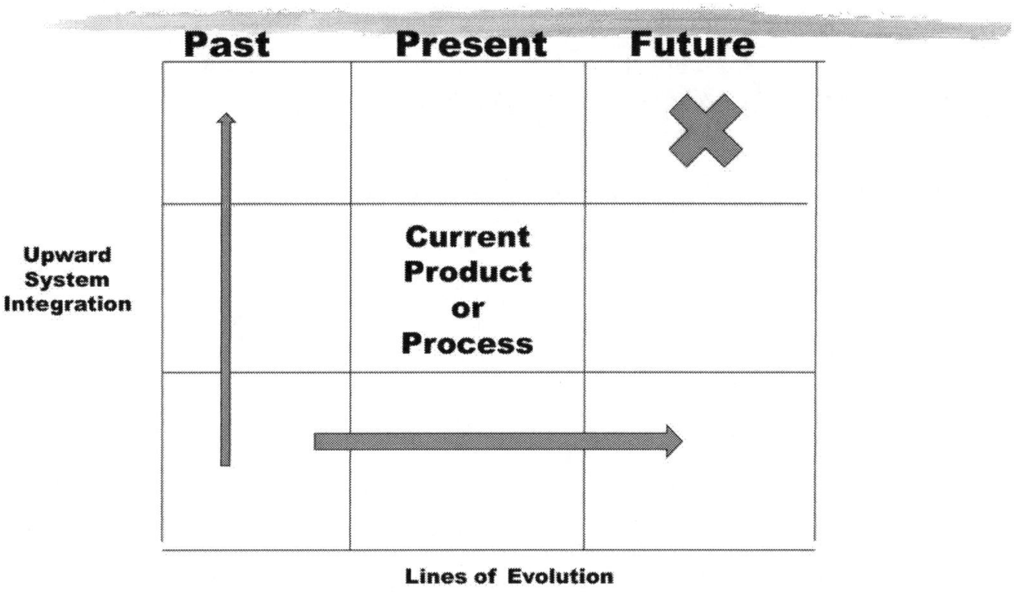

Evolutionary Potential: (Product and Process)

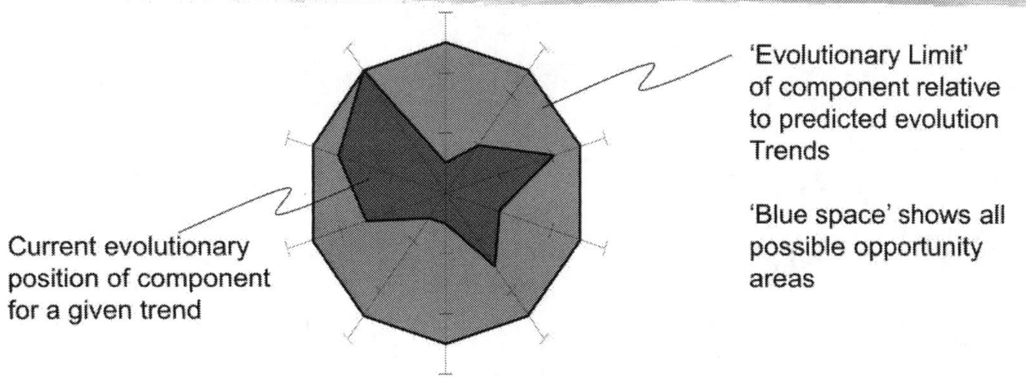

(Each spoke in the evolutionary potential radar plot represents one of the known technology trends identified by TRIZ researchers)

Summary

- Evolution and disruption are, in many respects, predictable
  - Ex: political, emotional trends and preferences
- Known lines and principles that can be applied
- Superior to conventional customer surveys
  - Typical interviewees are not aware of what is possible

Learning opportunities

- AIChE/ASME TRIZ classes
  - Minneapolis, April 28-30
  - Houston, June 2-4
  - Atlanta, October 20-22
  - San Diego, November 3-5

https://www.asme.org/products/courses/triz--the-theory-of-inventive-problem-solving

(or ASME TRIZ into search box!)

Mitsubishi Chemical Corporation has developed a novel on-purpose butadiene production technology. Main reaction is an oxidative dehydrogenation from n-butene to 1,3- butadiene.

It is a proven technology with excellent reaction performance, good operability, and superior product quality in the demonstration plant.
Mitsubishi proprietary catalyst's main characteristics are high butene conversion, high selectivity to butadiene, lower waste and long catalyst life.
All industrial C4 hydrocarbon mixtures such as Raf-2, FCC C4, and 1-butene can be processed by this technology.

Process Design Package for commercial plant has been completed. Whereas it is expected that for a medium-to-long term, butadiene supply will become very tight because gas cracker will replace steam naphtha cracker; Mitsubishi Chemical has regarded the on-purpose butadiene production technology as the key solution for this situation.
From 2014, Mitsubishi Chemical has started the worldwide licensing activities for this technology to all butadiene producers and consumers.

PAPER 26A

THINKING LIKE SHERLOCK HOLMES FOR PROCESS FILTRATION TECHNOLOGY SELECTION

Barry A. Perlmutter, President & Managing Director
BHS-Sonthofen Inc.
Filtration, Mixing & Recycling Divisions
14300 South Lakes Drive
Charlotte, North Carolina 28273-6794
Telephone: 704.845.1190; Fax: 704.845.1902
E-mail: barry.perlmutter@bhs-filtration.com

INTRODUCTION:

Sherlock Holmes and Dr. John Watson are fictional characters of Sir Arthur Conan Doyle. Process engineers who live in the real-world can learn many things from the two of them for solving process filtration problems. This paper will intertwine the detective techniques (mindfulness, astute observation, logical deduction and others) of Holmes and Watson with the problem solving skills required to select process filtration systems.

One example that Holmes proves time and again is that there is no benefit to jumping to conclusions. The paper begins with a discussion of the bench-top laboratory tests that are conducted for problem analysis, technology selection and scale-up. The tests include pressure/vacuum/centrifugation, filter media, cake thickness, temperature and viscosity concerns, filter aids and similar process parameters.

Another technique used by Holmes and Watson is "recreating events." Holmes talks through his theories out loud to Watson and only then do gaps and inconsistencies rise to the surface that were not apparent before. The paper continues with four case history examples discussing slurry testing, process analysis and then process filtration selection for pressure filtration, vacuum filtration, centrifugation and clarification.

Finally, the paper concludes with a general review of the problem-solving skills of Holmes and Watson such as the "occasional silence", "employing distancing" and "learning to tell the crucial from the incidental." These skills can be utilized by process engineers as a framework for "idea-generation" when analyzing an operating bottleneck issue or new process development problem. In all cases, by combining Holmes and Watson with accurate lab and pilot testing, the optimum filter selection can be realized.

LABORATORY TESTING AND WHY THERE'S NO BENEFIT TO JUMPING TO CONCLUSIONS:

According to Holmes and Watson, it is important to train yourself to be a better decision maker. For example, using checklists, formulas, structured procedures; those are your best bet. Figure 1 shows a typical Experimental Test Routine.

Overview of Bench Top Testing for Pressure and Vacuum Filtration

The BHS bench top testing is conducted using the BHS Pocket Leaf Filter, as shown in Figure 1. The test device is a BHS pocket leaf filter with a filter area of 20 cm^2 and a vacuum and pressure connection. The testing will analyze cake depths, operating pressures, filter media, washing and drying efficiencies and qualitative cake discharge. The data collection sheets are shown in Figure 2. The steps in filtration testing are as follows:

First, it is necessary to clearly state the process description. This includes the slurry characteristics (particle size distribution, particle shape, density, etc.), washing of the cake (i.e. number of washes and wash ratios), drying / pre-drying of the cake (vacuum, pressure blowing, and mechanical pressing) as well as the upstream and downstream equipment. With this definition, the type of samples that need to be collected and analyzed can be determined.

Secondly, it is necessary to know what are the requirements for the operation such as solids/hour and cake quality (percent moisture, percent contaminants, etc.).

Thirdly, with the above in mind, the testing much determines the following objectives:

- Choice of a suitable filter cloth
- Vacuum or pressure filtration
- Wash ratios for the washing of the filter cake
- Drying techniques

Overview of Bench Top Testing for Centrifugation

Centrifugation lab testing include static settling test, filtration rate test and spin settling rate test. The settling test will be able to determine the densities of the solid and liquid phases and if there are different densities then centrifugal forces can be applicable for separation.

The filtration rate tests are conducted with the BHS pocket leaf filter using vacuum. Depending upon the vacuum filtration rates, the type of centrifuge can be determined.

Finally, the spin rate test will determine the effect of G forces and the time to separate the slurry into distinct phases. A bench-top test tube spinner is used for these tests. The baseline testing is at a time of 90 seconds.

In summary, if none of the three lab tests produce a satisfactory separation, then another type of solid-liquid separation technology is required.

PROCESS FILTRATION SELECTION FOR PRESSURE FILTRATION, VACUUM FILTRATION, CENTRIFUGATION AND CLARIFICATION:

According to Holmes and Watson, it is easy to succumb to certainty but every time you find yourself making a judgment upon observation, train yourself to stop and repeat. Then go back and restate from the beginning and in a different fashion and most importantly, out loud instead of silently, as this will save you from many errors in perception. Process engineers can benefit from discussing options with technology suppliers that can provide different filtration solutions.

Case History: Pressure Filtration

Process testing was conducted at the site's laboratory and in the plant. For the bench-top lab testing, the BHS pressurized pocket-leaf filter (PLF) with 20 cm^2 of filter area. For the continuous pressure pilot testing, a pilot RPF with 0.18 m^2 of filter area is installed, as shown in Figure 4.

The objectives of the PLF testing are as follows:

Filtration time vs. filter media
Filtration time vs. slurry feed mass
Filtration time vs. differential pressure
Filtrate quality vs. filter media
Cake solids wash time and quality
Cake solids discharge characteristics
Production Scale-Up and Process Guarantee

The lab testing proved to be uniquely challenging both to feed the PLF as well as to maintain a pressure to keep the gas as a liquefied solvent. The plant engineers and BHS developed a confidential method to meet these challenges.

The PLF tests demonstrated that acceptable filtration and solids wash rates could be obtained for this product and acceptable solids levels were observed for the mother liquor filtrate. Washing results and drying quality parameters were also achieved.

Additional pilot plant tests with the BHS continuous pilot unit, RPF 0.18, are recommended to confirm the PLF lab tests. In these tests, BHS would be able to identify the necessary slurry solids percentage, cake solids thickness, solids wash time, solids drying time as well as cake discharge. Finally, the pilot testing will be the basis for the mechanical design of the RPF to ensure that the RPF can be designed for the process with a liquefied gas slurry.

While the actual data is confidential, the plant engineers and BHS process engineers gathered the following parameters from the pilot RPF 0.18 m^2 testwork.

Process Parameters:

Slurry Feed Pressure:
Slurry Feed Flow:
Wash Pressure:
Wash Flow:
Dry Pressure:
Drying Air Flow:

RPF Parameters:

Drum Speed:
Separating elements:
Cake blow back:
Cloth blow back:
Backpressure:
Cake Thickness:
Filter Cloth:

To fully evaluate the RPF performance, the site also compiled the following:

Slurry solids concentration
Filtrate quantity (mother liquor, wash, blow down, etc.)
Filtrate yield
Cake Moisture
Total Cake quantity

Scale-Up From RPF 0.18 M^2 Pilot Data

Calculate Specific Filter Performance from Pilot Testing
= kg of dry solids/m^2/hour
Calculate Production Area Required from Filter Performance and Client Required Production Rate
Using the drum speed, time for filtration, washing and drying and several other RPF factors, the specific filter area is calculated

Pressure Filtration and Typical Scale-Up Calculation-Example Only

The scale-up is based on 224 g slurry with 1:1 composition = 112 g dry solids:

Drum revolutions:

$$n = \frac{3600\frac{s}{h}}{(4+8+15)s} \cdot \frac{270°}{360°} \cdot 0{,}85 = 85\frac{\text{revolutions}}{h}$$

270°:	active angle
0.85:	factor for separating elements

Specific filter performance:

$$Q = 500\frac{1}{m^2} \cdot 85\frac{1}{h} \cdot 112g = 4760\frac{kg\ dry\ solids}{m^2 \cdot h}$$

Required filter area:

$$A = \frac{20000\frac{kgdrysolids}{h}}{4760\frac{kgdrysolids}{m^2h}} = 4{,}2\ m^2$$

Selected filter: BHS Rotary Pressure Filter, type B16 with 5.4 m² is sufficient to operate 20,000 kg dry solids per hour.

Case History: Vacuum Filtration

Bench top laboratory tests are valuable in selecting a solid/liquid separation device. The initial lab tests suggested a vacuum belt or rotary filter would achieve cake quality equal to or better than the current centrifuge with a major reduction in processing time. The footprint would be comparable to the current centrifuge and the unit would be suitable for conversion to a continuous process. After further discussions, the decision was to select a vacuum belt filter for pilot testing.

There are five objectives in running a pilot test filter:

1. To verify the time for formation of the cake and the initial saturation prior to dewatering of the cake
2. To evaluate the effect of cake thickness on the dewatering time
3. To investigate alternate ways to improve cake dryness (i.e. compression, gas blowing) that may eliminate the drying step
4. To evaluate the quality of the cake (dryness) and its effect on release from the filter media (Some initial tests would be required to make an initial selection of the cloth, but 2-3 cloths may need to be tested in the pilot unit to verify release characteristics)
5. To evaluate wash ratio needed to remove solubles and color bodies

The initial laboratory test data suggest a full-scale filter with from 0.5 to 1.5 m^2 of filter area would fit into the current process operations and reduce cycle time in half or better. A BHS 0.1 m^2 vacuum belt filter is available for test, which would allow for a feed rate of 0.5 gpm. Suggested testing order and condition changes:

1. Using a pocket filter and various samples of cloth, pull a vacuum of 20 in.hg. until no liquid is flowing. Invert the filter and observe the cake release. Describe it qualitatively (soupy, chunks, fine powder). Scrape out any remaining material and weight it separately from the material that was released. Select 2-3 cloths for the pilot testing from these tests. (optional) During the experiment measure how much time it takes for the cake surface to become dry and the dewatering time.
2. The estimated filtrate throughput for a 7 mm cake during cake formation and at the end of cake formation for vacuum filtration. Since there are 10 zones, in the BHS filter, samples from the second or third zone would be taken to evaluate the moisture after cake formation (dry surface). It may be necessary to stop the unit for this evaluation so it should only be done occasionally. Cake thickness can be checked at this time. The other zones can be sampled to determine the rate of dewatering after cake formation and wash ratio.
3. A wash ratio comparable to the centrifuge operation should be used for the previous tests. In the next series, the wash ratio could be varied to evaluate removal of solubles as well as the effect on cake stickiness.
4. While maintaining the same cloth rate, the feed rate can be increased and decreased to vary the cake thickness.
5. Throughout these tests the visual quality of the cake, especially at the discharge would be evaluated.
6. The test unit has an optional compression zone that could be employed. It is also possible to evaluate gas blowing with and without compression.

The results of the testing is that the BHS continuous-indexing vacuum belt filter will be able to produce a cake with better washing and drying compared with the existing centrifuge.

Case History: Centrifuge Selection

The choice of centrifuges (filtering or sedimentation) is dependent upon particle sizes, density of the solids and liquids and the process and application. Filtering centrifuges have a rotating perforated basket or bowl with filter media while sedimentation centrifuges have a rotating non-perforated bowl.

The initial testing is as follows:

- Preliminary data to determine centrifuge type and the initial parameters for pilot tests
  - Bench centrifuge tests
  - Filter bucket (specialized filtration bucket)
  - Vacuum filtration (Buchner or pocket filter)
- Cake moisture versus G-force and time
- Effect of cake thickness on time to reach moisture goal
- Effectiveness of wash
- Optimizing conditions (extensive pilot tests)
- Evaluate ways to avoid cake cracking
- Sliding and conveying properties of cake (shear)

In terms of filtering centrifuges, the choices are between batch and continuous feed. Continuous feed and liquid flow can be either continuous moving cake or intermittent moving cake. As for batch feed (fixed cake, batch liquid flow), the choices are the type of feed (Vertical, inclined, horizontal axis) and the type of discharge.

Major differences between the filtering centrifuges are

Batch

- Clear liquid
- Cake heel
- Wider size and feed concentration range (100 ppm to 50%)
- Can be inerted
- Can be used at high temperature

Continuous

- Liquid clarity poor
- Total discharge
- Feed > 15%
- Best above 10 micron
- Not suited for volatile liquids and hazardous or abrasive solids
- Operates at ambient temperatures

Why select a filtering centrifuge:

- Drier cakes than other centrifuges (generally)
- Compact relative to their throughput
- Fully automatic operation
- Particles from 1 micron to 2 mm (coarse – best)
- Fragile solids may have attrition issues
- Wide range of feed concentrations (non-abrasive solids)
- OK for moderately hazardous solids and volatile liquids
- When low moisture required
- Minimal washing required
- High in first cost; gives moderate clarity to the liquid

Why not select a filtering centrifuge:

- If the solids are only required in slurry form
- If coarse solids will screen or free drain to the necessary moisture content
- If fine compressible solids are to be filtered, washed and dried
- If coarse to medium sized solids have exacting washing requirements but require only moderate dewatering
- If feed solids content is low and the particles are very fine
- If the use of filter aid is contemplated

As for sedimenting centrifuges:

- Uses difference in density to separate solids
- Also used to separate liquids
- Has limited washing capability
- Wide range of feed concentrations
- Requires uniform feed
- Potential for particle breakage
- May have more wear than filtration devices
- Often wetter cake produced

In summary, the centrifuge selection depends upon the process characteristics and the laboratory testing to select the type and design.

Case History Clarification:
Replacing a Manual Plate Filter and Bag Filter Combination

This specialty chemicals manufacturer produces various resins that require filtration. Current production includes a neutralization step which yields metal salts. These salts are filtered out with a manual plate filter followed by a bag filter for polishing. Two solvent washes follow the filtration step to recover as much resin as possible. After washing, the filters are steamed and opened. The solids are disposed manually for each batch and the filter paper is replaced. The goals are to eliminate exposure to heptane, reduce the maintenance and operation on the two filters and to recover a dry catalyst. Current production is 3000 gallons in 4 – 5 hours.

The results and conclusions showed that the filtration flux rate from the BHS laboratory tests ranged between 10-30 $L/m^2/min$ at approximately 20 psi feed pressure. The sock filter cloth is polyester with an air permeability of 1.0 cfm/ft^2.

The tests showed that one BHS candle filter, Figure 5, with 10 m^2 of filter area can complete the cycle in 4.3 hours and replace the manual plate filter and bag filter.

The cycle time is as follows:

Cycle Times		
Filling	5	min
Filtration	10	min
Wash	4	min
Drain	10	min
Dry	5	min
Vent	2	min
Discharge	5	min
Reserve	9	min
TOTAL	**50**	**min**

CONCLUDING REMARKS & TAKEAWAYS:

Holmes and Watson provide a unique view of problem solving. The world of a process engineer is a distracting place and Holmes and Watson know that without the occasional silence, as in *The Hound of the Baskervilles*, there can be little hope for success. Engineers can benefit from conducting lab testing at the technology supplier's site to have time to think about the process issues, at hand. Finally, Holmes and Watson excel at "deduction from facts and deduction difficulties." All that matters is what the premises are (process definition, requirements and testing objectives) and how the testing unwinds the crucial from the incidental (what is the critical process parameter) and finally ending up in the logical conclusion (optimum process filtration solution).

In summary, it is important to put the project all together from many different prospectives. These include know the process, observe the testing, deduce the solution only from what is observed (and nothing more) and learn from your colleagues and the technology supplier's successes and failures. It is always difficult to apply Holmes' logic but as Holmes' states "you know my methods, now apply them." Engineers must practice these habits such that even under stress to solve a process problem, these stressors will bring out the very best thought patterns that are needed.

References:

1. All information about Holmes and Watson are taken from Maria Konnikova's *Mastermind-How to Think Like Sherlock Holmes* (Viking Penguin (USA) Inc. 2013).

2. Testing: The logical way to select a centrifuge, Mike Vastola, TEMA Systems.

3. Davies, E. (1965): Selection of Equipment of Solid/Liquid Separations.

4. All information about centrifuges is taken from Healthsite Associates, Tom Blackwood, presentation, St. Louis Engineers Club, 14 November 2013.

Figure 1: Davies' Experimental Test Routine

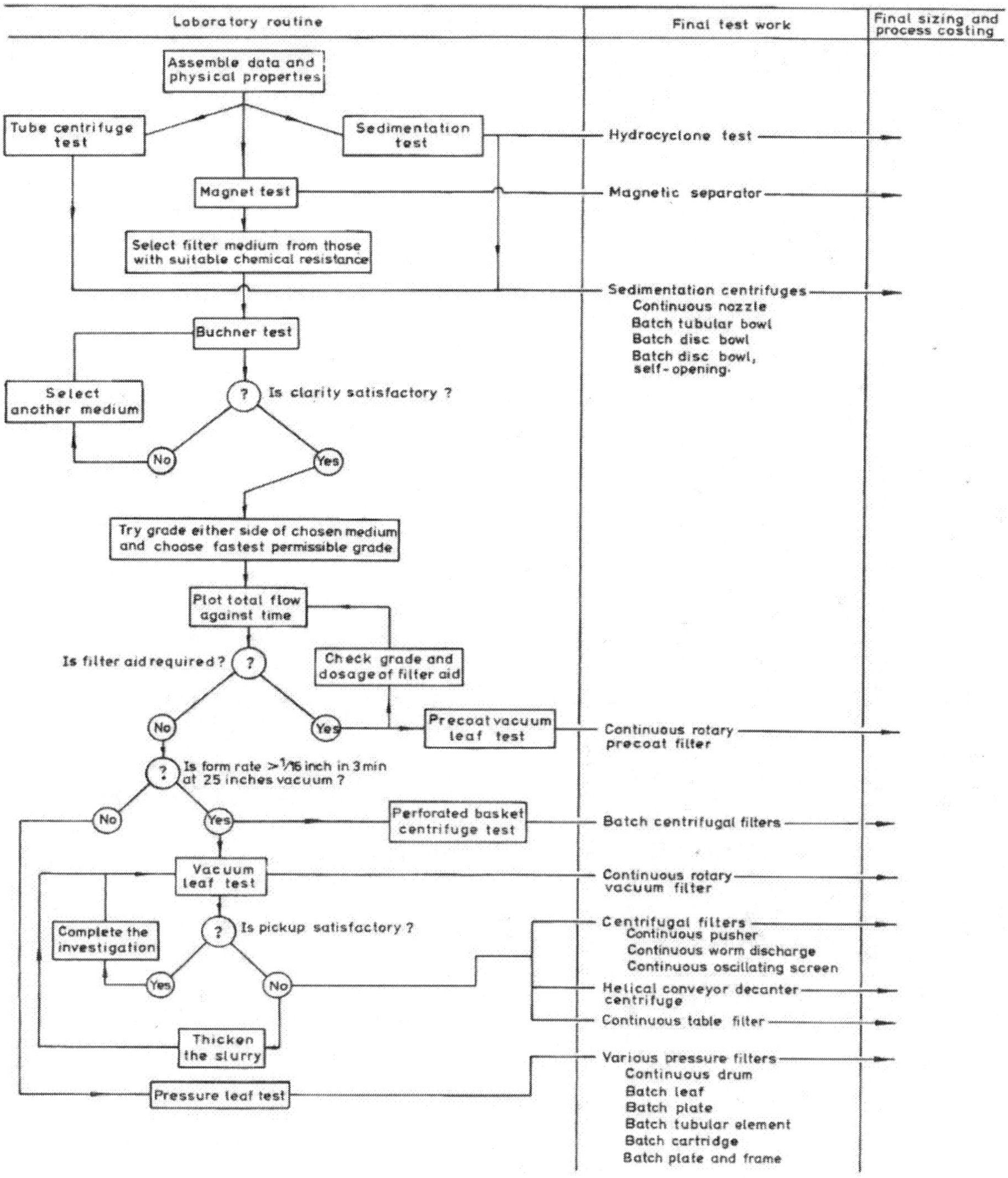

FIG. 4.1 DAVIES' EXPERIMENTAL TEST ROUTINE

Figure 2: BHS Pocket Leaf Filter

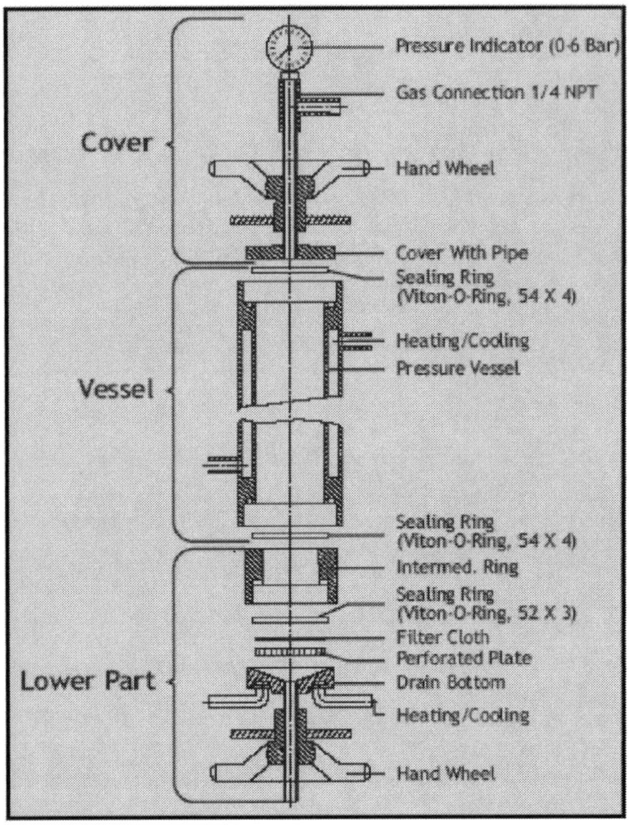

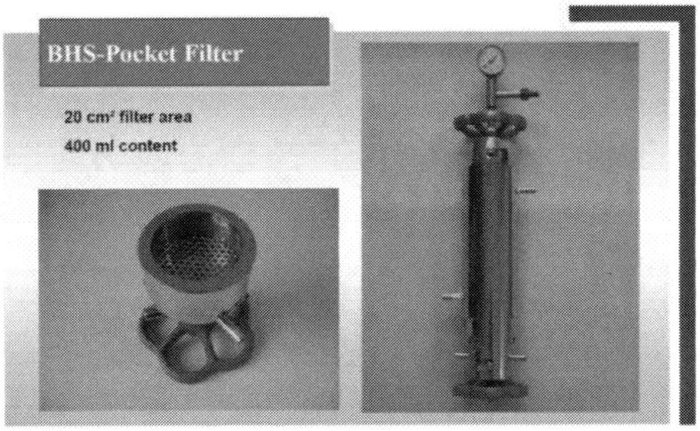

Figure 3: Data Collection Sheet for BHS Pocket Leaf Filter

Customer:		Test Number:
Date :		Run #
	Filter Media	
	Suspension	
Filling	Volume of Slurry	
	Density of Slurry	
	% Solids in Feed	
	Temperature	
Filtration	Pressure/Vacuum	
	Volume of Filtrate	
	Time for Filtration	
	% Solids in Filtrate	
Wash 1	Wash Material	
	Pressure/Vacuum	
	Volume of Filtrate	
	Time for Filtration	
Wash 2	Wash Material	
	Pressure/Vacuum	
	Volume of Filtrate	
	Time for Filtration	
Wash 3	Wash Material	
	Pressure/Vacuum	
	Volume of Filtrate	
	Time for Filtration	
Drying	Pressure/Vacuum	
	Temperature	
	Flow Rate	
	Time for Drying	
	Pressing Pressure	
Cake	Weight	
	Thickness	
	% Residual Moisture	
	Dry Cake Weight	
	Cake Discharge OK?	

Figure 4: BHS Rotary Pressure Filter, RPF 0.18 M^2, Pilot Filter

Figure 5: Pilot Vacuum Belt Filter Layout

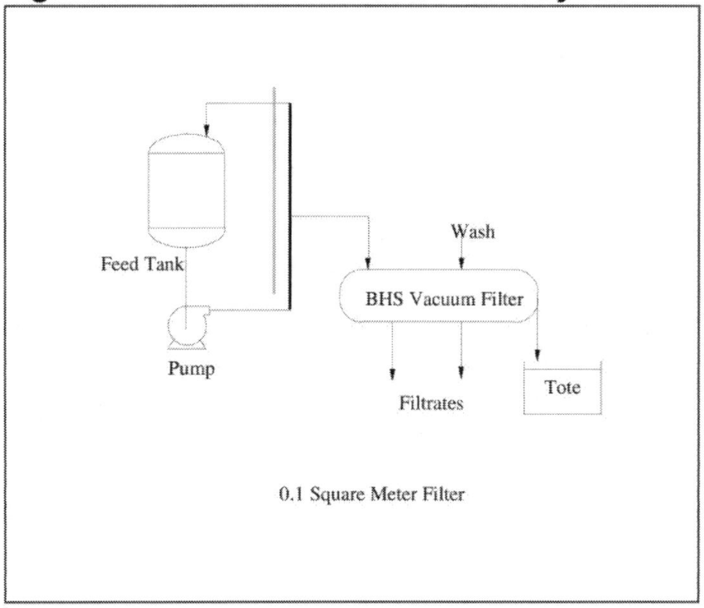

Figure 6: Dry Cake after Discharge

352728 Managing Process Safety in Pilot Plants and Bench Scale Units

Abstract:

Today's standards for safe execution in pilot plants and bench scale units are high. Increasingly, management of change (MOC) procedures for pilot plants resemble those of larger commercial units. The pilot plants can no longer be considered "tinker toys" for rapid testing of new configurations. This change can be accommodated via increased use of multi-throughput experimentation and numerical modelling for simulation of process configurations, prior to pilot testing.

This paper will present approaches to optimizing safety and productivity in execution of R&D programs.

Mechanism-free Chemical Engineering Kinetics

Blake Brousard[1], Maxwell Smith[1],
Jessica Orsak[1], and Jonathan H. Worstell[1,2]

[1] University of Houston
[2]Worstell and Worstell, Consultants

Overview

Review the Traditional Procedure for
Chemical Kinetics/Reaction Engineering

Review how rigorous is the Traditional Procedure

Propose a new Procedure

Challenge you

Reaction Engineering

We all took Reaction Engineering at University

Taught to postulate a 1st order reaction mechanism for

$A \rightarrow B$

and a 2nd order reaction mechanism for

$A + A \rightarrow B$

and a mixed order reaction mechanism for

$A + B \rightarrow C$

Postulated Rate Laws

For the reaction

$A \rightarrow B$

$$\frac{d[A]}{dt} = -k[A]$$

and for

$A + A \rightarrow B$

$$\frac{d[A]}{dt} = -k[A]^2$$

Postulated Rate Laws

And, for a mixed order reaction

$$A + B \rightarrow C$$

$$\frac{d[A]}{dt} = -k[A][B]$$

But, if [B] >> [A], then

$$\frac{d[A]}{dt} = -k_{Psuedo}[A]$$

where

$$k_{Psuedo} = k[B]$$

Integrate Rate Equation

Then taught to

Integrate the rate equation.

Plot the result.

If result is a straight-line, then we call it
a proof of our postulated mechanism.

Plot Evaluation

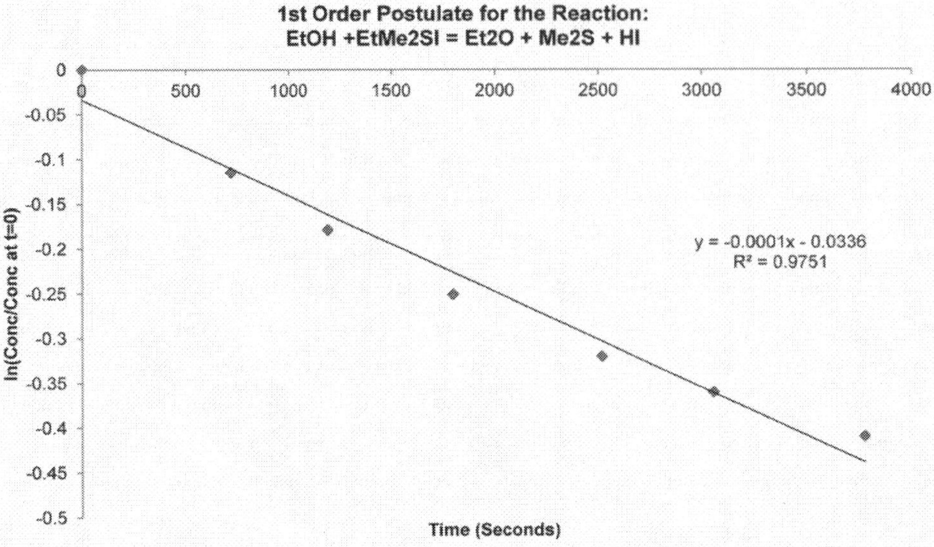

Plot Evaluation

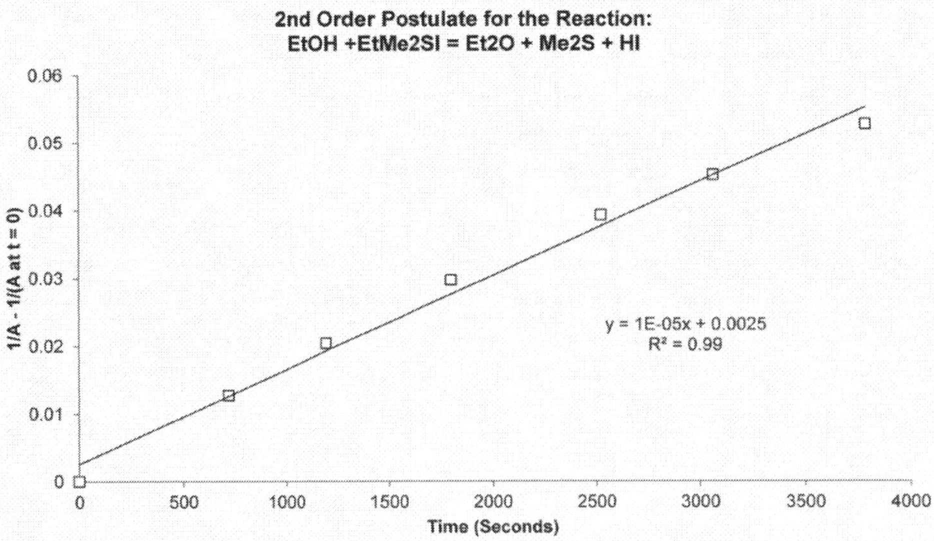

Issues

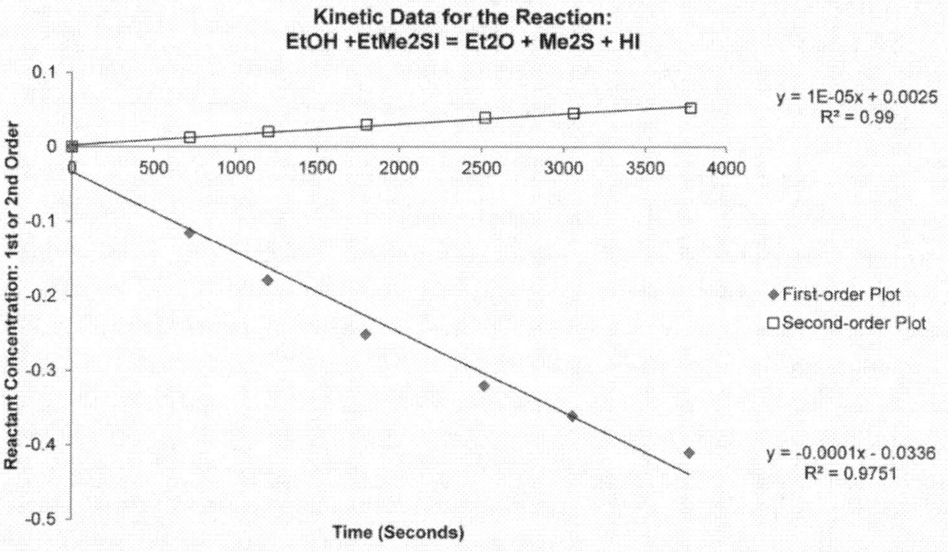

Questions

How rigorous is this proof ?

Not very

How did these mechanisms develop ?

From 19th Century experimental studies

Why are they so popular ?

Easily visualized

1st order is probability equation

2nd order is collision of two molecules

Easily integrated

Are these mechanisms justified ?

Well --- they work

Reaction Engineering Procedure

Review of Current Procedure

Postulate a reaction mechanism

Integrate the postulated rate equation

Compare experimental data to theoretical result

Calculate the rate constant

Use the rate constant to design a commercial plant

Should we continue to follow this procedure ?

Albert Einstein

Thermodynamics

Most stable science
Will not change in the foreseeable future

Why ?

Based on bulk physical properties
Based on physical "states"
Not based on mechanism for achieving particular state

Mechanism-free Chemical Kinetics

Can we build a mechanism-free chemical kinetics ?

Yes

It is possible with today's computing power and software.

Typical Course of Reactions

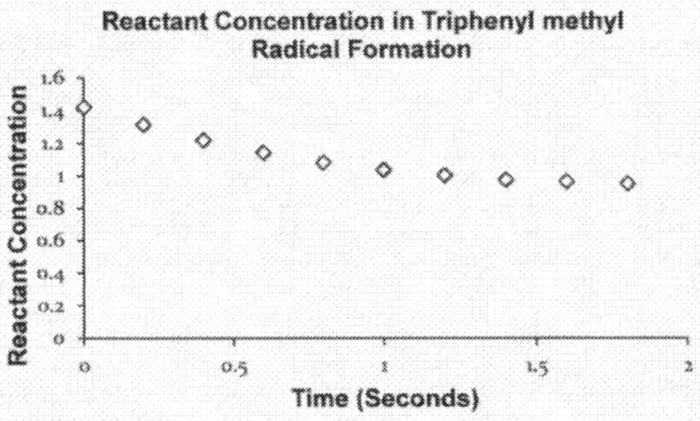

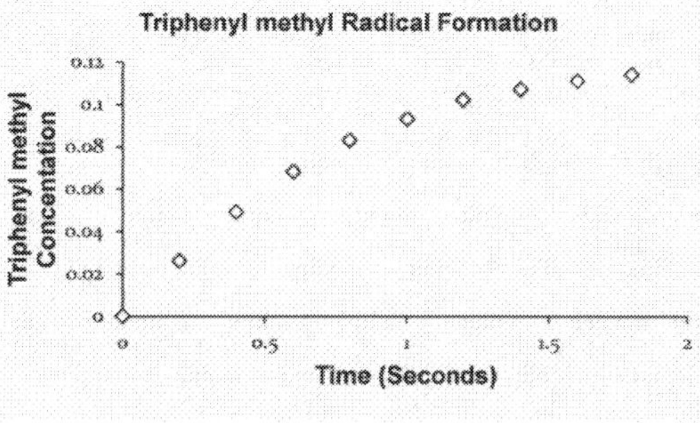

Current Data Fitting Capability

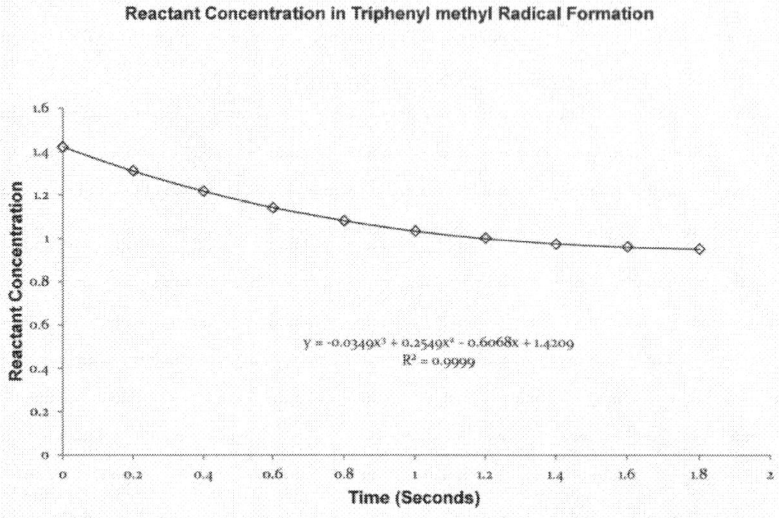

Current Data Fitting Capability

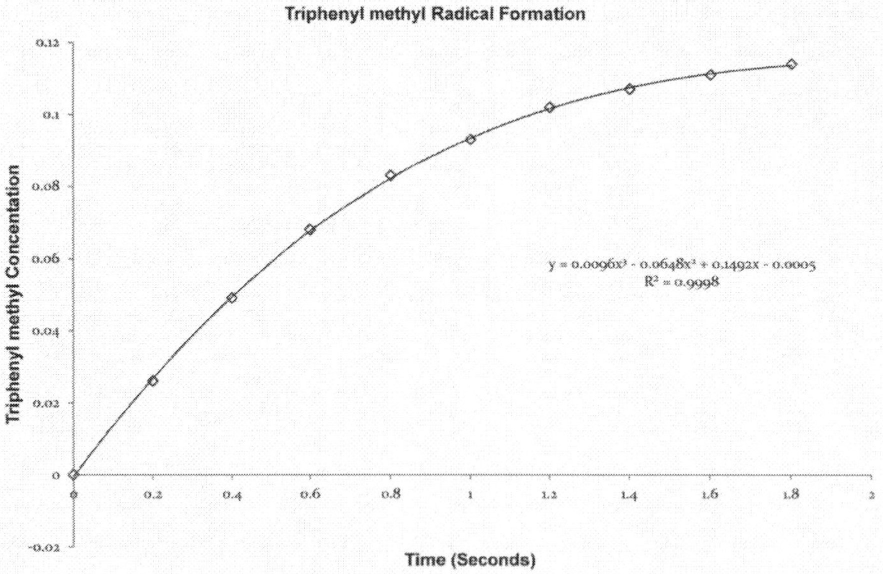

Fitted Concentration vs Time Data

What does a concentration versus time data fit give us ?

A concentration as a function of time polynomial.

For reactant

$$C_R = \alpha - \beta\, t + \gamma\, t^2 - \ldots$$

For product

$$C_P = \alpha + \beta\, t - \gamma\, t^2 + \ldots$$

What are these coefficients ?

α is the initial concentration of reactant or product in the reactor.

Interpreting Correlation Coefficients

Differentiating C_R or C_P gives

$$\frac{dC_R}{dt} = -\beta + 2\,\gamma\, t - \ldots$$

At t = 0

$$\frac{dC_R}{dt} = -\beta$$

Thus, β is the initial rate for the reaction.

Interpreting Correlation Coefficients

γ indicates the separation of molecules in the reaction:
separation grows between reactant molecules;
reactant consumption → - γ
separation shrinks between product molecules.
product formation → + γ

How do these constants change with volume ?

How do these constants change with agitation ?

How do these constants respond to temperature ?

Can we use this knowledge for reactor scaleup ?

Sucrose Inversion

Consider sucrose inversion.

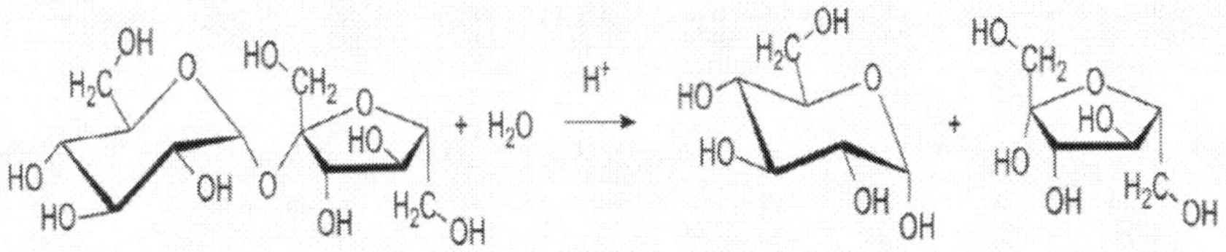

Sucrose Glucose Fructose

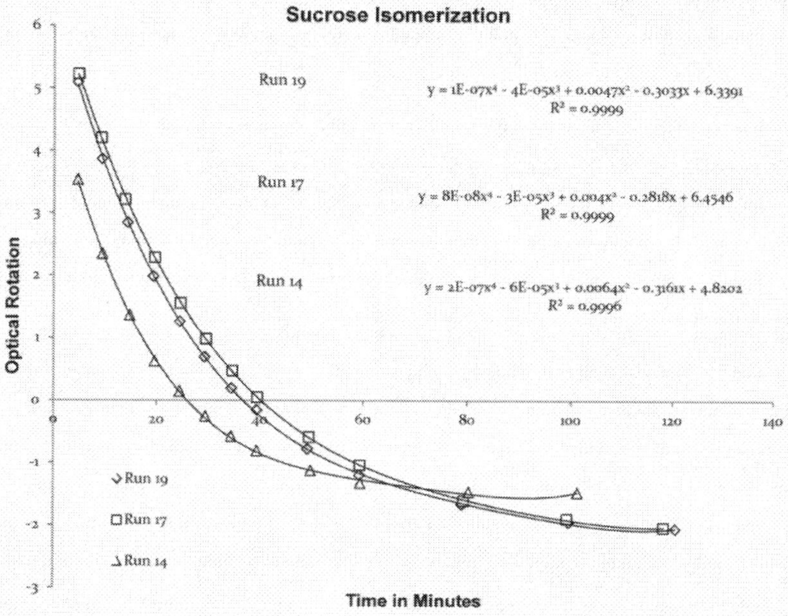

Volume Dependence of Polynomial Coefficients

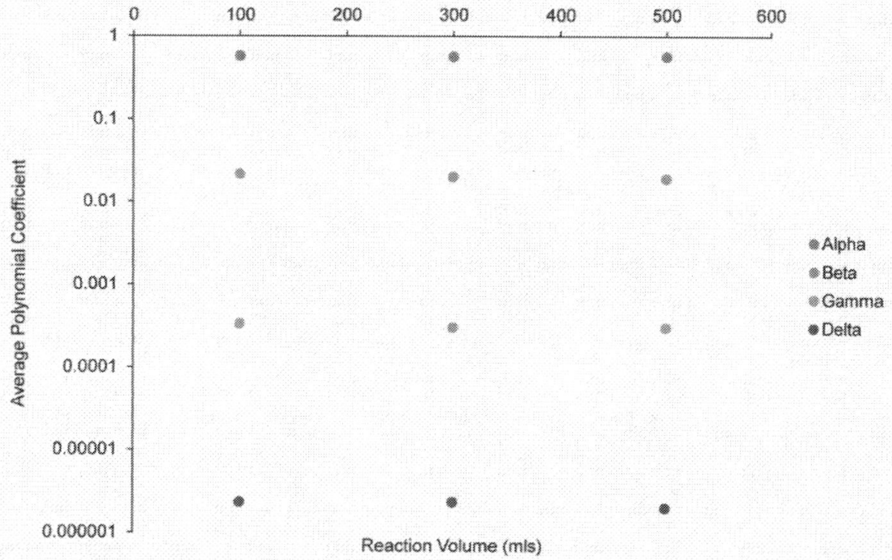

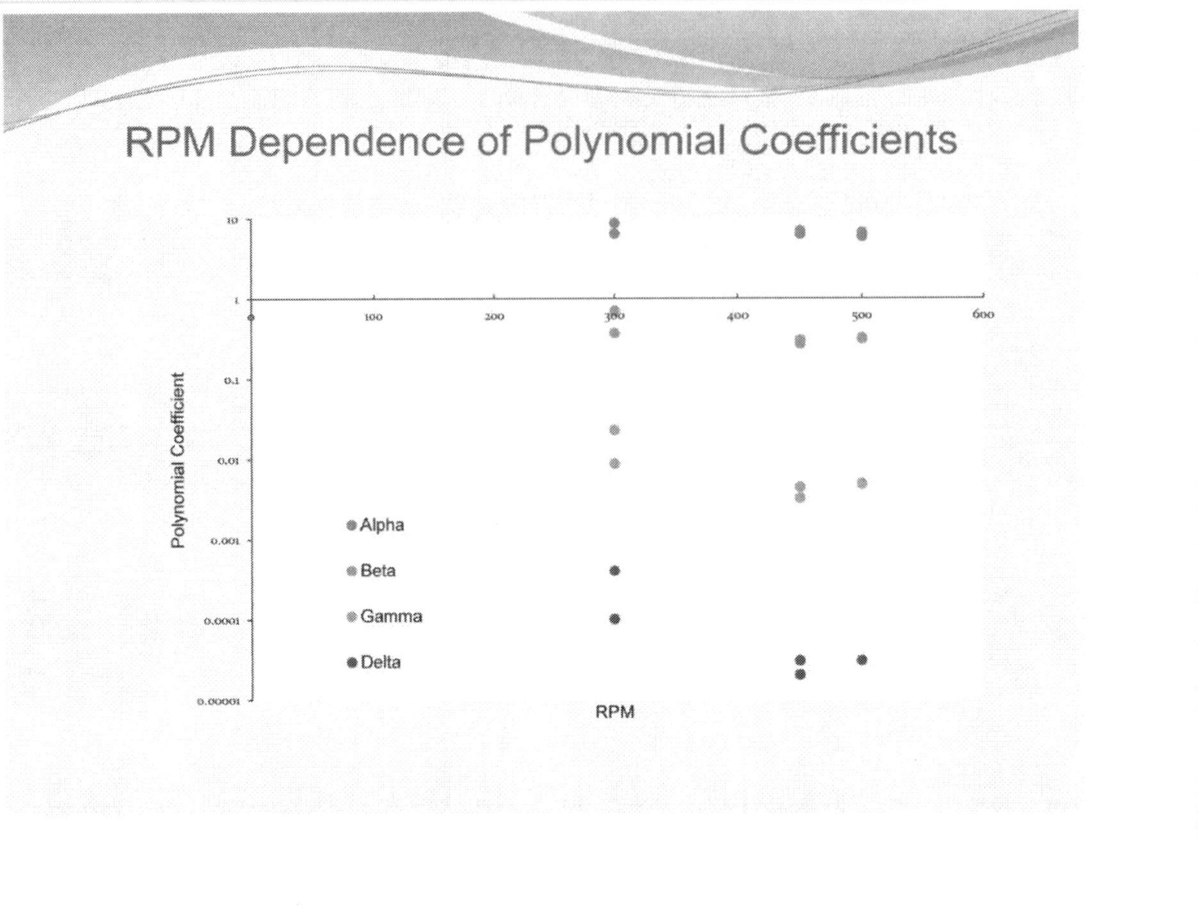
RPM Dependence of Polynomial Coefficients
Polynomial Coefficient
RPM
10
1
0.1
0.01
0.001
0.0001
0.00001
0
100
200
300
400
500
600
Alpha
Beta
Gamma
Delta

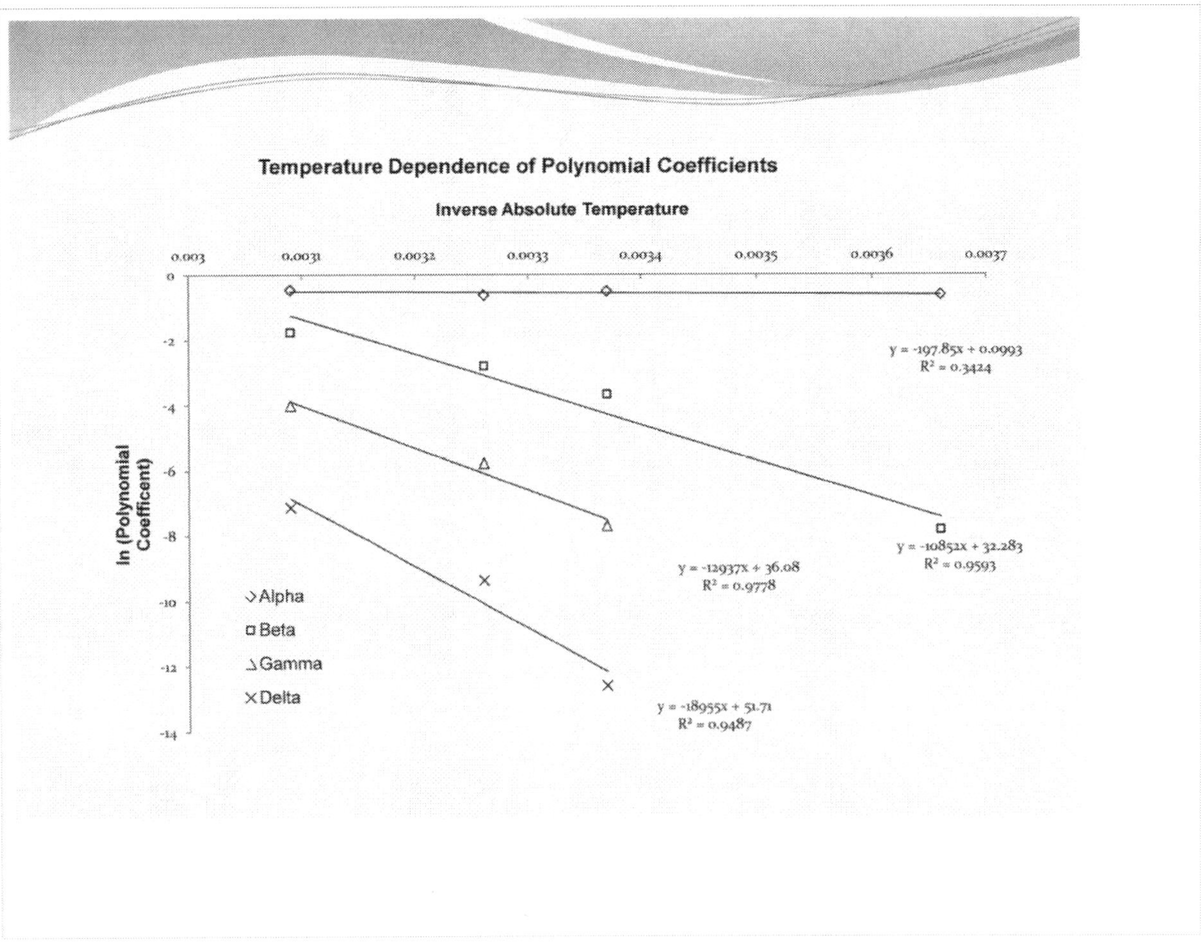
Temperature Dependence of Polynomial Coefficients
Inverse Absolute Temperature
0.003
0.0031
0.0032
0.0033
0.0034
0.0035
0.0036
0.0037
0
-2
-4
-6
-8
-10
-12
-14
ln (Polynomial Coefficent)
y = -197.85x + 0.0993
R² = 0.3424
y = -10852x + 32.283
R² = 0.9593
y = -12937x + 36.08
R² = 0.9778
y = -18955x + 51.71
R² = 0.9487
Alpha
Beta
Gamma
Delta

What Have We Shown Here ?

Laboratory batch reactors yield for reactant concentration

$$C_R = \alpha - \beta\,t + \gamma\,t^2 - \ldots$$

and, for product concentration

$$C_P = \alpha + \beta\,t - \gamma\,t^2 + \ldots$$

All coefficients are temperature dependent, except α.

Coefficients β, γ, and δ respond to temperature as Arrhenius functions.

All coefficients are volume independent.

Coefficients β, γ, and δ respond to agitation quality;
α appears independent of agitation quality.

Conclusion

Mechanism-free chemical kinetics is possible for batch reactors.

Batch reactors can be scaled using polynomial regressions for concentration versus time data.

Polynomials, such as

$$C_R = \alpha - \beta\,t + \gamma\,t^2 - \ldots$$

and

$$C_P = \alpha + \beta\,t - \gamma\,t^2 + \ldots$$

are easily obtained using today's available software.

Path Forward

Test this concept for other reactions in batch reactors.

Test this concept for recycle batch reactors using heterogeneous catalysts.

Link this concept with the energy balance.

Challenge to You

Test this method using your laboratory batch reactors.

Test this method using your pilot plant batch reactors.

Test this method using current plant batch reactors.

Reaction Engineering

We are taught to

Postulate a reaction mechanism

Integrate the postulated rate equation

Compare experimental data to theoretical result

Calculate the rate constant

Use the rate constant to design a commercial plant

351786 Dynamic Kinetic Test Unit for Determination of Kinetic Parameters Under Prototype Conditions

Abstract:

The Dynamic kinetic test unit (DKTUnit) is an innovative approach for conducting catalytic reaction tests to determine kinetic parameters in a prototype environment of single-phase and two-phase flow reactors that are typical of reactive distillation (RD) and other heterogeneous catalytic reactors. DKTUnit focuses on a kinetic test unit in which process conditions can be closely simulated, thereby reducing the uncertainty associated with scaling laboratory test data to a pilot plant RD column and other catalytic flow reactors. Once validated using the pilot plant, this kinetic test unit can even bypass further pilot plant testing by applying the data, with appropriate design tools, to commercial units with high degree of certainty.

Currently, batch kinetic experiments are typically carried out using a stirred reactor in which catalyst particles are allowed to move freely within the reactive liquid media. The rate of reaction as a function of time is monitored by taking and analyzing liquid samples using Gas Chromatography (GC) or other analytical methods. The major problems with this approach are that the fluid dynamics associated with catalyst loading in side reactors and structured packed columns – the porosity and catalyst effectiveness – cannot be simulated. As a result, there is unknown degree of uncertainty in applying kinetic parameters obtained with batch experiments to performance prediction of flow reactors with heterogeneous catalysts. Employing test data from a pilot-scale RD system, an adjusting parameter for catalyst effectiveness is commonly used for validating the performance prediction models or simulation codes, such as ASPENPlus.

DKTUnit is based on a patented High Temperature Fouling Unit (HTFU) for petroleum fouling that has been validated in refinery service. The unit consists of an autoclave with internals designed for simulating prototype flow conditions of catalytic reactors. DKTUnit consists of four configurations: a) flow tube equipped with impeller to induce flow across catalyst bag mounted in the flow tube; b) a similar configuration with structured catalyst packing mounted in the flow tube; c) catalyst bags mounted on a rotating frame for simulating liquid cross-flow across catalyst bag; and d) catalysts bags mounted on a rotating frame for simulating parallel-flow of catalyst bags. DKTUnit uses a variable speed impeller and frame mounted on a shaft with variable liquid flow across catalyst so that kinetic data can be determined.

A preliminary analysis is carried out for developing a methodology for applying the kinetic parameters determined by the DKTUnit to flow reactors as well as RD. The results from this design study will be presented and plans for applying the DKTUnit to RD processes will be discussed.

351788 Rapid Process Scale-up through Innovative Design Methodologies

Abstract:

Rapid Process Scale-Up Through Innovative Design Methodologies

John Prindle[1], Richard Doctor[1], Aspi Kolah[2], Carl Lira[2], Dennis Miller[2], and C.B. Panchal[1]

[1]E3tec Service, LLC, Clarksville, MD 21029, USA

[2]Department of Chemical Engineering and Materials Science,
Michigan State University, East Lansing, Michigan 48824, USA

Abstract

Scale-up of chemical processes is an expensive, time-consuming component on the path to commercialization. While many new lab-scale technologies have been created to improve specific aspects of scale-up, additional improvements can be achieved by making detailed performance prediction tools an integral part of the process development. This integration should be seamless and is particularly necessary when scaling up a multiphase reaction system or systems involving process intensification.

Questions involving commercial process robustness and sensitivity to operation/feed parameters are compounded when process units, such as distillation and reactors, are integrated as part of process intensification. Conservative approaches dictate the need to move incrementally from lab-scale to small pilot-scale to large pilot-scale and, then finally, to commercialization, involving tests at each stage. Even with this conservative approach, design uncertainties persist going from one step to the next. Seamless integration of detailed modeling tools and novel lab-scale technologies can allow the elimination of one or more of these scales and/or reduce design uncertainties, while simultaneously focusing efforts at the remaining scales to obtain only those data required for validating the design tools. This over-arching methodology increases the speed to commercialization and can lead to the development of innovative, reliable continuous processes where conventional scale-up approaches would point toward batch.

Our presentation will discuss efforts in developing and applying detailed performance prediction tools and integrating them with novel experimental techniques during the scale-up process to rapidly identify regions in the operating space where efficient, robust commercial operation can be achieved. Our particular focus is on Heat-Integrated Reactive Distillation (HIRD) with the incorporation of side-reactors to provide further operation flexibility. Examples will be presented for the scale-up of a continuous HIRD process with and without side-reactors for the production of tri-ethyl citrate from citric acid and ethanol.

351274 Energy Systems Model: Development of a Discrete-Event Simulation Tool to Facilitate Operating Decisions

Abstract:

The Dow Energy business is responsible for managing and providing power, steam and other utilities to some of Dow's largest sites located around the world. At certain sites, 100% of the steam requirements are generated on-site, and power is additionally supplied to the external grid and any disruption to the steam and energy supply would severely impact operations. An energy plant typically contains gas and steam turbines, boilers, synchronous condensers, and transformers. Each of these assets has the potential to experience unplanned shutdowns, in addition to scheduled maintenance shutdowns. It is important to ensure that steam and power requirements are still met, despite the uncertainties associated with asset shutdowns. Interrupted steam and power supply can cause short-term problems in meeting customer commitments and damage customer confidence in Dow's supply chain .This drives the need for a model that effectively captures energy system operations so that data based decisions on effective and robust operating strategies and capital investments can be made. We present an approach that would enable evaluation of the reliability of steam and power supply under various conditions (e.g. the removal/addition of new assets) through computationally inexpensive simulations.

Our model development includes (1) extracting asset reliability information from asset specific historical data and used in the model to accurately depict unplanned shutdowns; and (2) understanding interactions between the various energy assets in order to accurately represent the energy plant from a steam and power perspective. The model is based on discrete event simulations (DES) and has been developed in ExtendSim®. Our model output focuses on statistics that would be pertinent to operating decisions and includes: equipment reliability, steam and power supply reliability. We additionally focus on developing a tool that is user-friendly to promote its use, longevity, and ability to easily leverage to other sites in the energy business. We then demonstrate the use of the tool on a hypothetical case study that aims to drive key business decisions.

In summary, the modeling approach we present involves the development of a discrete event simulation-based tool that can be used to facilitate operating decisions by quantitatively characterizing a site's exposure to the risk of not obtaining the required steam and/or power supply. The tool would be used to drive key proactive energy decisions that include process development and enhancement related to steam and power assets.

Oxygenated Compounds as Reducing Agent for NO_x Removal by Selective Non-Catalytic Reduction (SNCR) Method

AIChE 2014 Spring Conference

Maoqi (Mark) Feng, Ph.D.

Southwest Research Institute®

March 31, 2014

Agenda

- Introduction
- Background
- Objectives
- Technical Approach
- Test Results
- Summary and Conclusions

2

Introduction

- NO_x are of regulated emissions from stationary sources
  - Clean Air Act (CAA), Section 111(b) requires to use "*best demonstrated technology* (BDT)"
  - California requires up to **75%** reduction in NO_x emissions
  - Texas Natural Resources Conservation Commission (TNRCC) adopted new rules for Houston-Galveston Area (HGA) - **90%** reduction for major industrial sources

- Examples of Stationary Sources
  - Coal-fired power plants
  - Waste incinerators
  - Process heaters

3

Technical Background (1 of 4)

- Commonly used NO_x reduction technologies
  - Flue Gas Recirculation (FGR)
    - ✓Similar to EGR in automotive NO_x control
    - ✓FGR $\Rightarrow O_2 \Downarrow \Rightarrow$ Combustion Temperature $\Downarrow \Rightarrow NO_x \Downarrow$
  - Selective Catalytic Reduction (SCR)
    - ✓$4NO + 4NH_3 + O_2 \Rightarrow 4N_2 + 6H_2O$ (Major Reaction)
    - ✓$NO + NO_2 + 2NH_3 \Rightarrow 2N_2 + 3H_2O$
  - Selective Noncatalytic Reduction (SNCR)
    - ✓$4NO + 4NH_3 + O_2 \Rightarrow 4N_2 + 6H_2O$

4

Technical Background (2 of 4)

➢ Comparison of selected technologies

In-combustion

Methods	Need a Catalyst?	Temperature, ˚C	NOx Efficiency	Relative Cost
FGR	No	N/A	10~25%	0.85
SCR	Yes	300~450	70~90%	3.5
SNCR	No	800~1050	25~50%	1.0

Post-combustion

5

Technical Background (3 of 4) – SNCR Overview

➢ **Key Terminologies**
- NH_3-to-NO_x Ratio (ANR): **Typically 2**
- NH_3 slip

$$4NO + 4NH_3 + O_2 \Rightarrow 4N_2 + 6H_2O$$

➢ **Operating Conditions**
- Temperature (**850~1100°C**)
- Residence time

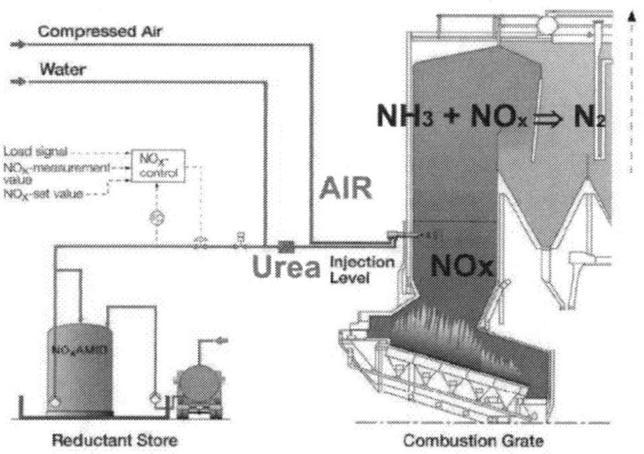

6

Technical Background (4 of 4)

– Remaining Issues with SNCR

- Relatively low NO_x conversion efficiency
  - Flow distribution
    - ✓Reaction efficiency
    - ✓NH_3 oxidation
  - Limited residence time
    - ✓Narrow temperature window
    - ✓Typically less than 1.0 second

7

Objectives

- Improve the current SNCR technology
  - NO_x efficiency ↑ 10-15%
  - Reducing Agent to NO_x ratio = 2 - 4

- Use oxygenated compounds as reducing agent to replace urea (or NH_3)

8

Reaction Conditions

➢ Simulated flue gas and SNCR conditions

Parameter	Concentration
NO_x	200, 400, 600 ppm
O_2	10%
H_2O	6%
CO_2	8%
N_2	Balance
Reducing Agent	$RA/NO_x = 3$
(SO_2)	*0, 200, 500 ppm*
(HCs)	*0, 200, 400, 600 ppm*
Space Velocity:	Residence time 1 s
Temperature:	800, 900, 1000°C

9

Reaction Setup

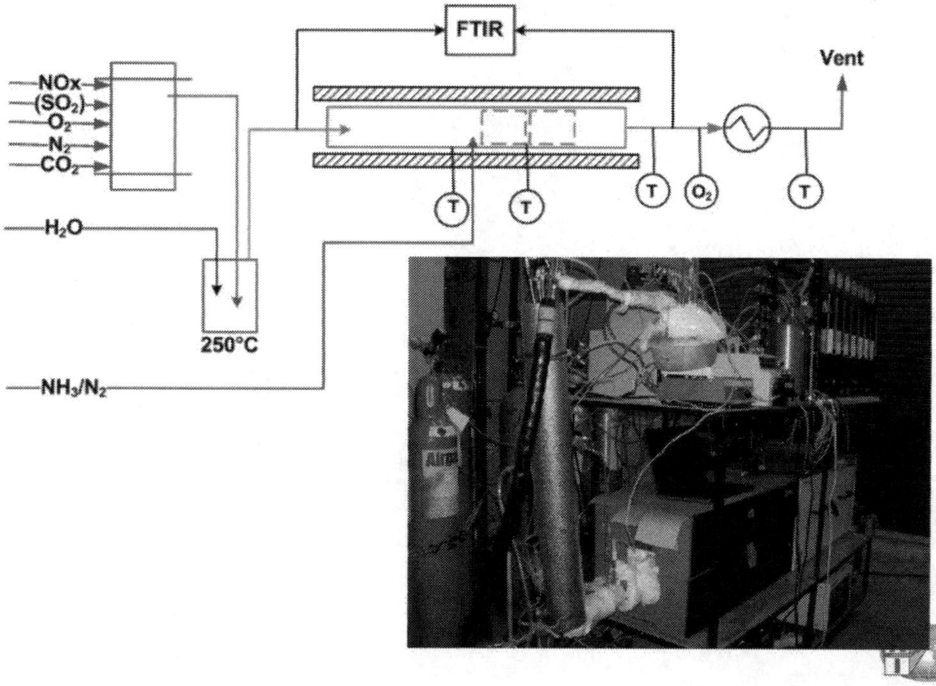

10

Advantages of Structured Packing

➢ A solution to combine two functions into one "structure"

- Improved flow distribution ⇒
  - ✓ NO_x efficiency ⇑
  - ✓ Reducing agent oxidation ⇓
- Increased surface area ⇒
  - ✓ NO_x efficiency ⇑

11

Structure Packing

➢ Three Types of "Structures"

- Ceramic packing
- Non-metalic structures
- Clay cakes

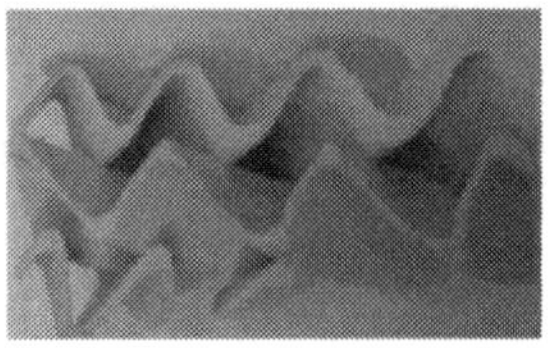

12

Pressure Drop

- Pressure Drop
  - <u>Issue</u>: High surface area may result in high pressure drop
    - ✓Generally, surface area ⇑ ⇒ pressure drop ⇑
    - ✓High surface area is desired (for NOx efficiency)
    - ✓Low pressure drop is required
  - <u>Mitigation</u>: CFD helps to balance pressure drop vs. surface area

13

CFD Modeling

- CFD modeling of structured packing for SNCR system
- ✓ Three dimensional model was created in SolidWorks
- ✓ Software FlowSimulation

- Compare experimental data with modeling data

14

CFD Modeling

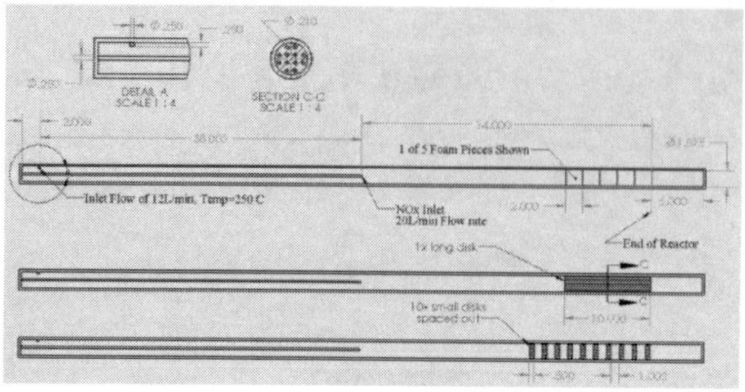

- ➢ A - five structured packing
- ➢ B - one single insert with channels
- ➢ C - 10x small inserts with channels

15

Mixing

- ➢ Air mass fraction contour plot of Configuration A and B

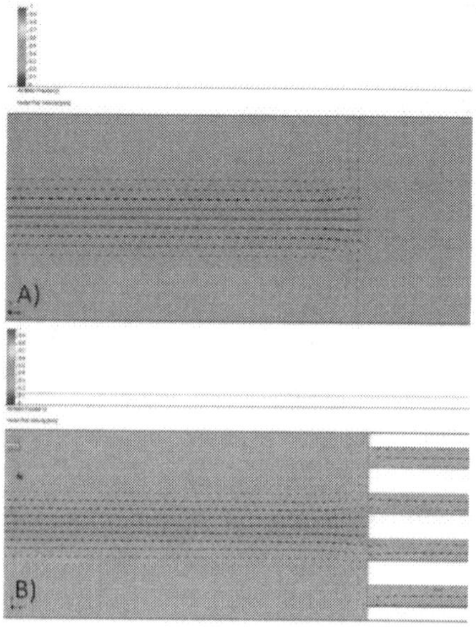

16

Mixing

- Air mass fraction contour plots with small ranges displayed for Configuration B and C
- Additional vortices created by the second inserted disk in Configuration C

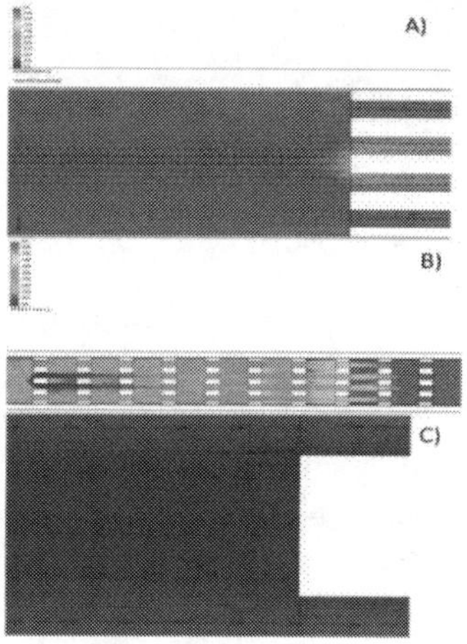

17

Pressure Drop

- Pressure contour plot of the reactor in Configuration A

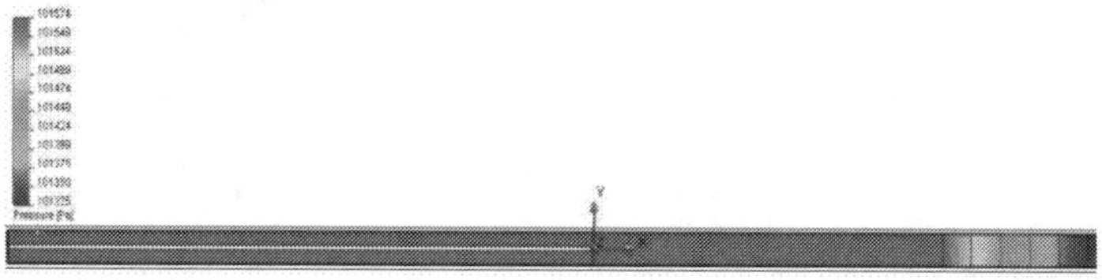

18

Gas Velocity

- Comparison of velocity contour plot of the three configurations

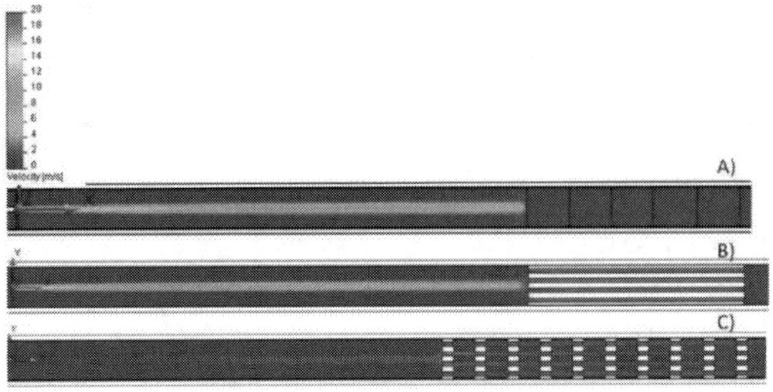

19

Test Results

- NOx removal efficiency was increased 10-15% by the structure packing
- CFD modeling predicted the pressure drop of different structure packings
- CFD modeling can be used to simulate the flow and temperature distribution
- Mixing of the reducing agent and NOx is key to SNCR efficiency improvement and preventing reducing agent oxidation

20

Test Results

- Propane was used as reducing agent for NO_x removal
- A mixture of C1-C4 was used as reducing agent for NO_x removal
- Proprietary oxygenated compounds as reducing agent (US Patent 8,535,627)
- Effect of hydrogen was tested – NO_2 reduction

21

Test Results

- Effect of CO was tested - NO_2 reduction
- SO_2 effect: Not directly related to NOx removal efficiency; corrosion
- Water effect: slightly decrease the NOx removal efficiency

22

Summary and Conclusions

- Introducing a well designed "structure" improved SNCR NO_x removal efficiency
- CFD modeling was useful in process intensification
- Oxygenated compounds reduced plugging due to NH_3 slip and ammonium sulfate formation
- Less corrosive

23

Acknowledgement

- This project was funded by the internal research fund of Southwest Research Institute (Project No. 01-R8213).

Thank you!

Questions?

24

353086 Flare Emission and Greenhouse Gas Reduction Study during an Ethylene Plant Startup on Recycle through Dynamic Simulation and Process Optimization

Abstract:

Currently, emission characterization and reduction have been gaining more attention in the Chemical Process Industry and playing a more prominent part in industrial sustainability. Flare emissions release volatile organic compounds (VOCs) and pollutants into the atmosphere, which react with nitrous oxide in sunlight to generate ozone and smog. Therefore, it is essential to limit this amount to improve local and regional air quality, especially around chemical plant locations. Besides, environmental performance also takes into account material usage, energy consumptions, non-product outputs and pollutant releases. From the economic point of view, reducing burned-off gases saves material, helps avoiding penalty and also creates a more effective way to distribute energy, thus lowering utility cost. Ethylene production process is one of the biggest and most complicated processes in petrochemical industry. A typical start-up of an ethylene plant flares approximately 5.0 million lbs of raw materials and generates at least 15.4 million lbs of CO_2, 40.0 Klbs CO, 7.4 Klbs NO_x, 15.1 Klbs hydrocarbons, and 100.0 Klbs HRVOCs. Hence, managing flare emission during startup, shutdown and malfunction of an ethylene plant are very important. Since flaring occurs in both normal and abnormal conditions, flare source reduction approach employs simulation and optimization, emphasizes the upstream section of flare system and the whole plant integration, in which more upstream modification and optimization contribute in limiting unwanted products.

In this study, rigorous steady-state and dynamic models of a front-end De-Ethanizer ethylene plant with sulfur recovery unit are constructed to serve as both the generic foundation to further explore causes roots of process abnormal behaviors and subsequently an optimization tool to provide guidance to both design and operational strategies. Rigorous steady-state and dynamic models are constructed and validated before case studies are carried out. Based on these models, possible flaring sources and incidents are identified, which are demonstrated extensively in this paper throughout three case studies of CGC startup, precooling chilling train, and withdrawal from full-reflux columns during an approach of flareless startup on total recycle.

Because charged gas is continuously produced from furnaces and sent to flare to retain compressors' inlet pressure, CGC startup is reportedly the most critical step and has largest amount of vent gas. Several scenario cases of charged gas at different flowrates and compositions are examined to find the least flaring one. Another case study focuses in recycling off-specs from full-reflux C2-/C3-supplier columns to CGC as partial substitute for furnace feed. In the chilling train section, specifications of equipment sizes and heat capacities are considered for simulating pre-cooling process. Optimal procedure is achieved for utilizing methane to cool down cold boxes at a reasonable temperature-dropping rate in order to protect these equipments. In dynamic simulation environment, all of the major steps and options are simulated and evaluated while flaring can also be predicted. These results serve as a basis for emission calculation and greenhouse gas study such as CO_2 control. The process can be considered from smaller scale (sectional model) to more comprehensive scale (whole plant model). The best results in these case studies are incorporated and validated in the integrated model of the whole plant in order to stabilize the system in the shortest time while minimizing material and energy loss.

Following this representative, utilizing dynamic simulation as a driving force for operation during abnormal conditions effectively provides an in-depth understanding and provides more detailed planning and expectations for all related personnel before the real action takes place. This procedure contributes a great deal in evaluating new strategies, predicting process dynamic behaviors, and achieving more precise timing to control commissioning and startup strategy.

351658 Chemical Plant Startup Simulations for Flare Emission Reduction

Abstract:

HRVOC (Highly Reactive Volatile Organic Compounds) emissions from industries are highly responsible for ground-level ozone pollution. Especially, HRVOC emissions from flares of MSS (maintenance, startup and shutdown) & events are the biggest. Thus, reduction of flare emissions could significantly reduce the total HRVOC emissions, and could reduce ozone pollution. On the other hand, reduction of flare emissions could save raw materials and intermediate materials, decrease startup/shutdown time, and thus reduce the total cost of startup and shutdown operations.

In this paper, startup procedures of an ethylene plant were simulated with dynamic simulation. The ethylene plant is a typical front-end demethanizer process. Simulation models have all the major process facilities, including cracking furnaces, quench towers, charge gas compressor (CGC) system, chilling train, demethanizer, deethanizer, depropanizer, debutanizer, C2 converter, C3 converter, C2 splitter and C3 splitter. Dynamic simulation covers all startup steps which involve flare emissions. The simulation results have been validated with real data from ethylene plants.

Three startup scenarios were simulated with dynamic models. The first scenario is a "baseline" scenario, which means no flare reduction methods are used in the scenario. The overall emissions of the baseline scenario are compared with emission data from actual startup events reported to local air quality authorities, and they are within normal range.

The second scenario simulates startup procedure with certain flare reduction methods, including commissioning distillation columns before CGC starting, using gas from columns to help CGC start, and precooling chilling train. An optimized startup schedule is built based on dynamic simulation. Comparing the baseline scenario, the second scenario reduces total startup time by 20%, and reduces total vent gas by 47%.

The third scenario simulates nitrogen startup procedure, which means use nitrogen to start CGC and precooling chilling train. The startup schedule is also optimized based on simulation results. Comparing the baseline scenario, the third scenario reduces total startup time by 36%, and reduces total vent gas by 58%.

352687 Dynamic Simulation for Flare Minimization Strategy in an Ethylene Plant Shutdown

Abstract:

Ethylene plant shutdowns generate a great amount of flare emissions that inevitably damage air quality of the surrounding environment. During an ethylene plant shutdown, plant generally cut off feed and flared off all off-spec products until the plant was de-inventoried. Here, an optimal shutdown strategy is designed for reducing flare emission and preventing losses of hydrocarbon materials. In this paper, dynamic simulations are employed to model the operations of shutdown for both the general strategy and the optimal strategy.

First, the general shutdown strategy is simulated and dynamic emission profile is presented. The shutdown of an ethylene plant would be executed following three steps: (i) Preparation of shutdown. Feed rates are reduced to minimize the liquid inventories in facilities ahead of the shutting down. (ii)Shutting down. The plant is shut down, beginning with shutting down Cracked Gas Compressor (CGC), followed by gradually cutting off the furnace feed. (iii) Decommissioning including liquid discharge, vapor discharge and N_2 purge.

Second, the optimal shutdown strategy is simulated to recover more liquid and vapor hydrocarbon materials in the facilities. Compared with the basic shutdown strategy, the optimal shutdown strategy includes certain flare reduction methods. For example, after shutting down the furnaces, gas from other facilities is used to keep CGC running to help recover leftover hydrocarbon materials in the flash drums and columns, rather than send hydrocarbon materials to flare. In the dynamic model, recycle lines are built and the operations are simulated to monitor the key process variables and study the feasibility. So, through the dynamic simulations, the shutdown strategy can be investigated to help the shutdown reduce the risks on safety, reliability, business and environment.

352087 Assessing the Use of Ceramic Membrane Bioreactor for Anaerobic Treatment of High-Load Food Wastewater at Bench Scale and Pilot Scale

Abstract:

Abstract: A novel type of ceramic membrane was tested in two anaerobic membrane bioreactors (AnMBR): a bench scale reactor (12 L) and a semi-industrial pilot plant (2.5 m^3). A highly loaded industrial wastewater from corn-processing was successfully treated, reaching COD reductions close to 95%, while up to 99% removal was achieved for a simulated wastewater. The performance of both AnMBRs is still being optimized for the industrial effluent; the digesters show a similar behaviour and the semi-industrial filtration module seems to have an improved permeability. This study confirms many of the benefits attributed to the AnMBR technology, combining the advantages of the anaerobic digestion with the efficiency and robustness of the ceramic filtration.

Keywords: Membrane bioreactor; ceramic membranes; anaerobic digestion; corn-processing wastewater

INTRODUCTION

The Anaerobic Membrane Bioreactor (AnMBR) combines the anaerobic digestion process with a membrane filtration to provide complete solid-liquid separation. AnMBR system makes possible to work with high biomass concentration to treat industrial complex wastewaters heavily loaded with solids, oil and grease, and/or toxic compounds (Futselaar, 2013; Skouteris, 2012). Anaerobic digestion offers several widely known advantages over conventional aerobic processes, since no oxygen is required and the biogas production reduces the associated energy cost of wastewater treatment and it compensates for high-energy requirements related to cross-flow filtration, which is a common drawback of membrane filtration-based technologies. Nevertheless, it presents some limitations that have limited its wide-spread use (lower purification efficiency, poor effluent quality and process instability). The AnMBR technology allows overcoming these limitations as it combines the advantages of the anaerobic process and MBR technology in a robust and compact solution, characterized by the higher biogas production and the excellent effluent quality due to the small pore size of the membranes and the total retention of the suspended solids. As a result, AnMBRs technology has gained popularity in recent years, and more attention has been focussed on the further development of this process for the treatment of high-concentrated wastewater, mainly food and beverage industry wastewaters (Fuchs, 2003) although they have not yet been employed widely at industrial scale (Grant et al., 2008, Christian et al, 2011). However, recent technical advances in membrane and system design are promising great reductions in operation costs, and thus, broader application of AnMBRs (Scott, 2008). As membrane fouling is a limiting issue for implementing AnMBRs, ceramic membranes may have a good opportunity, since the inorganic membrane materials are reported to present less fouling and to possess higher chemical and thermal resistance (Futselaar, 2013; Skouteris, 2012; Sutton, 2006; Wolff, 2013; Kroll, 2010).

In the present work two ceramic AnMBRs have been studied, a bench scale AnMBR fed with both synthetic wastewater and industrial corn-processing wastewater, and a semi-industrial plant fed with industrial corn-processing wastewater. The objective is to evaluate the maximum digestion capacity of the anaerobic reactor, to study the behaviour of the membrane and to optimize the filtration process to get the maximum permeate flow with an optimal transmembrane pressure (TMP).

MATERIAL AND METHODS

The AnMBRs consist on a CSRT digester coupled to an external ceramic membrane filtration module equipped with backwash and CIP devices. Both, bench and semi-industrial, reactors are operated in mesophilic conditions (35ºC) and with average MLSS of 20 g/L. The start-up was carried out with seed sludge from a municipal anaerobic digester.

Bench scale AnMBR

The bench scale AnMBR is a 12 L reactor, mounted with a single tubular ceramic membrane of 0.02 m^2. Cross-flow velocities of 2 and 3 m/s and different backwash sequences have been tested. The bioreactor was initially fed with a simulated wastewater with high starch content, increasing the organic load (OLR) gradually from 0.7 to 7 $kgCOD{\cdot}m^{-3}{\cdot}day^{-1}$. Then the substrate was replaced with the industrial effluent at an initial OLR of 6 $kgCOD{\cdot}m^{-3}{\cdot}day^{-1}$.

Semi-industrial pilot plant

The semi-industrial pilot plant with the capacity of 2.5 m^3 is equipped with a ceramic ultrafiltration system made up of 4 modules consisting of 7 ceramic membranes with 19 tubular channels (1.7 m^2 of filtration area per module). The MBR is operated at a CFV of 2 m/s and the reactor is fed with a highly loaded wastewater from the corn cooking process (COD ≈ 35 g/L). The OLR has been increased from 0.25, in the start-up, to 3.5 $kgCOD{\cdot}m^{-3}{\cdot}day^{-1}$.

RESULTS AND CONCLUSIONS

In the bench scale AnMBR, a 95-99% of both COD and TOC was removed from the simulated effluent at a gradually increasing OLR from 0.7 to 7.0 $kgCOD{\cdot}m^{-3}{\cdot}d^{-1}$. A high quality permeate with an output COD of 60-300 mg/L and TOC of 20-80 mg/L was achieved, whereby the average specific methane production was 0.34 $m^3CH_4{\cdot}kgCOD_{removed}^{-1}$.

The treatment of industrial corn-processing wastewater obtained about 90-95% of COD and TOC reduction at an OLR of 6.0 $kgCOD{\cdot}m^{-3}{\cdot}d^{-1}$, yielding an average specific methane production of 0.30 $m^3CH_4{\cdot}kgCOD_{removed}^{-1}$ and permeate of 300-400 mg/L in COD and 70-150 mg/L in TOC, respectively.

A permeate flux up to 50 LMH could be reached in the bench scale single channel filtration module, according to both the flux-step tests and the long-term studies.

Meanwhile, **in the semi-industrial pilot plant** the OLR was increased from 0.25 to 3.5 $kgCOD \cdot m^{-3} \cdot d^{-1}$ with COD reduction up to 95%, similar to the laboratory studies. The biogas production also increased with the OLR obtaining an average specific methane production of 0.32 $m^3CH_4 \cdot kgCOD_{removed}^{-1}$, close to bench scale results. Ceramic filtration worked satisfactorily, having a permeability over 120 $L \cdot h^{-1} \cdot m^{-2} \cdot bar^{-1}$ with permeate flux up to 45 LMH and average TMP of 0.25 bar.

Many of the AnMBR technology advantages were confirmed. The solids retention by membranes results in a high quality effluent, free of suspended solids, bacteria and reduced COD, which allows to reduce or, even, eliminate a post-treatment. Highly hydrophilic ceramic membranes maximized the performance.

REFERENCES

Fuchs, W., Binder, H., Mavrias, G. and Braun, R. (2003). Anaerobic treatment of wastewater with high organic content using a stirred tank reactor coupled with a membrane filtration unit. Water Research, 37, 902-908.

Futselaar. H., Rosink, R., Smith, G., Koens, L. (2013). The anaerobic MBR for sustainable industrial wastewater management. Desalination and Water Treatment, 51(4-6), 1070-1078.

Grant, S., Page, I., Moro, M., Yamamoto, T. 2008 Full-Scale Applications of the Anaerobic Membrane Bioreactor Process for Treatment of Stillage from Alcohol Production in Japan. Proceedings of the Water Environment Federation. WEFTEC 2008: Session 101 through Session 115: 7556–7570.

Kroll, S., Treccani, L., Rezwan, K., Grathwohl, G. 2010 Development and characterisation of functionalised ceramic microtubes for bacteria filtration. Journal of Membrane Science 365(1-2), 447-455.

Scott, C., Grant, S., McCarthy, P., Wilson, D., Mills, D. (2011). The First Two Years of Full-Scale Anaerobic Membrane Bioreactor (AnMBR) Operation Treating High-Strength Industrial Wastewater. Water Practice & Technology, 6 (2). doi:10.2166/wpt.2011.032.

Skouteris, G., Hermosilla, D., Lopez, P., Negro, C., Blanco, A. (2012). Anaerobic membrane bioreactors for wastewater treatment: A review. Chemical Engineering Journal, 198, 138-148.

Sutton, P. M. (2006), Membrane bioreactors for industrial wastewater treatment: applicability and selection of optimal system configuration, Proceedings WEFTEC 2006.

Wolff, C., Beutel, S., Scheper, T. (2013). Tubular membrane bioreactors for biotechnological processes. Applied Microbiology and Biotechnology, 97(3), 929-937.

353458 Common Issues that Can Limit Steam Trap Performance

Abstract:

The proper selection of steam trap type/technology for a given application is just one of the keys to achieve operational success. In this session we will discuss a variety of factors such as piping condition, piping arrangement, and steam trap installation/orientation/elevation that can also limit steam trap performance. Common application problems like steam locking and group trapping will also be discussed.

353457 Overview of Steam Trap Types and Their Operation

Abstract:

Steam traps are the quietly functioning devices underneath the pipe racks of our manufacturing facilities, giving the occasional breath of condensate/steam to the atmosphere. However they are not all created equally or function the same way. This presentation will give an overview of various types of steam traps, how they operate, and review steam trap selection for common applications.

A Practical Approach to Feedwater Quality Requirements in Boiler Systems

Paper No. 352255
Deborah Bloom, Principal Consultant, Nalco, an Ecolab company, Naperville, IL
David Puchan, Senior Industry Technical Consultant, Nalco Champion, an Ecolab company, Sugar Land, TX

ABSTRACT

The steam boiler system is a critical component of most manufacturing processes. Steam production reliability is often a key component in assuring product quality and overall production efficiency. Mechanical, operational, and chemical requirements of the steam will usually determine what boiler design and feedwater or makeup water quality is needed. If you define the steam requirements, you typically define the rest of the system.

This paper will discuss typical purity and quality requirements for some common steam users and show how that translates to boiler design, feedwater and makeup purity requirements, and condensate return. We will also discuss how typical treatment program choice, chemical control limits, and boiler blowdown rates are affected by steam purity requirements.

INTRODUCTION

Mechanical, operational, and chemical requirements for the steam will typically determine what boiler design and feedwater/makeup water quality are needed by the boiler system. Key questions to ask include, but are not limited to, the following:

- Are there steam purity or quality specifications that need to be met?
- How is the steam used? For example, does the system include turbines, superheaters, sterilizers, humidifiers, direct injection processes, or indirect heat transfer?
- What are the required temperature and pressure characteristics of the steam?
- How much steam will be needed and is the need continuous or intermittent?
- Are there government regulations or other regulations that need to be met?

Answers to these key questions will help define steam requirements. If you define the steam requirements, you will typically define the rest of the system - with one exception. Waste heat boilers may have more stringent feedwater quality requirements than are warranted by the pressure/temperature of the steam they produce.

Treatment program choice, chemical control limits, and boiler blowdown rates are also affected by steam purity requirements.

STEAM PURITY VS. STEAM QUALITY

Steam purity is defined as the amount of non-water components contained in the steam. These components can be dissolved in the steam, in water droplets entrained in the steam, or carried as discreet solid particles with the steam. Steam purity in industrial systems is normally measured in terms of parts per billion (ppb) sodium ion or silica. Gaseous components such as ammonia, amine, and carbon dioxide are not normally included in this expression, but there can also be limitations on these in the steam.

Steam quality is defined as the amount of moisture in the steam, with 100% quality equal to 100% dry steam, i.e. no moisture. Moisture from the boiler will contain boiler dissolved solids which can be harmful to the steam user. In addition, the moisture itself can be erosive to turbine inlet nozzles and blades, pipe elbows, and valves. For industrial boilers that typically operate at less than 2000 psig drum pressure, the primary route for inorganic contaminants, other than silica, getting into the steam is various forms of mechanical carryover - droplets of boiler water entrained in the saturated steam, including foaming. Therefore, the boiler water total dissolved solids (TDS) are directly proportional to the saturated steam TDS. Only silica has any appreciable vaporous solubility in saturated steam at these operating pressures, and its specific concentration in the boiler water must be appropriately controlled to limit its concentration in the saturated steam when drum operating pressures are above 600 psig. Typical desired steam silica concentrations are less than 10-20 ppb (as SiO_2).

Superheated steam is generally required for turbines and turbine drives to prevent damage by moisture and steam entrained contaminants. Superheated steam has been heated to a temperature significantly greater than the saturation temperature of the operating pressure and therefore by definition has no moisture. Appreciable dissolved solids entering the superheater in the saturated steam can cause a buildup of deposits in the superheater elements, often leading to overheating and tube failures. As a result, a low total dissolved solids content is desirable for steam going to an industrial superheater. Superheaters vary considerably in design but can nominally tolerate 100-200 ppb of total dissolved solids in the steam. This is equivalent to about 33-67 ppb of sodium (as Na) in the steam (1).

Superheaters, turbines and turbine drives normally have the highest steam purity restrictions among common steam users. Superheaters can tolerate higher levels of impurities in the steam prior to experiencing deposition problems. Acceptable steam purity for an industrial, condensing steam turbine is typically less than 10 ppb sodium (as Na) in the steam and less than 20 ppb sodium (as Na) in the steam for an industrial backpressure turbine. Some turbine designs are more sensitive to deposition than others and new turbine installations may have even more restrictive steam purity guarantee limits. Table 1 shows the chemistry limits from several manufacturers for condensing steam turbines operating under normal conditions. High purity feedwater/makeup is typically required to meet these steam restrictions, as well as the purity requirements for use as attemperation water prior to a turbine. High purity feedwater/makeup can usually be achieved by polished or mixed bed demineralization, polished reverse osmosis, or a high percentage of polished condensate (greater than 90%).

TEMPERATURE AND PRESSURE REQUIREMENTS AFFECT BOILER DESIGN

Temperature and pressure requirements for steam users affect boiler design and feedwater quality as well. Higher pressure steam users (greater than 600 psig) require higher pressure boilers, while turbines and turbine drives usually require superheated steam and must therefore be provided by a boiler with a superheater. Integral superheaters are typically limited to water tube boilers, heat recovery steam generators, and waste heat boilers.

Knowledge of the boiler design – firetube, water tube, heat recovery steam generator, or waste heat boiler, as well as direct fired vs. heat recovery – will help in understanding potential problems such as scaling and corrosion. By knowing the design of the boiler, one can anticipate where these problem areas are located and what may have caused them, such as:

- Water circulation issues
- Boiler blowdown location and limitations
- High heat flux areas

Table 1. Comparison of Manufacturers' Chemistry Limits for Condensing Steam Turbines Operating under Normal Conditions (2)

Steam (HP, IP, LP)	Alstom	GE[a]		Mitsubishi	Fuji	Siemens
		Reheat	No reheat			
Specific conductivity at 25°C, μS/cm	3-11	NS	NS	NS	NS	NS
Cation conductivity at 25°C, μS/cm	< 0.2	≤ 0.15	≤ 0.25	< 0.3	<0.3	< 0.2
pH at 25°C	9.0-9.6 [b]	NS	NS	NS	9.0-9.3	NS
Silica as SiO_2, ppb (μg/kg)	<20	≤ 10	≤ 20	< 10	<10	<10
Total iron as Fe, ppb (μg/kg)	<20	NS	NS	< 20	<20	<20
Total copper as Cu, ppb (μg/kg)	<3	NS	NS	< 2	<3	<2
Sodium as Na, ppb (μg/kg)	<10	≤ 3	≤ 6	< 5	<10 [c]	<5
Chloride as Cl, ppb (μg/kg)	NS	≤ 3	≤ 6	< 5	NS	NS
Sulfate as SO_4, ppb (μg/kg)	NS	≤ 3	≤ 6	NS	NS	NS
Oxygen as O_2, ppb (μg/kg)	NS	NS	NS	< 10	<10	NS
TOC as C, ppb (μg/kg)	NS	≤ 100	≤100	NS		NS

[a] Using phosphate treatment (greater than 2.6 Na:PO_4 molar ratio) or AVT
[b] pH 9.0 – 9.3 with copper-alloy condensers
[c] Na + K
NS is not specified

Temperature and pressure requirements also affect the boiler chemistry, which in turn affect feedwater and makeup quality.

BOILER SYSTEM CHEMISTRY AND CONTROL LIMITS

There are many water-related challenges encountered when operating a boiler. Two of the most common problems are scale or deposition and corrosion.

Scale is the result of dissolved minerals, such as calcium and magnesium, exceeding their solubilities and forming deposits (3). Boiler heat transfer efficiency can be negatively impacted by scale. In addition porous deposits formed by metal oxides (typically iron and copper) can also create serious under-deposit corrosion problems. Deposition, when not addressed in a timely fashion or when improperly dealt with, can result in ruptured boiler tubes from overheating or other boiler damage. This can cause boiler shutdown and lost or greatly reduced production.

Most boiler systems follow some sort of published guidelines to establish a suitable control strategy for their feedwater and boiler water chemistry. Equipment manufacturers, water treatment suppliers, American Boiler Manufacturers' Association (ABMA), and American Society of Mechanical Engineers (ASME) are among the more common sources for such guidelines. Table 2 shows an example of the recommended control limits provided by ASME (4). Note that chemistry limits generally become more restrictive for both feedwater and boiler water contaminants as system pressure increases. The table assumes that only demineralized makeup is used above 1000 psig. Waste heat boilers of unusual design and heat recovery steam generators (HRSGs) will frequently require more stringent control limits.

Table 2. ASME Suggested Water Chemistry Limits for Industrial Water Tube, High Duty, Primary Fuel Fired, Drum Type Boilers
Makeup water percentage: Up to 100% of feedwater
Conditions: Includes superheater, turbine drives, or process restriction on steam purity
Saturated steam purity target: See tabulated values below

Drum Operating Pressure[a,k]	psig 0-300 (MPa) (0-2.07)	301-450 (2.08-3.10)	451-600 (3.11-4.14)	601-750 (4.15-5.171)	751-900 (5.18-6.211)	901-1000 (6.22-6.89)	1001-1500 (6.90-10.34)	1501-2000 (10.35-13.79)
Feedwater[g]								
Dissolved oxygen ppm (mg/L) O_2 measured before chemical oxygen scavenger addition[h]	< 0.007	< 0.007	<0.007	<0.007	<0.007	<0.007	<0.007	<0.007
Total iron ppm (mg/L) Fe	≤0.1	≤0.05	≤0.03	≤0.025	≤0.02	≤0.02	≤0.01	≤0.01
Total copper ppm (mg/L) Cu	≤0.05	≤0.025	≤0.02	≤0.02	≤0.015	≤0.01	≤0.01	≤0.01
Total hardness ppm (mg/L)*	≤0.3	≤0.3	≤0.2	≤0.2	≤0.1	≤0.05	ND	ND
pH @ 25°C	8.3-10.0	8.3-10.0	8.3-10.0	8.3-10.0	8.3-10.0	8.8-9.6	8.8-9.6	8.8-9.6
Chemicals for preboiler system protection	NS	NS	NS	NS	NS	VAM	VAM	VAM
Nonvolatile TOC ppm (mg/L) C[f]	<1	<1	<0.5	<0.5	<0.5	<0.2	<0.2	<0.2
Oily matter ppm (mg/L)	<1	<1	<0.5	<0.5	<0.5	<0.2	<0.2	<0.2
Boiler Water								
Silica ppm (mg/L) SiO_2	≤150	≤90	≤40	≤30	≤20	≤8	≤2	≤1
Total alkalinity ppm (mg/L)*	<700[c]	<600[c]	<500[c]	<200[c]	<150[c]	<100[c]	NS[d]	NS[d]
Free OH alkalinity ppm (mg/L)*[b]	NS	NS	NS	NS	NS	NS	ND[d]	ND[d]
Specific conductance[L] µmhos/cm (µS/cm) 25°C without neutralization	5400-1100[e]	4600-900[e]	3800-800[e]	1500-300[e]	1200-250[e]	1000-200[e]	≤150	≤80
Total Dissolved Solids in Steam[i]								
TDS (maximum) ppm (mg/L)	1.0-0.2	1.0-0.2	1.0-0.2	0.5-0.1	0.5-0.1	0.5-0.1	0.1	0.1

*as $CaCO_3$
NS = Not Specified
ND = Not Detectable
VAM = Use only volatile alkaline materials up stream of attemperation water source

Maximum boiler water silica concentrations at pressures above 600 psig are selected so that the volatile carryover will not exceed 20 ppb silica (as SiO_2) in the steam according to well established silica volatility data (5). At lower operating pressures ranges, the boiler water silica concentrations are selected to avoid internal silicate deposits.

Recommended boiler water alkalinity and conductivity values also decrease as boiler pressure increases. Recommended boiler water total alkalinity values were set based on best correlation with boiler water pH, corrosion inhibition, and carryover tendency. Hydroxide alkalinity values are usually set for a specific system based on boiler design, pressure and treatment program.

Boiler water conductivity is the typical trigger for a blowdown control event, although silica and alkalinity should be considered reasons to blow down as well.

Feedwater total hardness, iron, and copper are specified to minimize potential boiler deposits. The decrease in recommended concentration relative to increased boiler pressure reflects the corresponding potential impact of the resultant deposits on heat transfer at higher pressures. Feedwater quality, and therefore makeup quality, dictates what the boiler blowdown rate must be to maintain the required boiler water chemistry limits. Higher blowdown rates result in increased water and energy loses, but are necessary to prevent deposition and corrosion. Some of the more common ways to prevent deposition from occurring in boilers include ion exchange (softening or demineralization) and reverse osmosis. Residual dissolved minerals are prevented from forming deposits by the addition of chemical treatment.

Table 2 also provides a connection between boiler chemistry and expected total dissolved solids in the steam. As boiler pressure increases, the specific conductivity required in the boiler water to achieve acceptable steam purity decreases. What's acceptable depends on how the steam will be used, but is typically 0.1 to 0.2 ppm TDS for superheater steam. Even this, however, is usually considered too high for most modern industrial turbines. ASME is the process of updating Table 2.

WASTE HEAT BOILERS

The design for a waste heat boiler generally depends on the process operating parameters. It is generally a one-of-a-kind unit, designed to accommodate and facilitate a plant's process flows. In flue gas waste heat recovery, where the process streams are low pressure, dirty, and require supplemental firing, watertube designs predominate. In process cooling applications, where the processes are corrosive and at relatively high pressure, firetube or shell-and-tube waste heat boiler designs are common.

Waste heat boiler design can cause limitations in flow, higher heat flux areas, limited or ineffective blowdown from areas where solids collect, poor circulation, and inadequate moisture separation from the steam. These problems can then lead to concentration mechanism corrosion, such as caustic gouging and phosphate wastage, creep rupture from overheating due to scale/deposits, and poor steam purity. System chemistry limits may need to be even lower than those shown in Table 2. For example, Table 3 shows recommendations for an Ethylene plant TLX/TLE feedwater chemistry; boiler water chemistry is specific to the treatment program that's in use.

Table 3. Boiler Feedwater Recommendations for Ethylene Plant TLX/TLE Operations (6)

Contaminant	Recommended Control Limit
Cation Conductivity	< 0.5 μmhos
Silica (as SiO_2)	< 20 ppb
Sodium (as Na)	< 10 ppb
Chlorides (as Cl)	< 6 ppb
Sulfates (as SO_4)	< 6 ppb
Calcium (as Ca)	Non-detectable
Magnesium (as Mg)	Non-detectable
Iron (as Fe)	< 10 ppb
Copper (as Cu)	< 2 ppb
TOC (as C)	< 100 ppb
Oxygen (as O_2)	< 2 ppb
pH	9.0 – 9.3 systems with yellow metals
pH	9.2 – 9.6 systems without yellow metals
Oil & Grease	<0.2 ppm

A FEW MORE WORDS ABOUT WATER QUALITY

By now it should be apparent that steam purity and temperature/pressure requirements determine boiler design, pressure, and chemistry. This in turn affects feedwater quality requirements, which affect makeup and condensate quality requirements. So, as mentioned initially, if you define the steam requirements, you will typically define the rest of the system.

Improving makeup water quality reduces the amount of feedwater contaminants entering the boiler system. There are a variety of pretreatment technologies available for industrial plants that can significantly reduce the hardness present in makeup:

- Hot/cold lime softening
- Sodium zeolite softeners
- Reverse Osmosis
- Demineralization

Table 4. Reverse Osmosis vs. Competing Technologies

Technology	Driver	Competing technology preferred if:	RO preferred if:
Hot-lime softener, filtration, sodium softening	Water quality	Need to remove hardness, alkalinity and silica only	Need to remove NaCl, TDS, alkalinity and silica to low concentrations
Sodium softeners	Water quality	Need to remove hardness only	Need to remove NaCl, TDS, alkalinity and silica to low concentrations
Dealkalizers	Water quality	Need to remove alkalinity only	Need to remove NaCl, TDS, alkalinity and silica to low concentrations
Demineralization	Cost	Cost of acid and caustic is low	Waste treatment cost high - To avoid handling of acid & caustic

Each provides a different effluent quality. Table 4 provides a comparison of these methods relative to reverse osmosis, including the drivers for choosing one technology over another.

CHEMICAL TREATMENT

The goal of chemical treatment is to optimize system chemistry and control in order to save water and energy. Left untreated, water in a boiler system will scale and corrode - even with the perfect pretreatment and chemistry control limits. Effective chemical treatment assures that the boiler system operates at maximum efficiency and reliability.

Scale and Deposition

Scale and deposit control is focused on preventing the precipitation of residual dissolved minerals remaining in the water after pretreatment, such as calcium and magnesium, from forming deposits in the boiler. Typical chemical treatment consists of using either a precipitating type program, such as phosphate, or a solubilizing type program, such as all-polymer. Both program types are fed relative to the amount of hardness still present in the feedwater. Phosphate programs can be used with very high feedwater hardness, but since the amount of precipitate formed in the boiler is in direct proportion to the hardness concentration, they also typically require higher blowdown rates to purge the sludge and minimize boiler scale formation or periodic mechanical or chemical cleaning to remove accumulated deposits.

There are many forms of all-polymer treatment in the market today. All can be effective at scale prevention when fed at the proper dosage relative to hardness. Newer versions of all-polymer programs tend to tolerate hardness upsets better than the original acrylate all-polymer programs. Boiler blowdown rates will be lower on all-polymer than with phosphate treatment, saving energy and water; and since all-polymer is a solubilizing type treatment, internal boiler surfaces will be cleaner which may result in additional fuel savings from improved heat transfer.

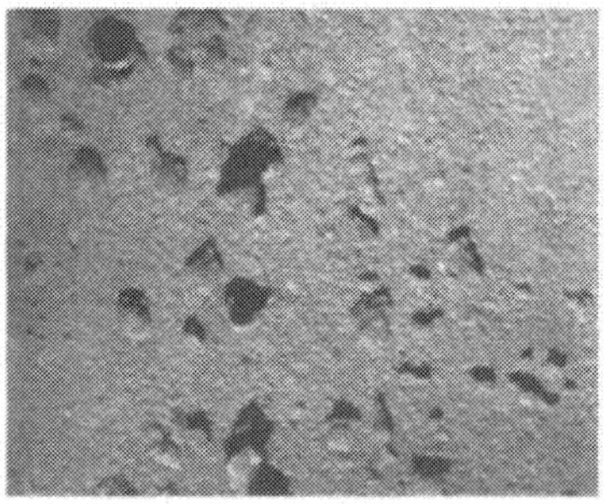

Figure 1. Oxygen attack, characterized by its classic pitting appearance

Corrosion

Corrosion is a serious concern in the pre-boiler and boiler system for two reasons. First, it shortens the asset life of the boiler and pre-boiler system. Second, the corrosion by-products (iron and copper oxides) resulting from unchecked corrosion can deposit in high heat transfer areas of the boiler as deposition. Oxygen corrosion is a specific and particularly devastating type of corrosion, Figure 1. It is a common problem that adversely affects the life of the key steam system components. Pitting attack from oxygen corrosion can cause premature metal failure.

Iron oxide deposits, just as the earlier described calcium and magnesium scales, reduce heat transfer efficiency and can result in under deposit corrosion. In uncontrolled scenarios, blown tubes (or other extensive boiler damage), boiler shutdowns and production losses can result.

Historically, the best practice for preventing oxygen corrosion in industrial boiler and pre-boiler systems has been mechanical deaeration for primary oxygen removal, followed by chemical oxygen scavengers to remove remaining trace quantities. Sulfite is typically the scavenger of choice in most low pressure systems, especially at pressures less than 250 psig. Sulfite adds solids to the system and is traditionally dosed to a prescribed residual in the boiler water. Non-sulfite scavengers also exist. In addition to scavenging oxygen, the non-sulfite programs will typically passivate metal surfaces further reducing corrosion and iron pickup, and do not contribute to boiler solids since most are volatile. As a result, non-sulfite scavengers may also reduce boiler blowdown requirements.

MONITORING AND CONTROL

For decades, the traditional practice employed to control many boiler water scale and oxygen control chemistries has been by sampling the blowdown. One primary reason for this was that the boiler concentrated the treatment chemicals sufficiently to enable them to be measured readily. Although modernization and improvement in test methods have enabled feedwater testing (e.g., cycling up is not required to run many of the tests), many operations hold onto the old practices of the 1930s through 1970s and test only in the blowdown. The major drawbacks resulting from this approach are related to the lag time between changes in feedwater conditions and those changes showing up in the blowdown.

In a blowdown based control scenario, the operator would sample the blowdown, perform an array of chemical tests, and adjust the feed rate of the scale inhibitor or the oxygen scavenger. This constant "add chemical - sample – test – adjust" control loop was the standard. This meant that if a test was missed, an opportunity for adjustment was also missed.

In addition to the challenge of the above testing loop, the physical capacity of the boiler often meant that a negative change in feedwater conditions might go completely undetected or not have a notable impact on the boiler blowdown composition for hours or even days. Feedwater conditions that were potentially adverse to system integrity (e.g., an oxygen excursion, a reduction in feedwater pH, or a badly contaminated condensate return stream) would only be seen after they had become history, if they were detected at all, because only the boiler water was being tested. What was missing was a proactive approach to protect the boiler as opposed to a reactive approach that adjusts after damage has occurred.

Ideally, one would want to measure and control boiler chemistry in the feedwater, because that would enable one to make adjustments to changing conditions in real-time, rather than chasing events or missing them entirely and suffering the outcome. The Nalco boiler feedwater treatment automation system allows us to do this. It is a combination of two different technologies. First is the patented solid state fluorometer, which is used to control boiler internal treatment (7, 8, 9), Figures 2 and 3. Nalco has used inert fluorescent tracers for system diagnostics and dosage control of boiler system treatment chemicals for many years. The fluorometer allows us to measure these materials at the part per billion level of detection.

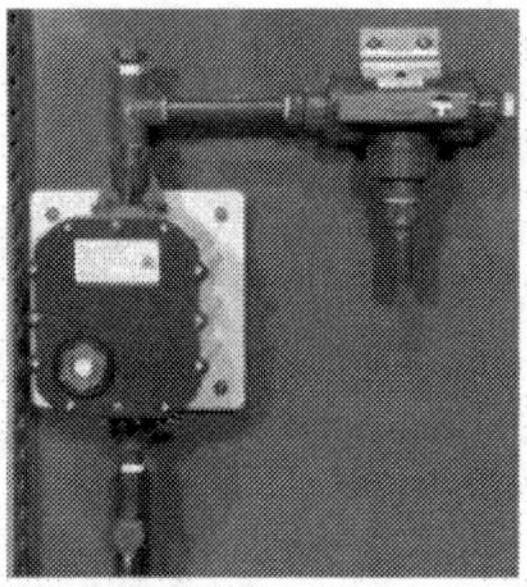

Figure 2. LED fluorometer provides continuous monitoring

Figure 3. Pen fluorometer used with grabs samples

This unique tracer technology makes it possible to determine the mass balance of a boiler system with greater accuracy, Figure 4. It also makes it possible to accurately and automatically control and monitor system chemistry and operating conditions, respond to boiler system changes, and maintain optimum treatment levels of highly effective scale and deposit inhibitors to keep boilers clean. As an example, this technology has been used in conjunction with soluble iron (Fe^{+2}) monitoring in a boiler system to quantify flow-assisted corrosion rates.

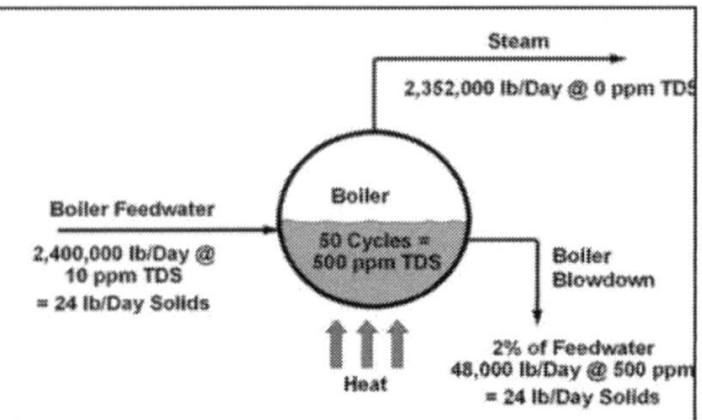

Figure 4. Boiler cycles of concentration

The second automation technology is the patented Nalco Corrosion Stress Monitor™, Figure 5. The Nalco Corrosion Stress Monitor instrument works by measuring the oxidation reduction potential at the temperature and pressure of the boiler feedwater. The At Temperature Oxidation Reduction Potential, or At Temperature ORP, detects changes in the corrosivity of the feedwater and can be used to control the oxygen scavenger feed. The AT ORP has a design rating up to 500°F and 2,800 psig. A uniquely designed controller controls both the fluorometer and AT ORP. Monitoring, On/Off Control, and PID (Proportional-Integral-Derivative) Control are all possible with this controller. These technologies are capable of communicating via web or DCS for monitoring and alarming existing conditions.

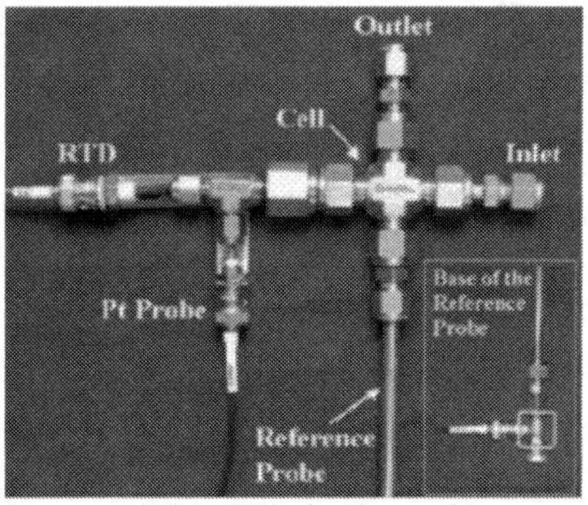

Figure 5. AT ORP Probe Assembly

OPPORTUNITIES FOR ENERGY SAVINGS

Controlling boiler deposition can have real energy savings benefits. There are many different sources of boiler deposition – two of the most common are feedwater hardness and dissolved iron entering the boiler feedwater.

The new control technologies can have profound impact on minimizing the potential for deposition, as well as mitigating its impact. Since the deposit inhibitor is measured in the boiler feedwater and adjusted in real time, regardless of changes in steam load, underdosing and overdosing can be almost completely eliminated. Boiler feedwater corrosivity is likewise monitored and controlled in real-time, so feedwater corrosion excursions, and the resultant deposit-forming dissolved iron can also be greatly reduced.

Other opportunities for energy savings spring from these technologies. It is important to remember that sulfite, one of the most common oxygen scavengers, contributes dissolved solids to the boiler feedwater. These solids increase boiler water conductivity. In a scenario where excessive levels of sulfite oxygen scavenger are being fed, improved control of oxygen scavenger feed can result in reduced boiler feedwater dissolved solids, higher cycles of concentration in the boiler, and reduced blowdown, all with associated water and fuel savings.

This technology also has the capability of monitoring and controlling feedwater pH, so energy and water savings from reduced feedwater dissolved solids, higher boiler cycles, and reduced blowdown can also be obtained by facilities that use caustic (sodium hydroxide) for controlling feedwater and boiler water alkalinity.

CASE HISTORY - A WESTERN U.S. REFINERY OPTIMIZES PERFORMANCE with 3D TRASAR™ BOILER AUTOMATION

Situation

A western U.S. refinery was experiencing deposition in their 600 psig boiler superheater tubes. This deposition forced operations to conduct frequent water washes, limit steam production, and operate the boiler at only 5 cycles resulting in an increase in energy, water, and maintenance costs. The superheater tube deposition led to very high skin temperatures and ultimately, tube failure.

Refinery management asked Nalco to help them understand their boiler system better. Nalco performed an Engineering Approach Analysis to determine system gaps and opportunities for improvement.

A Nalco Engineering Approach Analysis

Boiler Feed Water Quality - The 600 psig boiler system receives a combination of deaerated condensate return and reverse osmosis water that passed through a Graver Hot Process Softener used only for deaeration. Figure 6 shows a schematic of the boiler feedwater system.

Figure 6. Schematic of Boiler Feedwater System

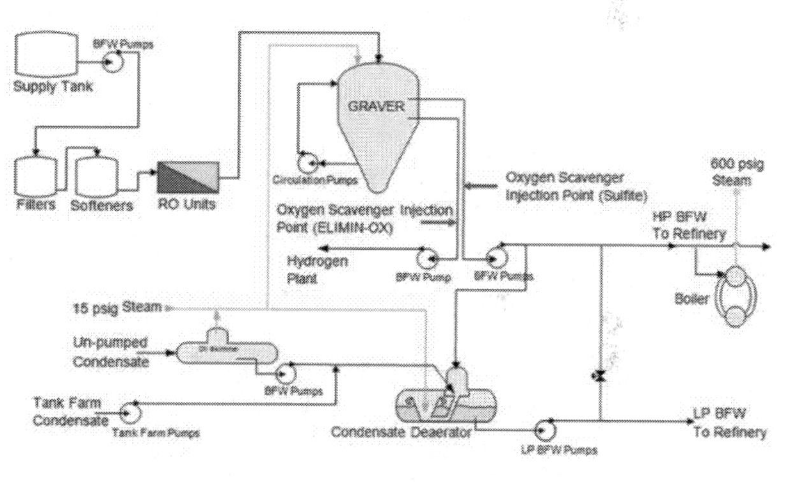

The feedwater from the Graver unit splits into two streams -one stream supplies feedwater to the refinery while the other stream supplies feedwater to a hydrogen plant. At the time of the initial audit, two different oxygen scavengers were fed into each of the boiler feedwater lines up-stream of the boiler feed water pumps. Sulfite was added to the refinery's feedwater line and erythorbate was injected into the feedwater line supplying the hydrogen plant. Sulfite cannot be used in hydrogen plants due to the potential for the product to poison the process unit's catalyst.

Figure 7. Manual Oxygen Testing of the Feedwater

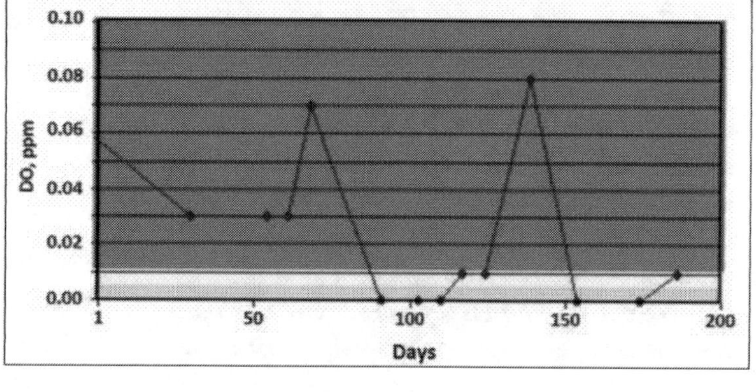

We suspected that the Graver was not mechanically removing the appropriate amount of dissolved oxygen. Manual testing confirmed this suspicion, Figure 7. Dissolved oxygen concentration varied, but at times was as high as 0.08 ppm or 80 ppb, which is well above industry best practices.

Plant operations fed tremendous amounts of sulfite to try to scavenge the high oxygen concentrations. Since the sulfite product is acidic, additional caustic was also required to help maintain feedwater pH. As more and more sulfite and caustic were added to the feedwater, the boiler water began to carryover. The solids in the boiler water deposited on the superheater tubes, increasing the outside metal tube temperature, and resulting in overheating and tube failure, Figure 8.

Figure 8. Superheater Tube Temperature

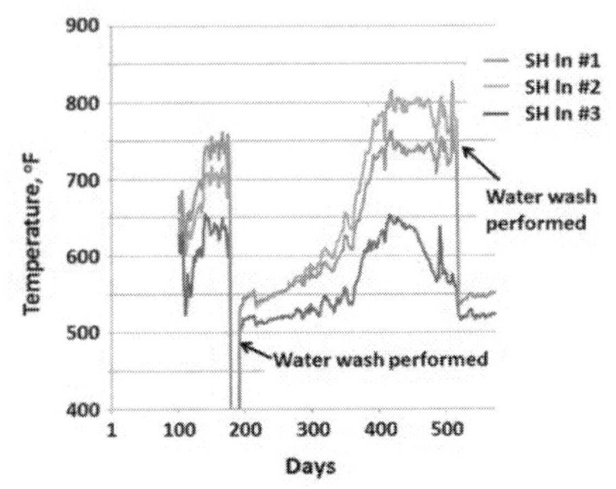

Recommendations

After careful review of the customer's system, chemical treatment program, and daily operations via the Engineering Approach Analysis, Nalco made several recommendations:

- Inspect and evaluate the boiler's internal steam separation equipment.
- Consider replacing the Graver with a traditional deaerator capable of mechanically reducing dissolved oxygen levels to less than 10 ppb.
- Evaluate whether a non-sulfite oxygen scavenger could be used.
    - Switching to a non-sulfite scavenger would reduce solids contribution to the feedwater and the potential for boiler carryover, thereby reducing the potential for deposition in the superheater tubes.
    - Switching scavengers would also potentially allow higher cycles of concentration in the boiler, reducing the energy and water usage.
    - Using an oxygen scavenger compatible with the hydrogen plant's requirements would allow the plant to use a single oxygen scavenger, which could then be injected into the Graver unit allowing longer reaction time.
- Utilize Nalco 3D TRASAR Boiler Automation to continuously monitor and control the chemical treatment and other critical steam system parameters based on system stress and demand.

Results

The decision was made to implement the Nalco recommended plan using a gradual approach. Soon after several of these actions were taken, the refinery experienced dramatic improvements.

Boiler Feed Water Quality - The 3D TRASAR Boiler technology confirmed that very high concentrations of dissolved oxygen were present in the feedwater, Figure 9.

Figure 9. Feedwater Dissolved Oxygen vs. Corrosion Potential

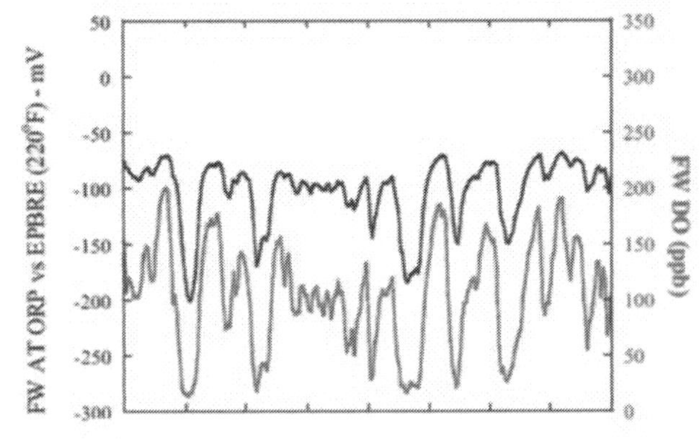

The refinery agreed to replace the Graver unit with a traditional deaerator as a long term solution, however in meantime, they needed a plan to reduce corrosion potential.

Erythorbate and carbohydrazide scavengers were tested as a replacement the existing sulfite application. Both met Hydrogen plant requirements, so the scavenger injection point was moved inside the Graver unit to improve mixing and residence time. Both scavengers performed well at removing dissolved oxygen, but carbohydrazide was the preferred chemistry because it had less effect on feedwater pH at high dosages and does not contribute dissolved solids like sulfite.

Once the best oxygen scavenger and injection point were in place, the next step was to put scavenger feed into automatic control mode via the Nalco Corrosion Stress Monitor. Figures 10 and 11 show time periods from both monitoring and control mode. Although oxygen concentrations and corrosion potential still spikes occasionally due to a sudden increase in dissolved oxygen, the system is quickly brought back into control range.

The AT ORP set point was -350 mv, reflecting system conditions. Once in control mode, ***control of corrosion potential improved from 50% on target to 91%.***

The real proof that 3D TRASAR Automation was improving boiler system reliability was the in superheater tube temperature data, Figure 12. Remember that boiler water was carrying over into the steam and depositing in the superheater tubes. The refinery had to shut down the boiler to water wash the superheater tubes to prevent failure from overheating. However with Nalco's action plan in place, the superheater tube temperatures have remained on target.

Summary

The refinery's utility area did not have a traditional deaerator, but rather a poorly operating Graver unit that resulted in excessively high feedwater dissolved oxygen concentrations. The 3D TRASAR Boiler Automation with the Nalco Corrosion Stress Monitor detected and responded immediately to changes in feedwater oxygen and corrosion stress conditions caused by the poorly operating deaerator. Manual feed of an oxygen scavenger provides poor feedwater corrosion stress control. The use of 3D TRASAR Technology to automatically control the feed of scavenger significantly increased percent on target control and reduced feedwater corrosion. Feeding carbohydrazide, a non-sulfite oxygen scavenger, reduced the overall solids loading on the boiler, eliminated boiler water carryover, and enabled the plant to maintain safe superheater tube temperatures. Because of the reduced solids loading, the refinery was also able to safely

Figure 10. Feedwater Corrosion Control using the Nalco Corrosion Stress Monitor

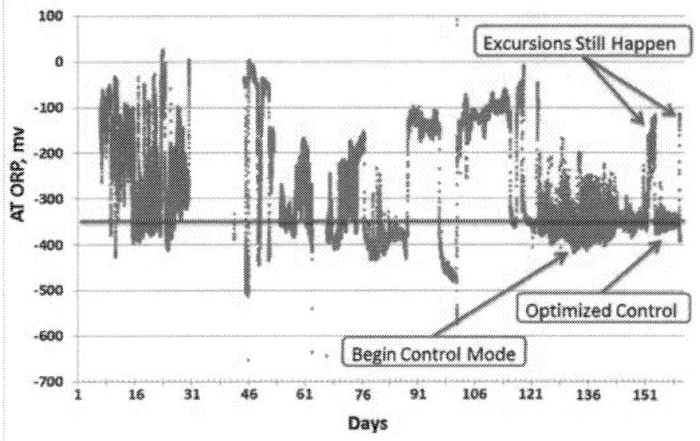

Figure 11. Feedwater Corrosion Control Showing Dissolved Oxygen

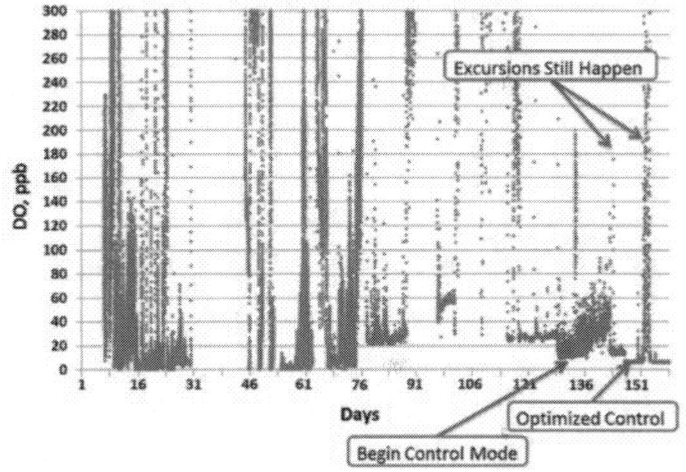

Figure 12. Nalco's Engineered Approach Analysis and 3D TRASAR Automation resulted in stable superheater tube temperatures and improved system reliability

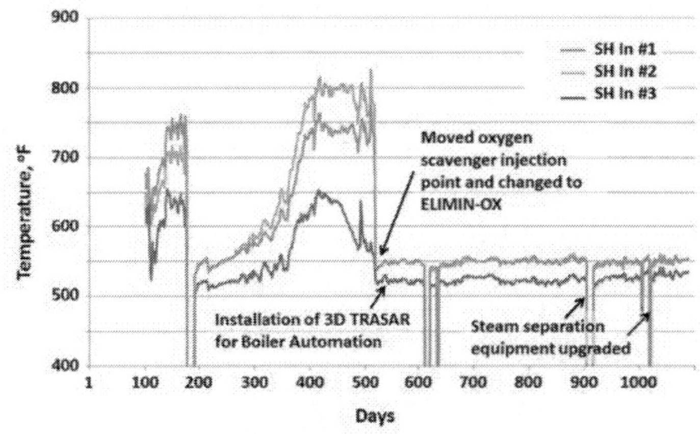

increase the boiler cycles, yielding $211,000 in annual savings.

CONCLUSIONS

Mechanical, operational, and chemical requirements of the steam will usually determine what boiler design and feedwater or makeup water quality is needed. If you define the steam requirements, you typically define the rest of the system. Typical treatment program choice, chemical control limits, and boiler blowdown rates are also affected by steam purity requirements. New control technologies can have profound impact on minimizing the potential for scale, deposition, and corrosion in the boiler system, as well as mitigating their impact. Water and fuel savings from reduced feedwater dissolved solids, higher cycles of concentration in the boiler, and reduced blowdown can be achieved.

REFERENCES:

1. "Industrial Steam Purity: Requirements, Proper Sampling and Practical Considerations", A Banweg, Paper IWC-08-28, Proceedings of the International Water Conference (2008).
2. ASME, "Consensus on Operating Practices for Control of Water and Steam Chemistry in Combined Cycle and Cogeneration Power Plants" (Jan 2012).
3. Nalco Water Handbook, 3rd Edition, 2009, McGraw Hill Publishing
4. ASME, "Consensus on Operating Practices for the Control of Feedwater and Boiler Water Chemistry in Modern Industrial Boilers" CRTD-Vol. 34 (1994).
5. "Selective Silica Carryover in Steam," E. Coulter, E. Pirsh, and E. Wagner, Jr., Transactions of ASME, 78:869-873 (1956).
6. "Boiler Technology to Monitor and Control Corrosion In Ethylene Plant Steam Generation Process", D. Puchan and M. Gerty, AIChE Paper Number 49 G, AIChE Spring National Meeting, San Antonio, TX (2010).
7. United States Patent #6,255,118 "Method for Using an All Solid-State Fluorometer in Industrial Water Applications"
8. United States Patent #5,736,405 "Monitoring Boiler Internal Treatment with Fluorescent-tagged Polymers"
9. "Monitoring and Control in Boilers With All Solid-State Fluorometers", Martin Godfrey, Corrosion / 2002, Paper No. 02012, NACE International Conference, Houston, TX (2002).

For presentation at the AIChE 2014 Spring Meeting; New Orleans, LA; March 31- April 3, 2014

353051 Steam Traps: Audience Survey of Operation and Maintenance

Abstract:

All chemical plants have steam traps. However, steam trap operation and maintenance is not identical from one chemical plant to another chemical plant. Their design and operation is not proprietary since they are standard purchased items at all chemical plants. Therefore, this presentation will be conducted in an "open" fashion; i.e., the audience can and will be encouraged to participate freely in it.

The presenters will moderate an interactive discussion with the audience pertaining to the operation and maintenance of steam traps. This discussion will delve into steam trap management and operation on a company-side basis and, more likely, on a site-basis. The presenters will also lead a steam trap maintenance discussion. In this latter discussion, an attempt will be made to establish a "best practices" approach to steam trap maintenance.

This session, in general, and this presentation, in particular, will provide chemical plant engineers with pertinent information regarding steam traps.

PROSYS™ Maximize Your Plant.

Operator Errors and What Can be Done to Minimize

Global Congress on Process Safety

March 31, 2014

10TH GLOBAL CONGRESS ON PROCESS SAFETY

PROSYS™ Maximize Your Plant™

10TH GLOBAL CONGRESS ON PROCESS SAFETY

Presenter

Tom Nolan

- **Graduated from Ohio University with B.S. in Chemical Engineering**
- **24 years experience in chemical process industry in a variety of roles**
- **MAIC and DFSS Six Sigma Black Belt**

PROSYS Maximize Your Plant™

10th GLOBAL CONGRESS on PROCESS SAFETY

The Cost of Errors

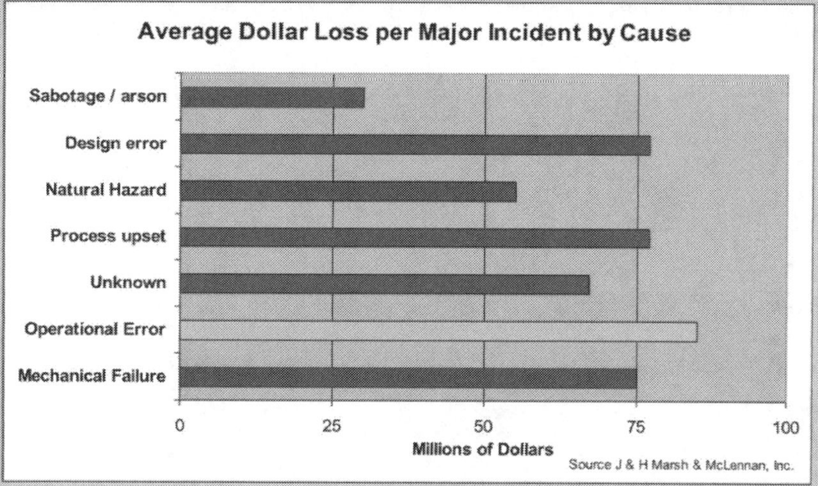

PROSYS Maximize Your Plant™

10th GLOBAL CONGRESS on PROCESS SAFETY

The Cost of Operator Errors

- **ASM estimates total loss due to operator error is $8B per year**
- **Chemical Safety Topical Committee - average of one chemical incident per day - cost of over $2 million per incident to comply with requirements**
- **Errors cause 42% of unscheduled shutdowns**
- **70% of process incidents occur during start-up or shutdown**

PROSYS Maximize Your Plant™

10th GLOBAL CONGRESS ON PROCESS SAFETY

Reasons for Errors

- Lack of Skill
- Lack of Knowledge
- Carelessness
- System Design
  - Operator set up to make errors by inappropriate design – built in errors
  - Errors that are predictable are preventable by better design

PROSYS Maximize Your Plant™

10th GLOBAL CONGRESS ON PROCESS SAFETY

Focus of this Discussion

- System Design
  - **Alarm Management**
  - **Operator Graphics**
  - Operator set up to make errors by inappropriate design – built in errors
  - Errors that are predictable are preventable by better design

PROSYS Maximize Your Plant™

10th GLOBAL CONGRESS ON PROCESS SAFETY

Alarm Floods

In a number of industrial incidents, alarm floods were identified as a significant contributing cause to the incident...

As found by EEMUA in 1999 and CSB

Alarm Flood defined by ISA 18.2 as -
"10 or more annunciated alarms in any 10 minute period per operator"

PROSYS Maximize Your Plant™

10th GLOBAL CONGRESS ON PROCESS SAFETY

Why do Alarm Floods Occur?

One reason is not providing dynamic alarm management

- **Alarms need to indicate abnormal situations that require operator action**
- **Processes do not operate in one state**
- **What is normal vs abnormal changes with state**
- **Alarms are typically configured for run therefore many alarms are triggered upon a Change of State: Run to Shutdown**
- **Many of the alarms are not applicable or actionable for the new state impleading the operators ability to act quickly on what is important**

PROSYS Maximize Your Plant™

10th GLOBAL CONGRESS on PROCESS SAFETY

What is Impacted by Alarm Floods?

- **Product quality**
- **Operability or profitability of the process**
- **Loss of equipment**
- **Loss of containment – environmental releases**
- **Injury and loss of life in plant or community**

PROSYS Maximize Your Plant™

10th GLOBAL CONGRESS on PROCESS SAFETY

What Makes Alarm Floods so Dangerous?

Can be a problem for three reasons:

- **A deluge of alarms can cause critical alarms to be missed**
- **Floods can be a significant distraction when dealing with process upsets**
- **Can be an indicator of larger systemic safety issues**

PROSYS Maximize Your Plant.™

10th GLOBAL CONGRESS ON PROCESS SAFETY

Impacting Alarm Management Design

- Alarm rationalization is <u>not</u> a process to eliminate alarms - its about quality
- Good rationalization will add alarms when appropriate
- Typical (Static) rationalization is only for run mode
- Dynamic rationalization considers all plant modes

PROSYS Maximize Your Plant.™

10th GLOBAL CONGRESS ON PROCESS SAFETY

Dynamic Alarm Management

- Dynamic rationalization does everything a static rationalization does plus asks "when" for each alarm
- Requires dynamic software to make changes based upon operating mode of the plant
- Eliminates redundant alarms and lowers operator loading during transitions
- Only alarms what is abnormal and actionable for the given state

PROSYS Maximize Your Plant™

10th GLOBAL CONGRESS on PROCESS SAFETY

Actual Performance Metrics vs ISA 18.2

Type	Avg Alarm Rate/hr	Low/Hi of Avg Rate/hr	Peak Alarm Rate/hr	Low/Hi of Peak Rate/hr	% Time in Flood	Standing Alarms >24 hrs
Before Rationalization	30	Low=6.2 Hi=61	638	Low=152 Hi=2402	17.2%	9
ISA 18.2 Target Metrics	6	---	<60	---	<1%	<5
After Dynamic Alarm Management	2	Low=0.09 Hi=2.5	25	Low=13 Hi=42	0.25%	4

PROSYS Maximize Your Plant™

10th GLOBAL CONGRESS on PROCESS SAFETY

Results After Dynamic Alarm Management

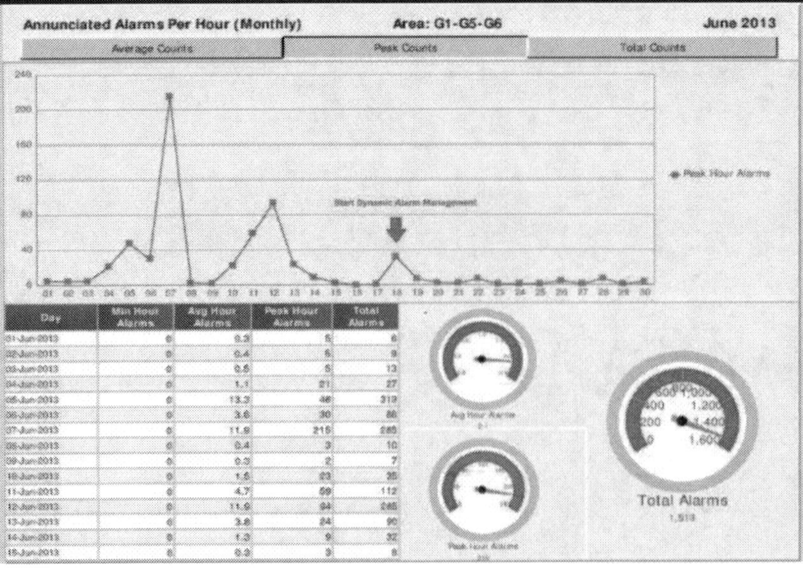

PROSYS Maximize Your Plant™

10th GLOBAL CONGRESS ON PROCESS SAFETY

Operator Graphics – The Risk of Changing Focus

- Distraction of changing focus from process graphic to faceplate window
- Added workload from managing multiple open windows
- Increased probability of errors when changing values for un-intended tag with multiple faceplates open
- All of above become multiplied and more complex when process is transitioning from one state to another

PROSYS Maximize Your Plant™

10th GLOBAL CONGRESS ON PROCESS SAFETY

Faceplate Operation

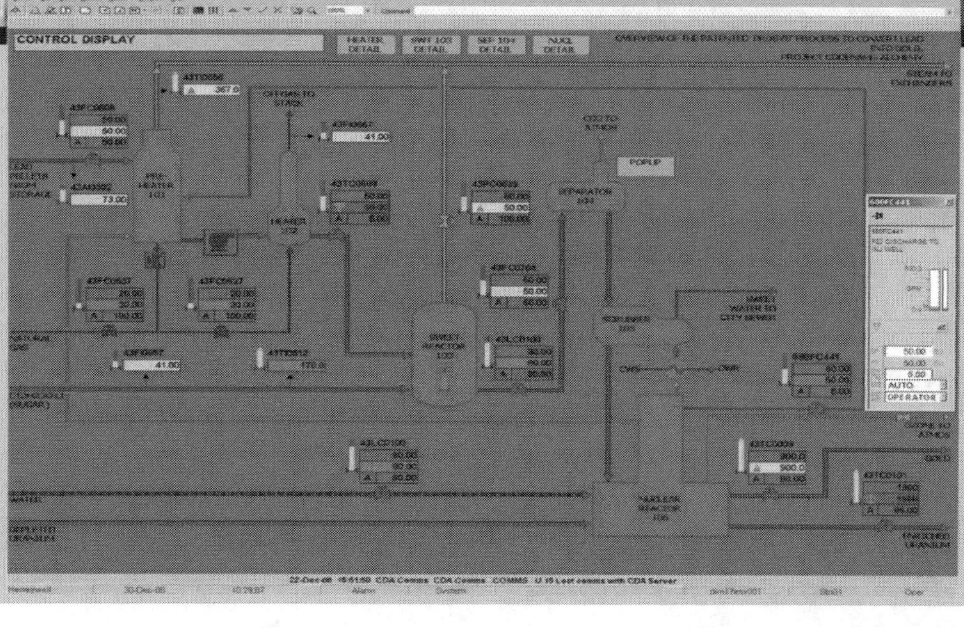

PROSYS Maximize Your Plant™

10th GLOBAL CONGRESS on PROCESS SAFETY

Potential for Errors

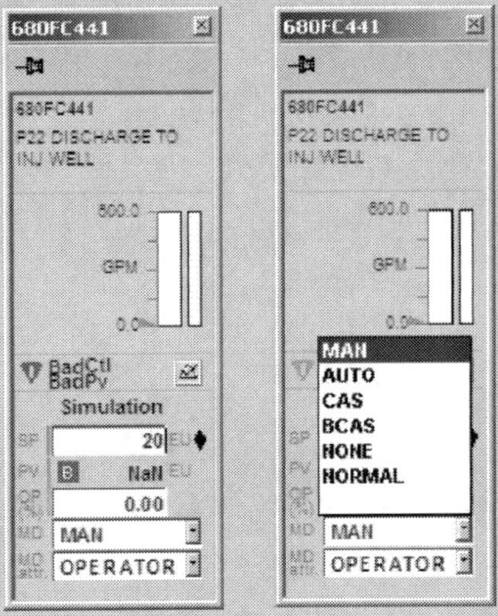

Faceplate Design -can introduce additional errors

- Mode drop down list covers SP, PV and OP values
- Operator may select wrong Mode if SP, PV and OP values are hidden from view
- Mode list offers more options than necessary

PROSYS Maximize Your Plant™

10th GLOBAL CONGRESS on PROCESS SAFETY

Direct Entry Operation

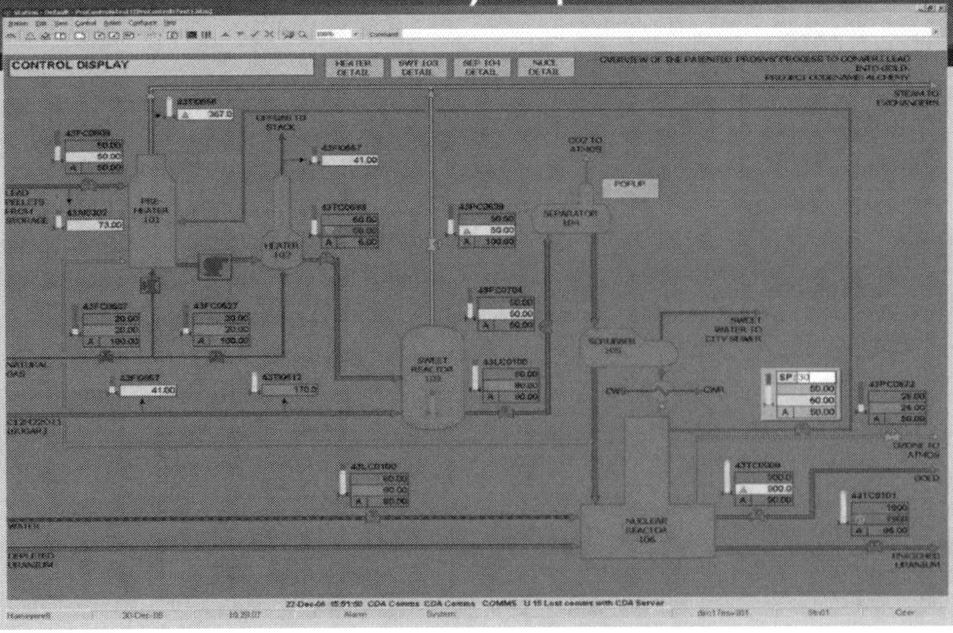

PROSYS Maximize Your Plant™

Direct Entry Fields

Values can be entered by:

- Select point
- Typing value via keypad
- Pressing enter

Key Factors:

- By maintaining focus of points after value change, it improves safety for input errors, multiple clicking functions and is more time efficient
- Easily recognizable operator enterable fields

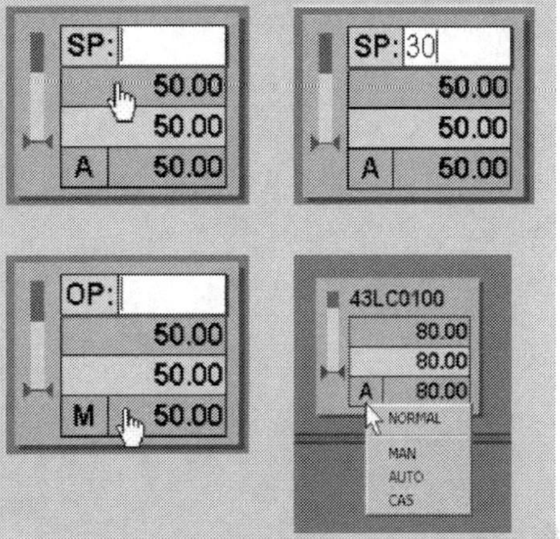

PROSYS Maximize Your Plant™

Evaluation of Methods

Comparison of Faceplate to Direct Entry

- **Keystroke Level Analysis**
- **Operator Loading Analysis**
- **Risk Analysis**

PROSYS Maximize Your Plant™ 10th GLOBAL CONGRESS on PROCESS SAFETY

Keystroke Level Model

Symbol	Time (s)	Description
K	0.28	Keystroke
P	1.1	Point to object
BB	0.2	Click on object
H	0.4	Home hands on keyboard or mouse
M	1.2+	Mental act or routine thinking

PROSYS Maximize Your Plant™ 10th GLOBAL CONGRESS on PROCESS SAFETY

Estimate Execution Time Analysis

- **Evaluates execution time by an operator comparing faceplates to direct entry for SP/OP and Mode Changes**

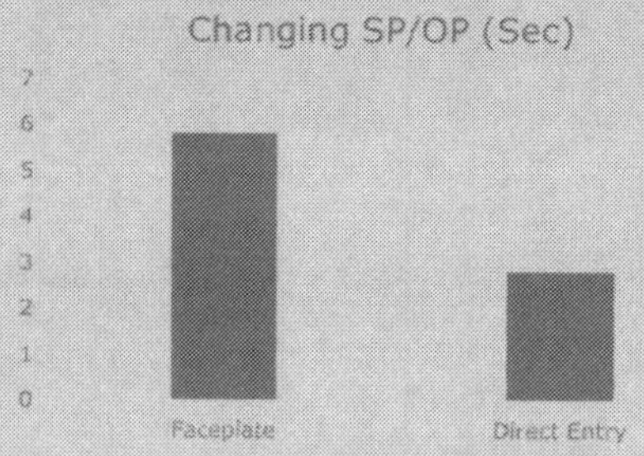

- **50 % Reduction**

PROSYS Maximize Your Plant™

10th GLOBAL CONGRESS on PROCESS SAFETY

Estimated Operator Loading Analysis

- **Evaluates Mental and Physical Load on Operator**
  - Tasks that require very little thought are removed – such as (BB), (H)

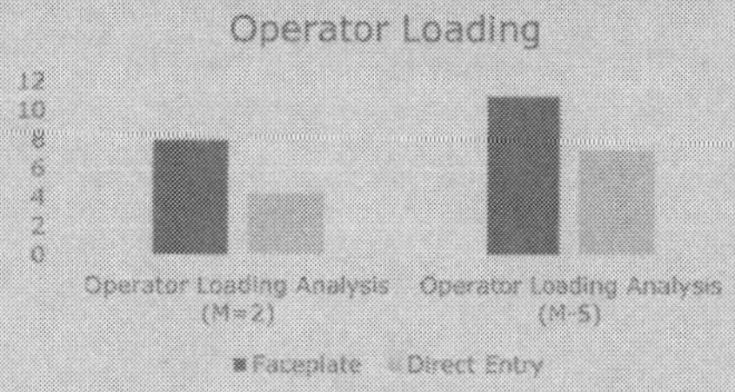

  - Mental and Physical tasks are weighted based on the amount of fatigue they cause – assigned a weight of M=2.0 to 5.0

PROSYS Maximize Your Plant™

10th GLOBAL CONGRESS on PROCESS SAFETY

Risk Analysis

- **Identifies actions where operator entry errors can occur**

PROSYS Maximize Your Plant™

10TH GLOBAL CONGRESS ON PROCESS SAFETY

Risk Analysis

Faceplate operation KLM for changing SP/OP	Potential Error
Move to shape (P)	Select wrong point
Click on shape (BB)	
Move to faceplate (P)	Point in faceplate is previous point
Click on SP/OP field (BB)	Change wrong parameter
Move hands to keyboard (H)	
Type in value and press enter (4K)	Mistype value and press enter
Move hands to mouse (H)	
Move to close faceplate (P)	
Click to close faceplate (BB)	
Potential error count	**4**

PROSYS Maximize Your Plant™

10TH GLOBAL CONGRESS ON PROCESS SAFETY

Risk Analysis

Direct Entry operation KLM for changing SP/OP	Potential Error
Move to SP/OP (P)	Select wrong parameter or point
Click on SP/OP field (BB)	
Move hands to keyboard (H)	
Type in value and press enter (4K)	Mistype value and press enter
Potential error count	**2**

PROSYS Maximize Your Plant.™

10th GLOBAL CONGRESS ON PROCESS SAFETY

Direct Entry Vs Faceplate

Summary of Benefits

- Execution Time SP/OP Changes - 51.5% less
- Execution Time for Mode Changes – 50% less
- Operator Loading – 33 % to 47% less
- Risk of Errors – 50% less

PROSYS Maximize Your Plant.™

10th GLOBAL CONGRESS ON PROCESS SAFETY

Conclusion

- **Many industrial errors are a result of operators using control systems with flawed designs**
  - Poor design and performance by alarm management distracts operators and/or occludes critical alarms
  - Operator graphics with multiple open faceplates can cause a change intended for one controller to be entered into the faceplate of another
- **The two mechanisms listed above are predictable and therefore preventable through better design available today**

PROSYS™ Maximize Your Plant.

Problem Statement

visium KMS
Process Safety Management Solution

- Any organisation can improve by leveraging knowledge and lessons learned.
- Knowledge can be externally or hidden locally, in a global enterprise, where lessons learned are difficult to share beyond their initial impact.
- There is so much data, information and knowledge (opportunities) coming at your company that can make you stronger or make you look like your not paying attention.
- Not paying attention or managing the available knowledge, can sink your ship… nobody wants that!

Expert Insight Delivered 2

OSYS
Part of the Rolls-Royce Group

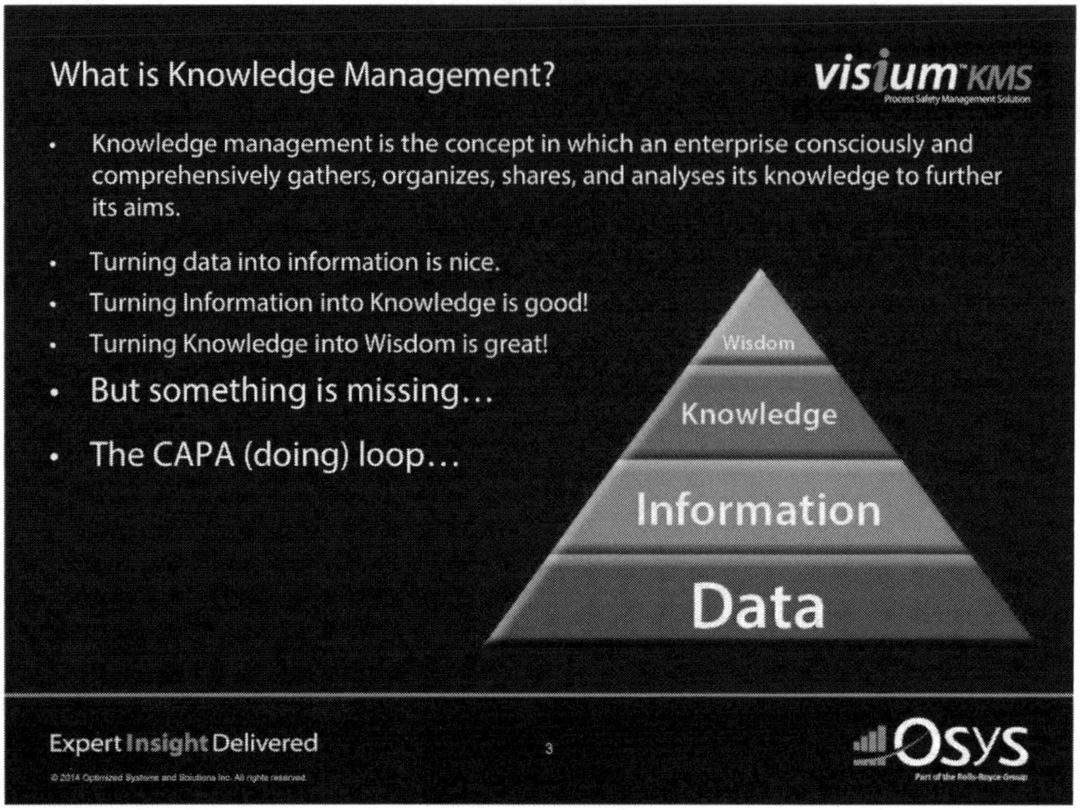
What is Knowledge Management?
visium KMS
Process Safety Management Solution
• Knowledge management is the concept in which an enterprise consciously and comprehensively gathers, organizes, shares, and analyses its knowledge to further its aims.
• Turning data into information is nice.
• Turning Information into Knowledge is good!
• Turning Knowledge into Wisdom is great!
• But something is missing…
• The CAPA (doing) loop…
Wisdom
Knowledge
Information
Data
Expert Insight Delivered
3
© 2014 Optimized Systems and Solutions Inc. All rights reserved.
OSyS
Part of the Rolls-Royce Group

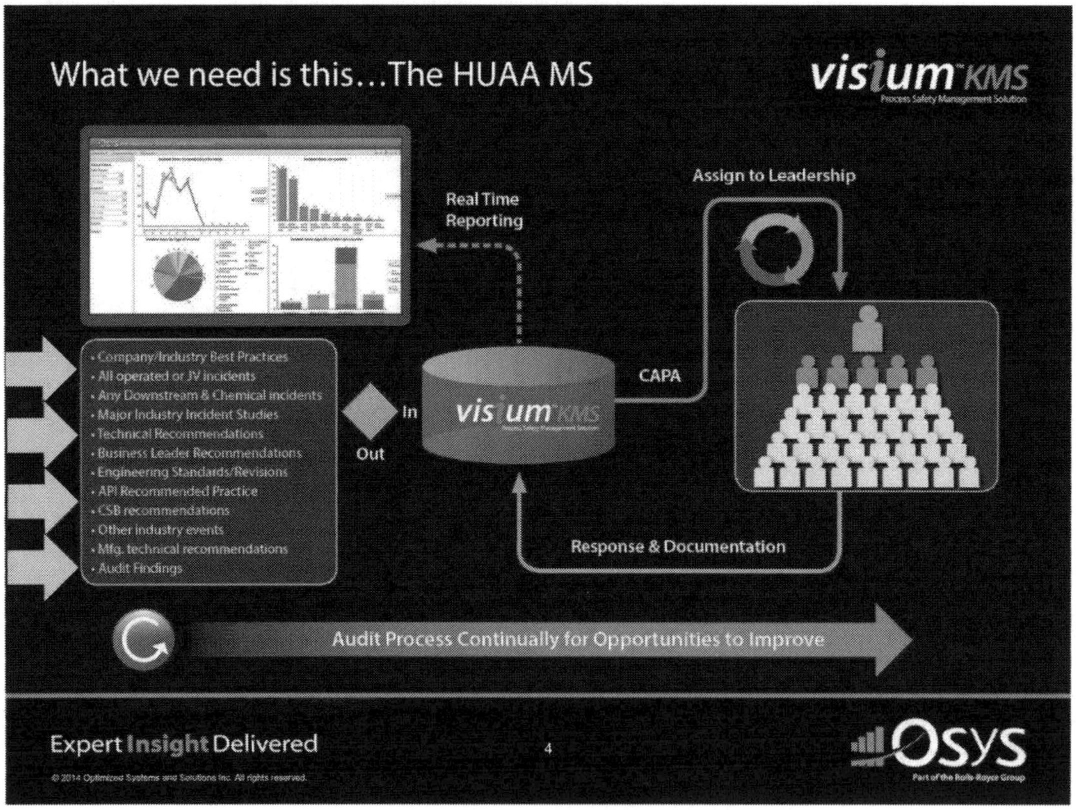
What we need is this…The HUAA MS
visium KMS
Process Safety Management Solution
Real Time Reporting
Assign to Leadership
• Company/Industry Best Practices
• All operated or JV incidents
• Any Downstream & Chemical incidents
• Major Industry Incident Studies
• Technical Recommendations
• Business Leader Recommendations
• Engineering Standards/Revisions
• API Recommended Practice
• CSB recommendations
• Other industry events
• Mfg. technical recommendations
• Audit Findings
In
Out
visium KMS
CAPA
Response & Documentation
Audit Process Continually for Opportunities to Improve
Expert Insight Delivered
4
© 2014 Optimized Systems and Solutions Inc. All rights reserved.
OSyS
Part of the Rolls-Royce Group

What is the HUAA Management System…

visium KMS
Process Safety Management Solution

HUAA – Heard, Understood, Acknowledged, Actioned.

1. Opportunities identified (heard) by your experts,
2. Captured by an electronic system,
3. Expertly Reviewed (understood),
4. Accepted or rejected (acknowledged) and
5. Assigned (actioned) to Sr. Managers for resolution and tracked to closure.

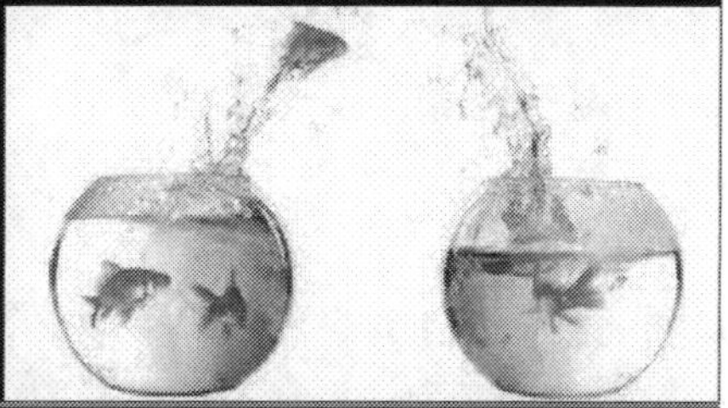

Expert Insight Delivered 5

OSyS
Part of the Rolls-Royce Group

HUAA – Management Systems - Detail

visium KMS
Process Safety Management Solution

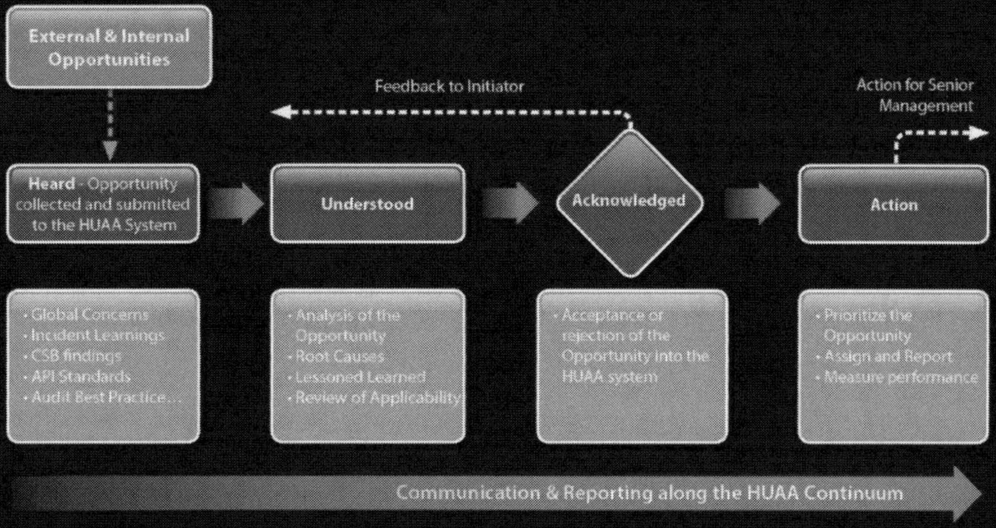

Expert Insight Delivered 6

OSyS
Part of the Rolls-Royce Group

HUAA Results

visium KMS
Process Safety Management Solution

- Efficient Collection - Important Opportunities are identified and collected from many sources (internal and external), and entered into the HUAA management system.
- Expert Vetting - These opportunities are vetted, parsed and resolved for inclusion or exclusion; documenting & communicating both.
- Accountable Assignment - Included opportunities are systematically assigned to Sr. Leaders and managed across the enterprise at SVP and C levels.
- Action - Improvement will be assured through the accountable actions of these leaders.
- Results - The organization will gain competitive advantage by active listening, learning from their experience, as well as the experience of others, and by taking ACTION!

Expert Insight Delivered 7 OSYS Part of the Rolls-Royce Group

Sharing is not enough to keep your ship afloat!

visium KMS
Process Safety Management Solution

Expert Insight Delivered 8 OSYS Part of the Rolls-Royce Group

352631 Autocad - Smart Grid – Agriculture - Architecture and the United States Government

Abstract:

The computational methods, process control, process design, process safety, public policy of the smart chip for agriculture industry, and architecture industry in the United States utilizing the computational methods of AutoCAD in the structural development stages of the smart grid for device/component/equipment and geographical development and public policy for grant funding.

352747 Production, Characterization and Catalytic Studies of Biobased Carbon Materials

Abstract:

Carbon-based nano materials were produced from wood chips, lignin and bio-char. The influence of reaction conditions on the production of nano-carbons was investigated. Carbon-based nanomaterials were characterized using scanning electron microscopy (SEM), transmission electron microscopy (TEM), high resolution transmission electron microscopy (HRTEM), and X-ray diffraction (XRD). The catalytic performance of these nanomaterials as syngas cleaning adsorbents and chemical conversion catalysts for bio-fuel processes has been evaluated.

347835 Uncertainty in Sour Gas Viscosities Estimation, What Is the Effect on Your Reservoir Inflow and Tubing Performance

Abstract:

In light of the need to find an applicable, consistent and promising correlation to accurately calculating sour gases viscosity under several reservoir conditions, this study aims to compare the existing correlations and other industry standard methods, check their accuracy and range of applicability for sour/acidic gases. Moreover, investigating the impact of uncertainties in estimating sour gas viscosities on reservoir flow performance relationship (IPR), reserves and resources estimations, tubing performance, pipeline and facilities design, the design of compressors and more importantly the need to find such correlation to be used in further advancements of reservoir simulation.

354052 Chromonic Nanocarriers for Chemotherapeutics: Size Distribution and Control Release Studies

Abstract:

Conventional chemotherapy fails to release the drug in the controlled manner; rather they release the drug instantly due to which re-administration is required as therapeutic index falls. Moreover, they target both the cancer cells and non-cancer cells which make the treatment of the cancer cells highly ineffective due to excessive toxicities. Recently, the nanoparticle mediated drug delivery for chemotherapeutics has gained much attention. The present study reports the use of a novel class of materials – *Chromonics* in cancer drug delivery. Folic acid as a chromonic material has been explored to develop nanocarrier for sustained release of chemotherapeutic drugs *in-vitro*. With the help of a biocompatible polymer as a tool, folate nano-domains were formed from self-assembled liquid crystalline folate solutions. These nano-domains were crosslinked to form a stable nanoparticle system. The effect of different factors like crosslinking agent, relative polymer and folate concentration, on the size distribution was investigated in detail. Based on the observations, we have reported that not only the synthesis but also the size of the developed nanoparticles can be controlled by varying the folate: HPMC weight ratios. The present study then explored the release of drugs from these nanocarriers. We have investigated the release of folic acid as a model drug from these nanoparticles as well as release of methotrexate and cytarabine as an anticancer drug for the period of 60 days. It was observed that release of drug was nanoparticle's size-dependent, release medium dependent, pH of the release medium dependent, crosslinking agent and its concentration dependent. The significant part of the current study is the synthesis of chromonic nanoparticles, how the size can be controlled, how controlled release can be obtained as well as how the release profiles can be modified depending on the dosage forms. Therefore, we can say that folate nanoparticles can be a potent nanocarrier for the sustained release of cancer drugs.

IPEX
Committed to Excellence

Double Containment Piping Systems: A Fail-Safe, Leak-Free Solution

Darin Johnson, P.Eng.
Applications Engineer
IPEX USA LLC

AIChE – 2014 Spring Meeting
New Orleans, LA

Monday, March 31, 2014

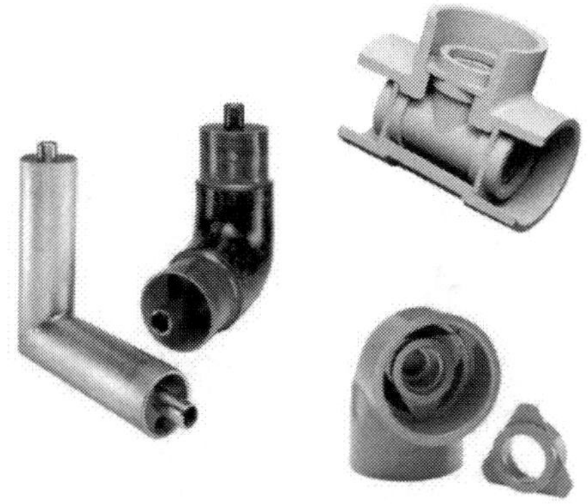

IPEX
Committed to Excellence

Why Double Containment Piping Systems?

- December 1988 – law change within the ***Resource, Conversation, and Recovery Act*** in the United States regarding underground piping.
- U.S. Environmental Protection Agency (EPA) requirement: Code of Federal Regulations (CFR), Title 40, Part 280.
  - Petroleum Products
  - Hazardous Substances (over 3200, CFR Title 40 Part 302.4)
  - Hazardous Wastes

Why Double Containment Piping Systems?

- Buried tank - all buried piping connected to it must be double contained.
- Above ground tank – associated buried piping consists of 10% or more of the tank volume, it also must be double contained.

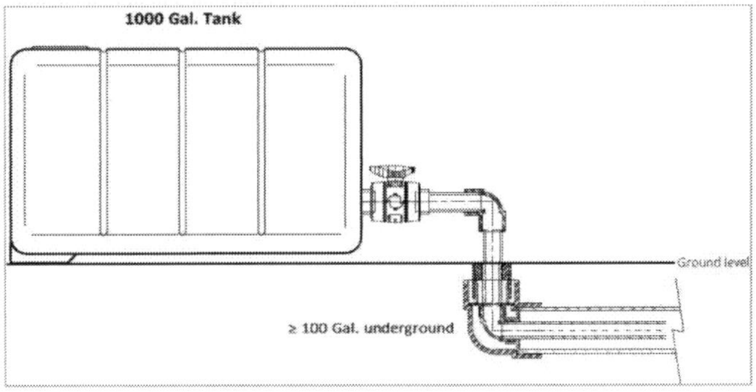

Why Double Containment Piping Systems?

- If the piping is less than 10% of the volume of the aboveground tank, and you choose not to double contain the buried piping, ***responsibility for any clean-up or remediation resulting from a release will still fall on the Client.***

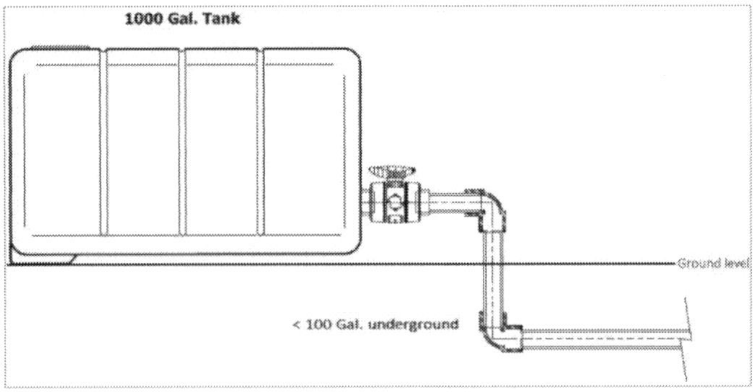

Why Double Containment Piping Systems?

- **Above Ground Piping Systems**
  - Different jurisdictions have to ability to require chemicals be double contained.

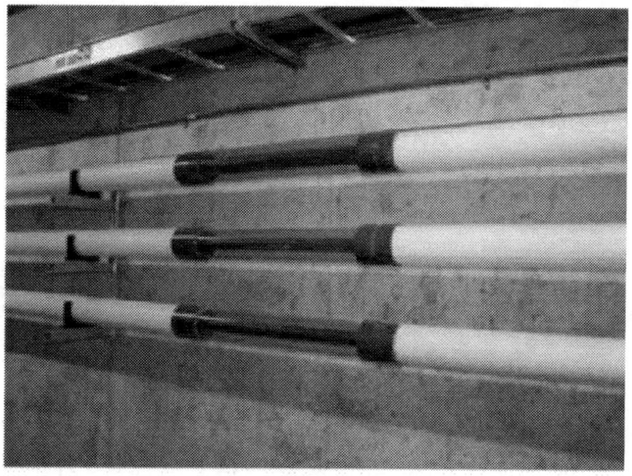

Nomenclature of Double Containment Piping

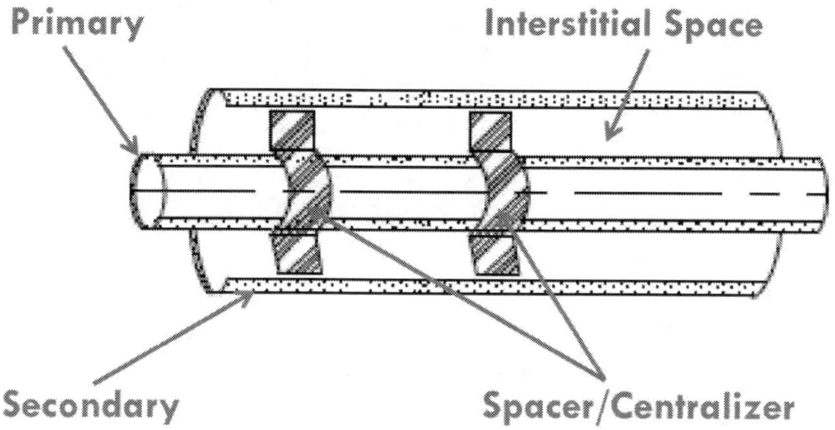

Double Containment Piping Systems: Material Combinations

SYSTEM Carrier	Containment	Carrier SIZE inches
PVC 80	PVC 80	1/2 - 12
PVC 80	PVC 40	1/2 - 12
PVC 40	PVC 40	1/2 - 12
PVC DWV	PVC DWV	1-1/2 - 12
PVC 80	CS SCH 10	1/2 - 12
CPVC 80	CPVC 80	1/2 - 8
CPVC 80	PVC 80	1/2 - 12
CPVC 80	PVC 40	1/2 - 12
CPVC 80	CS SCH 10	1/2 - 12
NAT PP 80	PVC 80	1/2 - 2
NAT PP 80	PVC 40	1/2 - 2
RED PVDF 80	CPVC 80	1/2 - 2
RED PVDF 80	PVC 80	1/2 - 2
NAT PVDF 80	PVC 80	1/2 - 2
ENFUSION PP	PVC DWV	1/2 - 6
S40 A53B CS BW	PVC 80	1/2 - 8
S40 A106B CS BW	PVC 80	1/2 - 8

SYSTEM Carrier	Containment	Carrier SIZE inches
S40 A53 BLK SW	FRP	1/2 - 2
S40 A53 BLK SW	PVC 80	1/2 - 2
S80 A53B CS SW	PVC 80	1/2 - 2
S10 T304L SS BW	PVC 80	1/2 - 8
S10 T304L SS BW	PVC 40	1/2 - 8
S10 T0304L SS BW	CPVC 80	1/2 - 8
S10 T304L SS BW	FRP	1/2 - 12
S10 T304L SS BW	S10 T304L	1/2 - 12
S40 T316L SS BW	PVC 80	1/2 - 8
S40 T316L SS BW	CPVC 80	1/2 - 8
S40 T316L SS BW	FRP	1/2 - 12
S40 T304L SS SW	PVC 80	1/2 - 2
S40 T304L SS SW	PVC 40	1/2 - 2
S40 T304L SS SW	CPVC 80	1/2 - 2
S40 T304L SS SW	FRP	1/2 - 2
S10 T316L SS SW	PVC 80	1/2 - 2
S10 T316L SS SW	CPVC 80	1/2 - 2

Double Containment Piping Systems: Design Considerations

- **Material Selection**
  - Primary Pipe
    - Chemical Compatibility
    - Pressure Capabilities
    - Thermal Properties
  - Secondary Pipe
    - Chemical Compatibility
    - Loading (Static & Live) for buried applications
    - Hanging requirements for above ground
    - Leak detection requirements
    - UV requirements

Double Containment Piping Systems: Design Considerations

- **Thermal Expansion & Contraction**
  - Two basic thermal relationships may be present:
    - When only one of the two systems experiences thermal expansion/contraction.
    - When both systems experience different magnitudes of thermal expansion/contraction.
  - The Floating System
  - The Restrained System

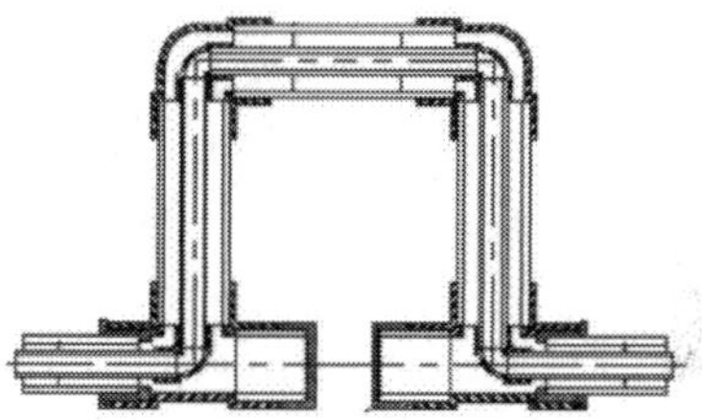

Double Containment Piping Systems: Leak Detection Requirements

- EPA requirement: (CFR), Title 40, Part 280
  - Pressurized systems require electronic leak detection that must be checked once per year.
  - Drainage systems (gravity flow) require visual leak detection with monthly manual inspection for a media leak.

IPEX
Committed to Excellence

Double Containment Piping Systems: Leak Detection Requirements

- The following forms of leak detection are acceptable as per regulations
  - Visual (drainage applications only)
  - Electronic low point probes
  - Continuous cable

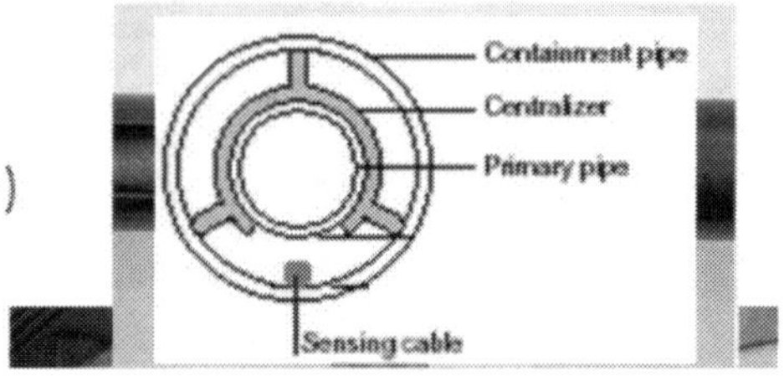

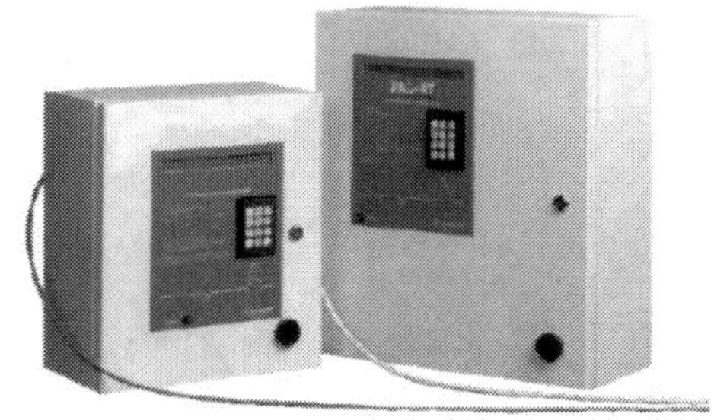

IPEX
Committed to Excellence

Double Containment Piping Systems: Applications

- Underground Applications
- WTP & WWTP
- Oil & Gas
- Hospitals
- Museums
- Food and Beverage
- Pharmaceuticals
- Laboratories
- Data Centers

IPEX
Committed to Excellence

Double Containment Piping Systems: A Fail-Safe, Leak-Free Solution

Questions?

Thank you!

DTP Process

"Now Ready to License"

JGC Corporation
Mitsuo Morita

2014

JGC CORPORATION

MITSUBISHI CHEMICAL

What is DTP Process ?

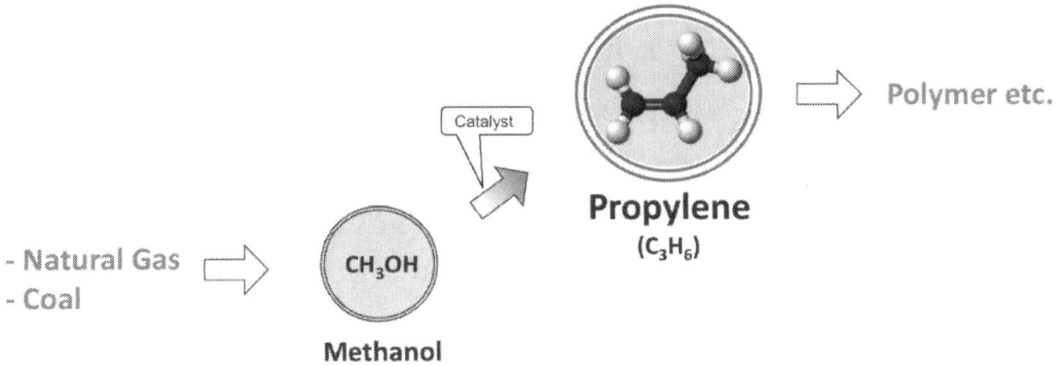

-On-Purpose Propylene Production from Methanol

JGC CORPORATION

- 2 -

MITSUBISHI CHEMICAL

What is DTP Process ?

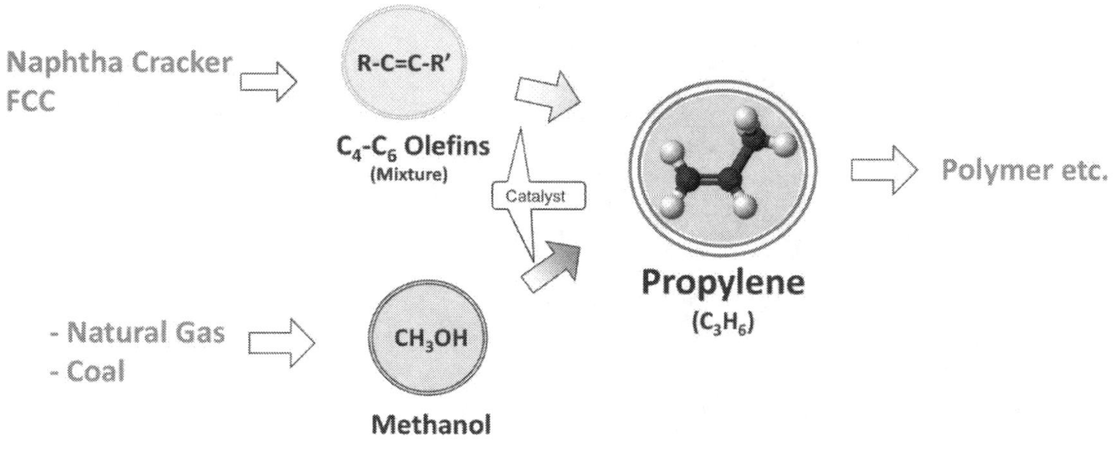

-On-Purpose Propylene Production from Methanol

- Additional Unutilized C_4-C_6 Olefins as Feedstock

JGC CORPORATION - 3 - MITSUBISHI CHEMICAL

Development History and Prospect

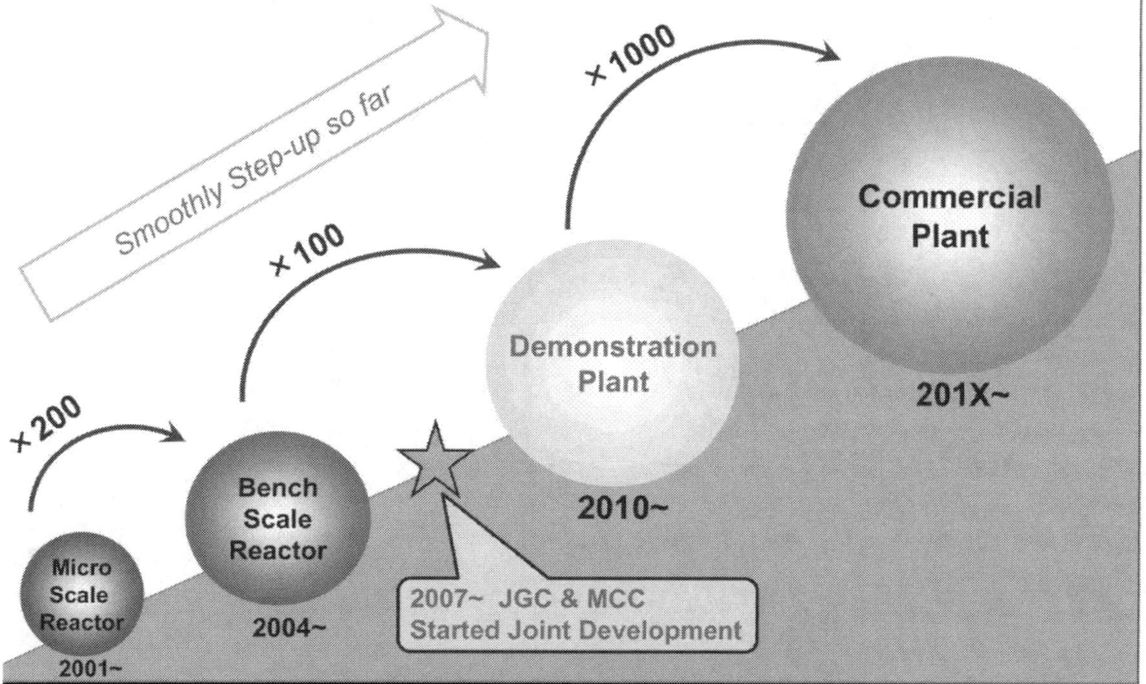

JGC CORPORATION - 4 - MITSUBISHI CHEMICAL

Demonstration Plant Overview

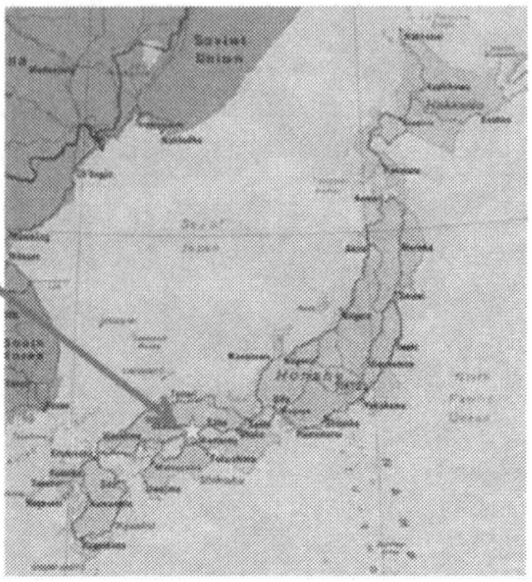

- **Located in Mitsubishi Chemical Mizushima Plant**
- **Successfully Start-up in August 2010**
- **Continuous Operation for Two Years**

Process Flow (Standalone Case)

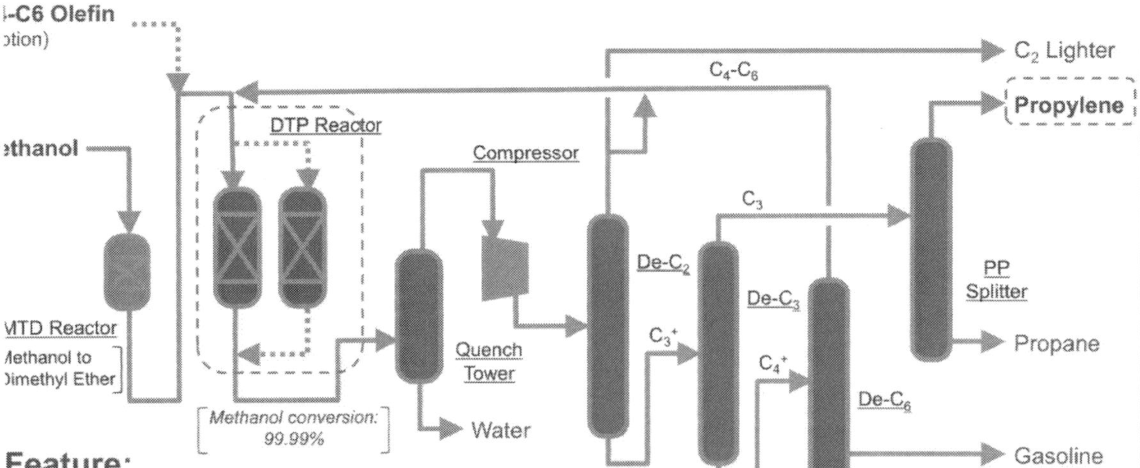

Feature:

High Propylene Yield: more than 70%

Applied Existing Proven Technologies

Process Flow (Integration Case)

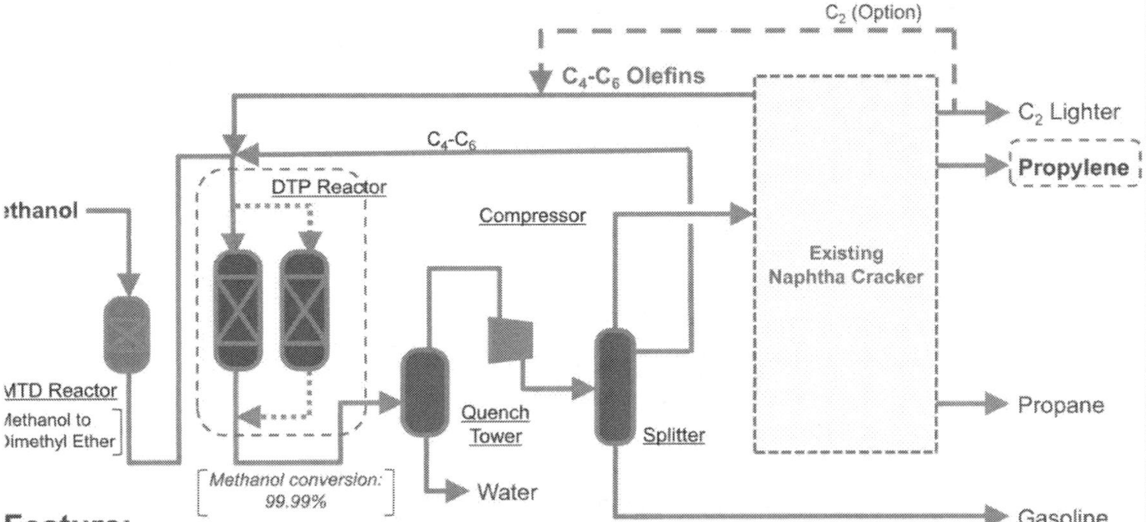

Feature:

High Propylene Yield: more than 70%

Applied Existing Proven Technologies

Low Investment Cost in case of Integration with Naphtha Cracker

Application

1) Naphtha Cracker Integration Case (Propylene Capacity: 20KTA – 200KTA)

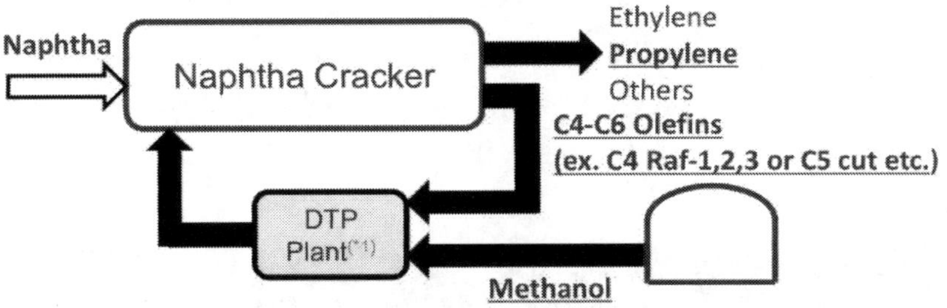

(*1) Utilizing facilities in Naphtha Cracker.

2) Standalone Case (Propylene Capacity: 250KTA – 500KTA)

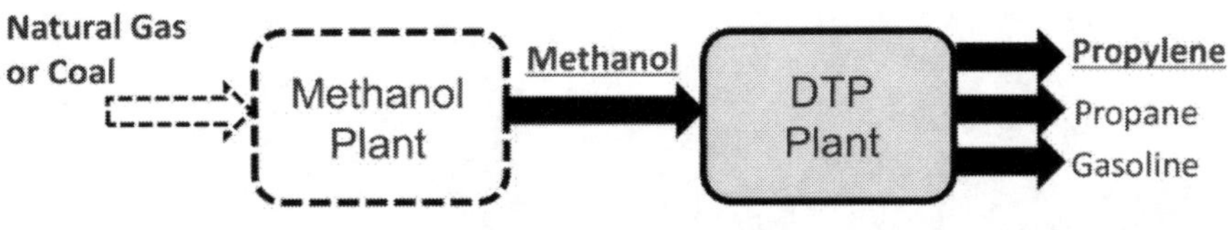

- Two Cases available depending on Client requirements !!

DTP is…

- *High Yield Propylene Production Process*
- *Flexible Feed Methanol to Olefin Process*
- *Process that increase P/E Ratio for Cracker*

Thank you very much !!

Contact:

Mitsuo Morita
Market Development Dept.
Global Marketing Division
2-3-1, Minato Mirai, Nishi-ku,
Yokohama 220-6001, Japan

Phone: 81-45-682-8371
FAX: 81-45-682-8183
E-Mail: morita.mitsuo@jgc.co.jp

352533 Self-Cleaning "Bernoulli" Type Filters Used in Onshore and Offshore Applications

Abstract:

Georg Schuenemann GmbH (SAB-BREMEN) has developed a new Filter Type F-480, dubbed the "Compact Giant" for applications in onshore water intake and cooling water systems, sea water desalination plants, and in offshore projects, such as FPSO's, Drillships, and Production Platforms.

This paper will describe the essential design features of the filters, based on the "Bernoulli principle" which result in a fully automated self-cleaning filter system, and choice of materials in a sea water service with high corrosion rates.

A case study will be presented in which stringent project requirements such as substantial flow rates, adaptibility to space limitations and pipe connections had to be observed. Reliability of the system, extended run times without the need for frequent replacement of filter elements, as well as ease of monitoring and operator involvement have been achieved.

352653 Effects of Support on Sulfur Tolerance and Regeneration of Pt Catalysts Measured By Ethylene Hydrogenation and EXAFS

Abstract:

We investigated the effect of support on sulfur tolerance and regenerability under reducing environments by measuring the rate of ethylene hydrogenation, combined with hydrogen chemisorption, and extended x-ray absorption fine structure (EXAFS). Alumina and titania (P25) supports were used to prepare Pt catalysts at 1 wt.%. The catalysts were tested after treatment in H_2S/H_2 at 250 °C followed by regeneration treatments in H_2 at 250 °C, 350 °C and 450 °C. The sulfided catalysts exhibited a 20-27 times decrease in the rate of ethylene hydrogenation and a 4-6 times drop in the Pt surface area. Regenerations up to 450 °C were unable to remove all the sulfur, as shown by the presence of Pt-S bonds by EXAFS. However, a partial recovery of the hydrogenation rate per mole of Pt was observed on sulfided Pt/Al_2O_3 after reduction at 450 °C, while the induction of strong metal support interactions (SMSI) at reduction temperature above 350 °C was observed on Pt/P25, regardless of the presence of sulfur. For Pt/P25, the reversal of the SMSI state together with sulfur removal by mild oxidation suggests that sequential reduction/oxidation treatments may be more effective in restoring the S-free state of TiO_2-supported catalysts.

352799 Improved Operational Efficiency and Reliability through Insulation Materials Selection

Abstract:

This paper will explore the implications of insulation materials selection on the energy efficiency and reliability for Ethylene production units. Considerations will include insulation properties, structural support, installation methods, installation and maintenance costs.

BOILER MACT COMPLIANCE: MERCURY IN FUEL GAS

PATRICK LAINE, WES YONGE, & VIC VICKERY
PEI, HOUSTON, TX 77043

Patrick Laine, Ph.D.
Chief Scientist
Houston, TX
850-866-6686
plaine@pei-tx.com

Wes Yonge, PE
Chief Engineer
Houston, TX
713-806-8396
wyonge@pei-tx.com

James 'Vic' Vickery
Chief Technology Officer
Houston, TX
281-536-0899
vvickery@pei-tx.com

1 Introduction & Objectives

For most ethylene producers the primary objective of this sampling and analysis effort will be to verify and "establish a mercury fuel specification through which gas-fired boilers that use a fuel other than natural gas or refinery gas may be considered Gas 1 unit" as identified in 40 CFR Part 63.7521.(f).

Measurement and Monitoring Requirements: Decision Rules

- ❑If Hg concentration ≤ 50% of Gas 1 specification* then no further testing required
- ❑If Hg concentration > 50% and ≤ 75% of the specification* then semi-annual sampling required
- ❑If Hg concentration > 75% of the specification* then monthly sampling required

*Specification is **40 µg/Sm³**

The goal will be to execute a sampling and analysis campaign that will produce compliance quality data. The key to producing and validating compliance quality data is a well-designed and executed QA/QC protocol. At a minimum, a performance based QA/QC protocol is recommended that includes the following components:

- Sorbent Trap Breakthrough Bed:
  - ❑ Sorbent traps consisting of a minimum of two sorbent beds in series so that the trailing bed can be analyzed to verify that no breakthrough occurred from the leading bed.
- Duplicate Samples:
  - ❑ The sampling system should be configured with two trains to accommodate the collection of simultaneous duplicate sorbent trap samples, in parallel, providing a means of verifying precision in the sample collection process.
- NIST (National Institute of Standards and Technology) Traceable Spike:
  - ❑ Sorbent traps with a third bed (downstream) that is spiked with a NIST traceable spiked mass of mercury serves as a matrix spike and provides a means of verifying that components in the matrix (the fuel gas) are not affecting the trapping efficiency of the sorbent trap.

2 Methods

Sorbent Trap Mercury Sampling Methods Comparison

Method	Description	Method Type: S = Sampling A = Analysis	Evaluation	Sample Collection Phase QA/QC
ISO 6978	Hg in Natural Gas by Thermal Desorption (CVAFS)	S & A	Method calls for low sample volumes can lead to non representative results.	No
EPA Method 30B	Determination of Gaseous Hg Emissions from Coal Fired Combustion Sources using Carbon Sorbent Traps	S	Requires modification (Modified M30B)	Yes
ASTM D6350	Hg S&A in Natural Gas (AFS)	S & A	Method calls for short term samples (i.e., low sample volume) which leads to poor accuracy.	No
ASTM D5954	Hg S&A in Natural Gas (AAS)	S & A	AAS is less sensitive than AFS.	No
EPA Method 1631	Measurement of Hg in Water	A	No	Yes

3 EPA Method 30B QA/QC Performance Criteria

QA/QC Test or Specification	Acceptance Criteria	Frequency
Analyzer calibration	Correlation Coefficient: 0.999	Daily
Transfer efficiency of analyzer	%Transfer Efficiency: 90%–110%	Once per sorbent trap
Initial calibration blank	< 0.5 ng	Once before each calibration
Sorbent Trap Section B breakthrough	≤ 10% of Section A Hg mass, for Hg concentrations > 1µg/dscm ≤ 20% of Section A Hg mass, for Hg concentrations ≤ 1µg/dscm	At each sample point
Spike Recovery Test	Average recovery between 75% and 125 % for Hg_0	Once per sample point
Paired Sorbent Traps Agreement	≤ 10% RD for Hg concentration > 1µg/dscm ≤ 20% RD or 0.2 µg/dscm absolute difference for Hg concentrations ≤ 1µg/dscm	At each sample point

4 Equipment

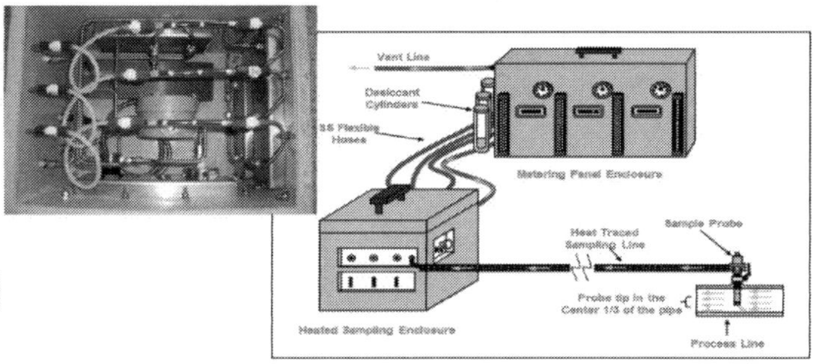

Flow →	Primary Bed	Breakthrough Bed	Spiked Bed
QA/QC Acceptance Criteria:	99% of Total Collected Hg	≤ 10% of Primary Mass	Recovery between 75% and 125%

Gold Trap:

5 Conclusions/PEI Recomendations

- ❑ There are five methods accepted by the Boiler MACT rule. PEI has extensive experience and expertise employing **all five** of these methods.
- ❑ Operators can achieve compliance without submitting a Fuel Analysis Plan by using ASTM D6350, however, we have noticed variable results with this method.
- ❑ The **PEI recommendation** for obtaining the most accurate compliance quality data is to use Modified EPA Method 30B in conjunction with EPA 1631 where traps should be analyzed in a certified ISO 17025 laboratory. The recommended pathway includes a rigorous, performance based QA/QC protocol that validates accuracy and precision of all sample collection procedures.

PEI

351376 Young Professional Simulation Tutorial

Abstract:

18C00 Young Professional Simulation Tutorial

Authors: Naomi Hua and Mike Donahue, Invensys

Process simulation is a complex skill to master. Even with years of experience, a typical user is constantly facing new problems that need to be addressed. There are many issues that can lead to unexpected and unnecessary errors with the use of process simulation tools. Regardless of what simulation program you are using, these issues are common from one program to another. In this presentation, targeted for the young professional, we will discuss the "Most Common Pitfalls in Process Simulation". For each topic, we will discuss various techniques and methods these troublesome problems and prevent their occurrence.

Topics to be discussed include:

1. Erroneous data entry
2. Simulation defaults
3. Bad plant data
4. Thermodynamic method selection

5. Simulation tolerances and convergence
6. Recycle identification and handling
7. Model complexity
8. Difficult or impossible specifications
9. Assay pseudo-component development
10. Assay property blending

1. Erroneous data entry is the most common mistake made when building a process simulation model. This presentation will cover methods that can help ensure proper data entry in your process simulation model, such as checking the data you entered, particularly the units of measure, using "alternate forms" of data input such as copy/paste and data import, and reviewing all warning and error messages.

2. Defaults are methods, values, and conditions that are automatically selected for you by the program when you begin a new simulation model. These defaults will affect the results calculated by your model and every simulation program has them. Most defaults within simulation programs fall under categories such as convergence criteria, thermodynamic methods, tolerance and units of measure. The problem with defaults is unintended results from not fully understanding the effects a particular default has on the model. Take the time to understand the defaults of the program and how these defaults will affect the results. If you do not agree with the defaults selected by the program, most simulation programs will allow the user to change the default.

3. Plant data changes over time. The feed information and process conditions of the unit operations may be different. Not only does plant data change over time, but measurement information is often very inaccurate. This could be due to faulty or poorly calibrated measurement equipment, leaks in lines, sensor failures, and other issues. Because of these changes and inaccuracies over time, you always have to validate the information that is entered in your model to ensure that you are obtaining the proper results. To do so, you can utilize programs that can help with data validation and reconciliation like SimSci's ROMeo online optimization program. With it, you can connect your model to your data historian to constantly import current data measurements into your model and then reconcile these measurements to determine when problems arise within your plant and where those problems are occurring. You will also need to use your engineering judgment to determine if the data is valid. More details will be provided during the presentation.

4. The proper selection of thermodynamic methods for your simulation is the most important decision that you make when modeling a process. The method you select could have a drastic effect upon your results. Often, any number of thermodynamic methods is suitable for a given application. It is up to the user to determine the best method to use based upon knowledge, personal experience, and best judgment. Some important factors that need to be considered are the components being used and the operating conditions.

Oftentimes, the selected thermodynamic method still needs to have additional inputs such as binary interaction parameters to simulate the process accurately.

5. Many of the calculations in process simulation are iterative. When these calculations iterate, the program compares the results calculated from one iteration to the results calculated in the next iteration in order to determine

whether it is converged to a solution or needs to continue calculating. This method of comparing results is where tolerances come into play and simulation programs use tolerances to determine when a solution is reached. One of the major mistakes when modeling a process is not understanding or using the correct tolerance values. A looser tolerance may make the file easier and quicker to converge, but it also reduces the accuracy of the model. A tighter tolerance may be more accurate, but might make it impossible for the model to solve or increase the convergence time exponentially. A common practice is that when you have reached a converged solution with your model, tighten the tolerances within the model and solve it again. If you notice that your answers have changed significantly, you need to continue tightening the tolerance values until your results change little from one iteration to another. You will then know that you have reached an accurate solution for your model.

6. Recycles in process simulation occur when you return to a previously calculated unit operation to recalculate the unit due to a change made elsewhere in the model. They are very common and can often be difficult to converge. Some suggestions on troubleshooting recycle convergence include initially solving your model without the recycles to see how it behaves, providing initial estimates for recycle streams, using tighter tolerances on unit operations within recycle loops than the total recycle tolerances themselves, and ensuring that all components have a way in and out of a recycle by performing mass balances and using purge or makeup streams where necessary.

7. The general rule of thumb when modeling is to start simple and add complexity as you go. Each time you add a few streams or unit operations converge and save the model before moving on. Make sure you justify any additions that you make to the model. You can most likely model your system without including every unit from your PFD in the simulation. When modeling, think about the goal of your simulation and not about literally translating every real-world piece of equipment into the simulation model. Make the judgment call on what is most important and what the purpose is of every unit you add to your simulation. For example, if your real-life process shows a heat exchanger, a pump and a splitter, but all that is necessary is that the temperature, pressure, and separation are achieved properly, consider using a flash drum to accomplish all three.

8. Specifications within process simulation are everywhere and every stream and unit operation you place in your model contains specifications. In its simplest form, a specification is a constraint that you place on the model that the model needs to meet in order to successfully converge. For streams, these are usually temperatures, pressures, flowrates and compositions. For unit operations, they range from process conditions, like temperatures and pressures, to product conditions, like an overhead composition in a column. There are some pitfalls, however, that you can run into with specifications such as azeotropes in columns, high purity specifications, and the possibility of multiple solutions.

9. Petroleum assays are sets of distillation data that are used to characterize all of the components that reside in a given crude sample. When you enter distillation data into a process simulator, the program converts it into components that it can use when modeling your process. The components that are generated are called pseudo-components, and each pseudo-component represents all of the unknown components that would be within a given boiling point range in the crude. Some of the things you will want to ensure when properly developing your assay components are did you use the correct distillation curve basis, did you enter the data correctly, did you use enough cuts and are you using the correct characterization methods .

10. When you have more than 1 assay in a simulation model, you need to be concerned with assay blending and how it will affect the results of your file. Assay blending is the mixing of 2 or more groups of components (i.e. "C sets of cuts/pseudo-components) in order to calculate averaged component properties for cuts that are in the same boiling point range as one another. Once these new averaged component properties are determined, the program will use these averages from this point on in the simulation model any time it needs to access component

properties for a given pseudo-component. All component properties are blended, including normal boiling points, specific gravity, molecular weight and refinery inspection properties, and usually the blended component properties are calculated using a weighted average

These topics are based on experience and feedback from end-users to technical support representatives for process simulators. They often represent the most commonly asked questions or areas that most users find to be problematic. While the specific examples highlighted in each topic come from SimSci software, the intent of this tutorial presentation is to be broad-based and is particularly applicable to young professionals seeking to develop their experience with simulation.

Polymeric Sulfonic Acid Composites for Fuel Cell Applications: Effects of Casting Solvent and Sonication Time on Membrane Performance

Jimoh K. Adewole[1], *Abdullah S. Sultan*[1, *], *Amir Al-Ahmed*[2], *and S. M. Javid Zaidi*[3]

[1]*Center for Petroleum & Minerals, King Fahd University of Petroleum & Minerals, Dhahran 31261, Saudi Arabia*

[2]*Center of Research Excellence in Renewable Energy, King Fahd University of Petroleum & Minerals, Dhahran 31261, Saudi Arabia*

[3]*School of Chemical Engineering, The University of Queensland, Australia*

Abstract

Polymer blends comprising sulfonated poly (ether ether ketone) (SPEEK) and 3wt% (2-acrylamido-2-methyl-1-propanesulfonic acid) (PAMPS) were prepared by solution casting method. Two solvents (N,N-Dimethylacetamide (DMAc) and N-Methyl-2-pyrrolidone (NMP)) were uesd separately. Residual solvents in the membrane samples after drying were quantified using thermo gravimetric analysis. Effects of periods of sonication (0 to 90 minutes) and types of solvent on the membrane performance were evaluated interms of morphology and water uptake. An attempt was made to correlate membrane performance with the time of sonication in order to select the most suitable solvent and membrane casting conditions for subsequent membrane preparations. For both solvents, the water uptake was observed to first decrease at 10 minutes of sonication and then increased as the sonication time was increased. The minimum water uptake recorded for DMAc samples was 19.23% at 20 minutes sonication time while that of NMP sample was 24.24% at 10 minutes. However membrane morphology was found to deteriorate with increase in sonication time. No correlation was obtained between the membrane properties and the sonication time. In general, membrane performance was better for membranes samples using DMAc than NMP solvent. Therefore, DMAc was selected as the casting solvent for membrane preparation on which further studies will be carried out.

Keywords: morphology, water uptake, residual solvent, thermogravimetric analysis

Introduction

Water soluble polymers have received academic and industrial interests due to their potential applications in fuel cells and other electrochemical devices (Ye et al. 2012). One of the water soluble polymers that have attracted publication interest is poly (2-acrylamido-2-methyl-1-propanesulfonic acid) (PAMPS) (Florjańczyk et al. 2003). Generally, some of the major performance criteria for fuel cell materials include proton conductivity, water and methanol uptake, good mechanical properties, chemical stability, thermal stability and methanol permeability (Gil et al. 2004; Swier et al. 2005; Ye et al. 2012). PAMPS was reported to posses higher proton conductivity than partially hydrated Nafion. In addition, it has been found to be more tolerant to water fluctuation in high temperature operations. Some of the major drawbacks in the use of PAMPS are its poor mechanical properties, thermal stability, water and methanol uptake and solubility in many solvents (Florjańczyk et al. 2003). One of the approaches that have been proposed to solve these problems is development of polymer composites.

This research paper investigates the synergistic effect of combining water soluble PAMPS with sulfonated poly (ether ether ketone) (SPEEK) which is non - water soluble polymer to prepare composite membranes for application in fuel cells. Specifically, preliminary results on the performnance of the composite were presented. Performance were evaluated based on the composite morphology and water uptake. Effects of sonication time and types of casting solvent on membrane properties were also evaluated.

Material and Method

Material

Sulfonated poly (ether ether ketone) (SPEEK) purcahsed from FuMA-Tech, Germany, and poly (2-acrylamido-2-methyl-1-propanesulfonic acid) (PAMPS) purchased from Aldrich Chemical Co., are the two major polymers used in this research. The SPEEK has a DS of 0.533 and solubility of 99% in DMAc. SPEEK is derived by sulfonation of Poly (ether ether ketone) (PEEK). PEEK is a commercially successful polymer due to its excellent thermal stability and solvent resistance (Mohr et al. 1991). For sometimes, further application of PEEK have been stalled due this solubility issues. By sulfonation, PEEK can be made soluble in most commercial solvents. Importantly, the solubility of SPEEK also depends on its degree of sulfonation (DS). Since, this work involves the use of two polymers, a solvent which will disolve both the PAMPS and SPEEK must be selected. Literature review showed that PAMPS dissolves in N,N-Dimethylformamide (DMF), N,N-Dimethylacetamide (DMAc), propylene carbonate (PC), and water (Florjańczyk et al. 2003). Initial solubility studies that was carried out in our laboratory was done for N,N-Dimethylacetamide (DMAc) and N-Methyl-2-pyrrolidone (NMP). The choice of NMP was due its unique solvent properties and its structtural distiction from DMF and DMAc (Figure 1). Moreover, no literature was found where NMP was used as casting solvent for PAMPS. Thus, DMAc and NMP were used as solvents in this work. All materials were used as received from the manufacturers.

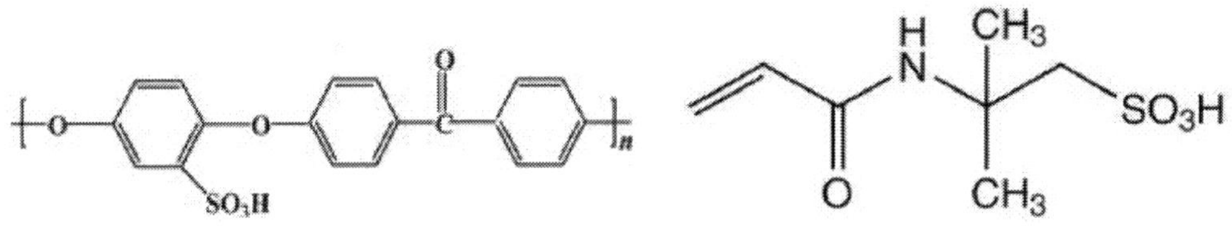

Figure 1 Structure of SPEEK and PAMPS

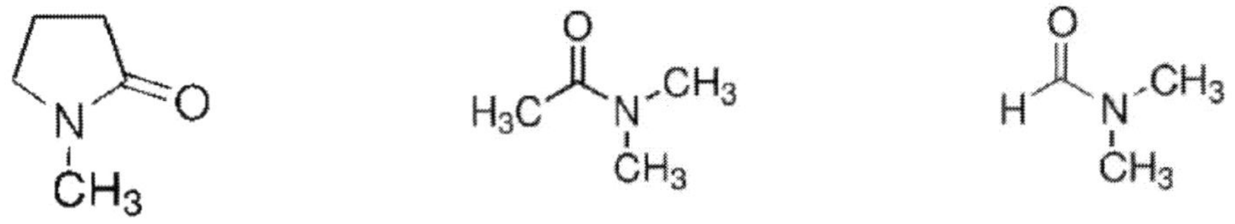

N-Methyl-2-pyrrolidone N,N-Dimethylacetamide N,N-Dimethylformamide

Figure 2 Structure of Solvents

Experimental Procedure

Membrane Preparation

SPEEK was first dissolved in solvents (DMAc and NMP) and stirred for 2hours at 70°C. PAMPS was then added and the blend was stirred for 24hours. The solution was sonicated for various periods of time ranging from 10 to 90minutes using an Ultrasonic bath. In untralsonic bath, the final temperature achieved for a given time of sonication depends on the tempearutre of the laboratory (summer or winter). It is therefoe genrally advisable that maximum temperature of the bath is determined and maintains (as the so - called equilibrium tempearture), when operating continuously under ambient conditions (Santos et al. 2009). During the sonication, equilibrium temperature was maitained by simply filling the bath with water that was preheated to sample temperature.

Membrane Characterization

Water uptake and morphological tests were performed. Water uptake was carried out by first vacuum drying all membrane samples at 100°C until constant weights were obtained. Dried samples were then immersed in deionized water for 24h at room temperature (25°C). Samples were taken out of the water after 24h, wipped to remove surface water and then weighed immediately. Water uptake was calculated using Eq. 1.

$$water - uptake = \frac{W_{wet} - W_{dry}}{W_{dry}} \qquad \text{Eq. 1}$$

Surface morphology of the membrane was done using scanning electron microscopy (SEM). Thermogravimetric analysis (TGA) was done from room temperature to 350°C using TA Q600 analyzer at heating rate of 20°C/min under nitrogen atmosphere with flow rate of 20mL/min using ~ 8 -10mg membranes samples.

Results and Discussion

Membrane Morphology by Scanning Electron Microspcopy (SEM)

Results of surface morphological analysis performed on membrane were shown in Figure 3 and 4. The overall pictures of this figure depicts a dense membrane with some surface defects observed for most of the sonicated samples. For example, Figure 3A showed some kind of layered arrangement with no visible pores at the surface. Also, the PAMPS seems to be well distributed throughout the SPEEK. The micrograph in Figure 3B can be observed to be rough and possesses some uneven ramdom distribution of PAMPS agglomerates indicating that the distribution of PAMPS have been disturbed by the 10minutes sonication. An increase in sonication time to 20 minutes was observed to improve the distribution of PAMPS. This was however accompanied by some cracks believed to be due to the slight increase in the viscousity of the blend solution. The vicousity increase was brought about by the excape of some fraction of the solvent during sonication. As the sonication time was increased, PAMPS was observed to become more evenly distributed and the cracks become more visible and larger in size (Figure 3D to F). SEM images of samples at sonication time of 10 and 30 minutes were reproduced in higher magnification of 10μm (Figure 4) in order to further revealed the effect of sonication on the morphology of the somposites. Similar morphological behaviour has been reported in literature, though for inorganic/organic clusters of nanoclay/epoxy composites membranes (Lam et al. 2005).

Water Uptake

Water uptake have been reported to be affected by the degree of sulfonation, membrane pretreatment, hydration, ambient relative humidity and water temperature (Xing et al. 2004; Wu et al. 2006). Figure 5 showed the effect of sonication time and solvent type on water uptake that was measured at room temperature (25°C). For both solvents, water uptake is same at zero sonication time. There was however a further decrease in the water uptake for NMP at 20 minutes sonication time before it beging to increase. Thus, the best sonication time for both solvent lied between 10 and 20 minutes with respect to water uptake. Interestingly, this result correspond to that reported by Lam (2005) on the effect of ultrasonication on microhardness of composites.

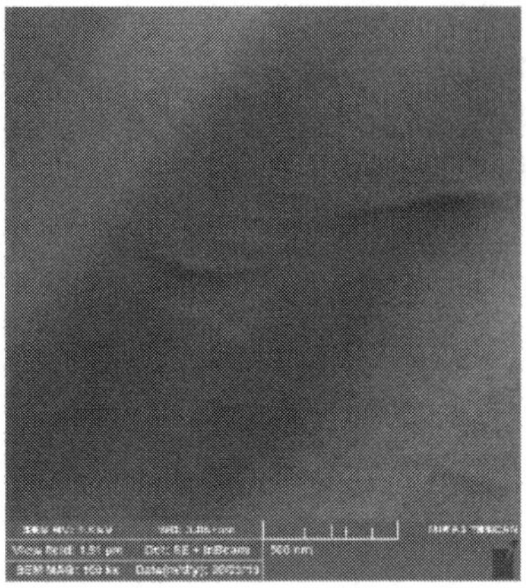

A) No sonication

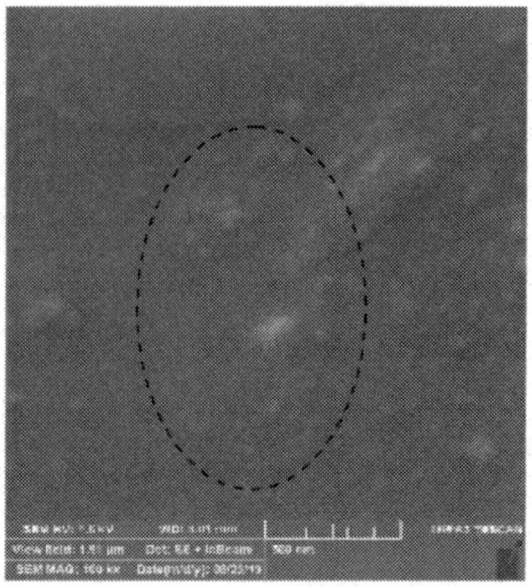

B) 10 minutes sonication

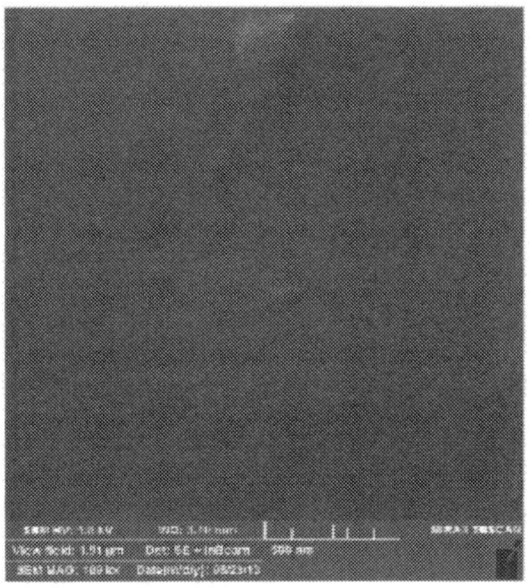

C) 20 minutes sonication

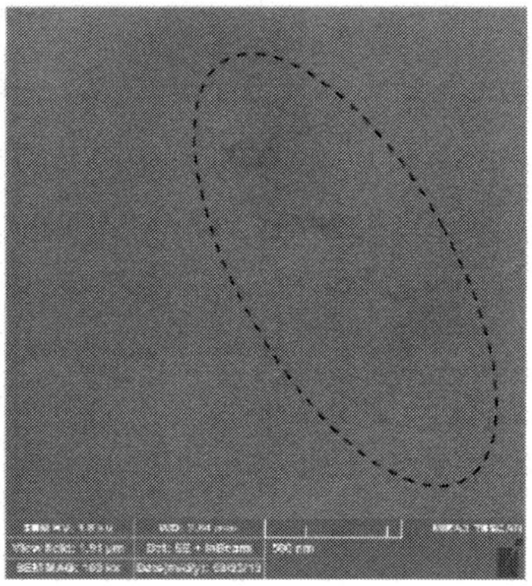

D) 30 minutes sonication

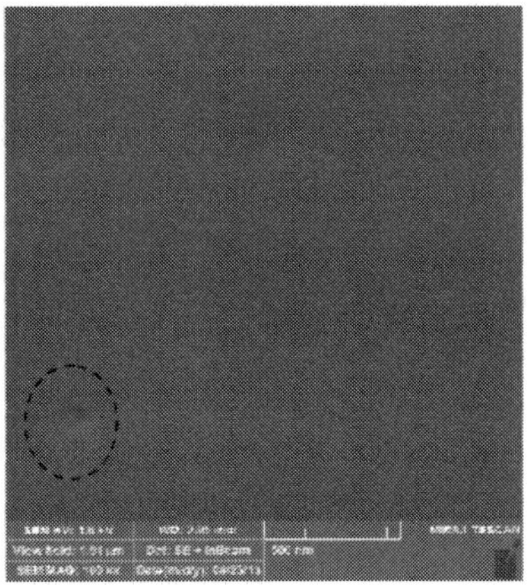

E) 60 minutes sonication

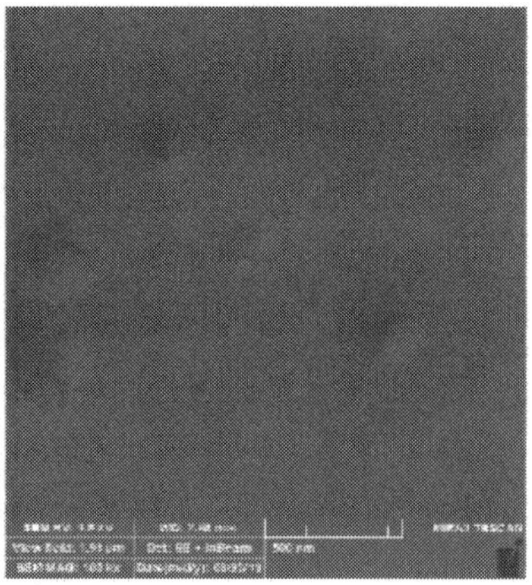

F) 90 minutes sonication

Figure 3 SEM micrograph of SPEEK-PAMS composite with various times of sonication with magnification of 500nm

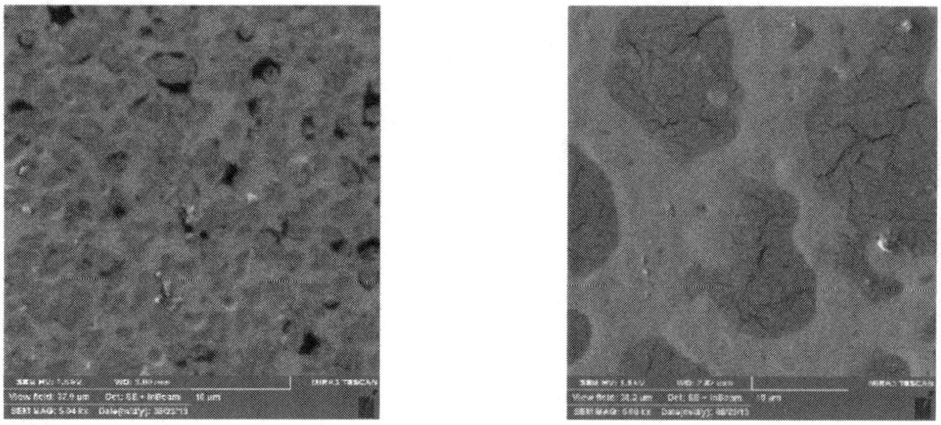

A) 10 minutes sonication B) 30 minutes sonication

Figure 4 SEM micrograph of SPEEK-PAMS composite with various times of sonication with magnification of 10μm

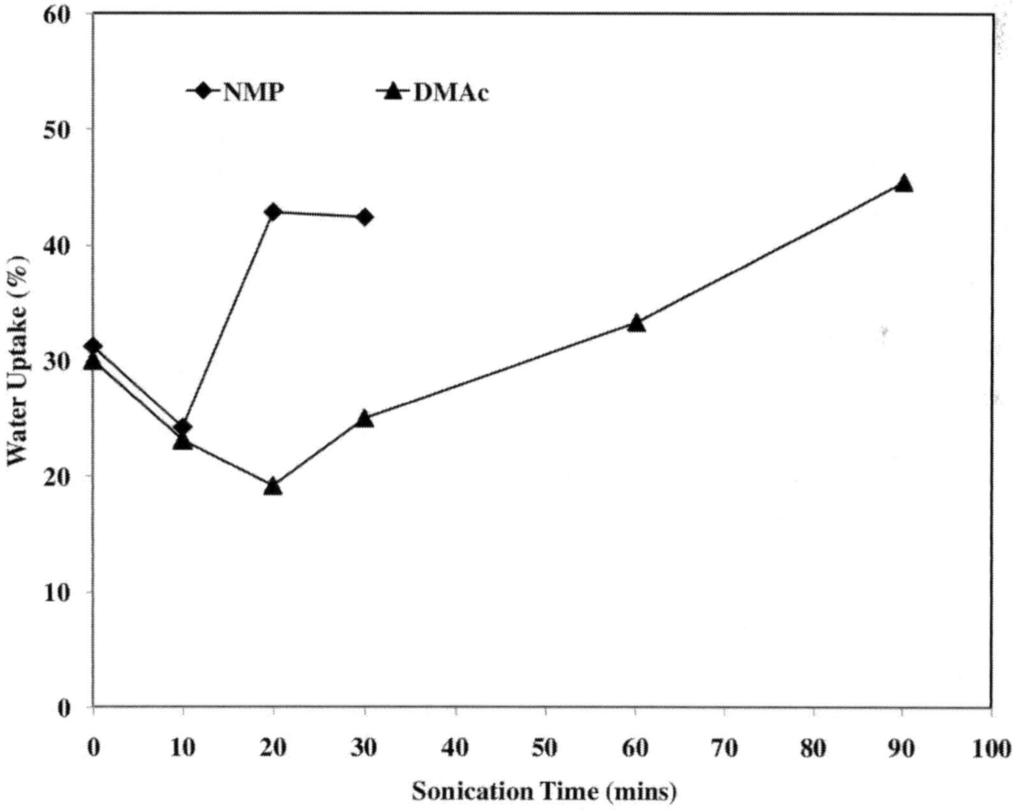

Figure 5 Water uptake of SPEEK-PAMS (3wt%) composite with variuos sonication times

Thermogravimetric analysis (TGA)

The difference in membrane performance resulting from the use different casting solvents have been attributed to the presence of residual solvents in the membrane samples after evaporation (Joly et al.

1999). TGA was thus used to quantify the residual solvent and evaluate its effect on the membrane performance as has been previously demonstrated in lieterature (Recio et al. 2007; Iqbal et al. 2008). Figure 6 shows the TGA curve for samples prepared from NMP and DMAc with different post vacuum heat treatment conditions. From this figure, weight loss can be attributed to evaporation of variety of substances, including solvents, from the samples (Rich et al. 1997). Table 2 contains the solvent and its boiling points, and weigth loss at the boiling point range that were extracted from TGA results. For sample heat treated for 24hrs, the overall percentage weight loss up to 204°C was about 8wt% of membrane sample for NMP while it is about 6.3wt% for DMAc. These will definitely include compnenents such as water, solvents and other unidentified substances. To quantify residual solvents, weight loss was estimated between 192 -214°C for SPEEK-PAMPS-NMP and between 154 and 176 for SPEEK-PAMPS-DMAc. The results obtained showed that the amount of the residual solvent in NMP is larger than that of DMAc by 1.25wt% (Table 1). To further investigate if this difference is responsible for higher value of water uptake for the NMP sample, membrane samples made from both solvents were further heat treated under vacuum for 72hrs. Residual solvent after this treatment was slightly higher in DMAc than NMP. Despite this, the value of water uptake of the NMP membrane is still higher than that of DMAc by approximately the same value as before the heat treatment (1.71wt%).

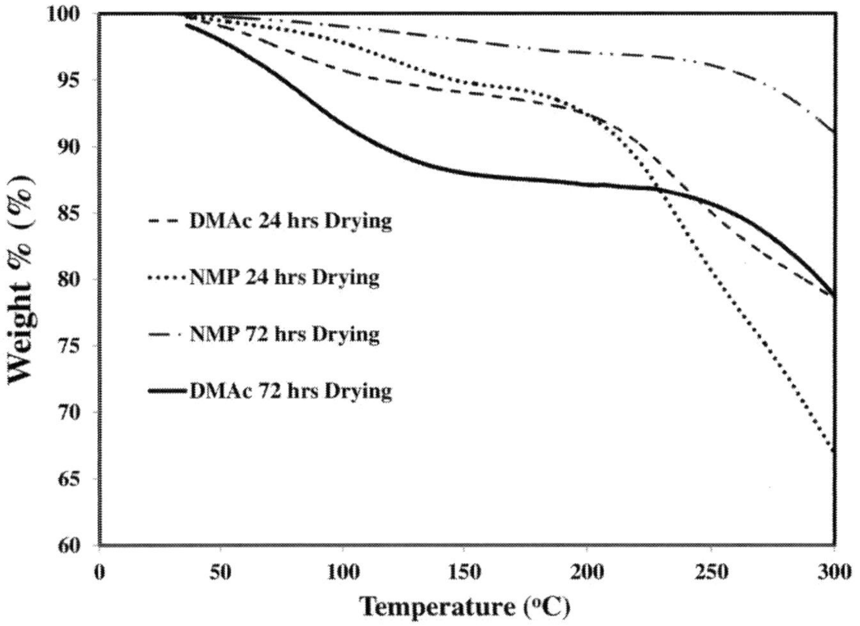

Figure 6 TGA of SPEEK-PAMPS-NMP and SPEEK-PAMPS-DMAc

Table 1 Residual solvent in SPEEK-PAMPS composites membrane samples

Membrane Sample	Solvent	B. Pt. (°C)*	Residual Solvent (wt%)	Water Uptake
DMAc 24 hrs Drying	DMAc	164.5 – 166	0.57	30.00
NMP 24 hrs Drying	NMP	202 – 204	2.85	31.25
DMAc 72 hrs Drying	DMAc	164.5 – 166	0.367	21.21
NMP 72 hrs Drying	NMP	202 – 204	0.217	22.92

*Boiling Points of Solvents

On a general note, results from Figire 1 – 5 showed that the best perfomance can be obtained for variety of sonication time depending the property of preference. Also, performance of DMAc solvent was observed to be better than that of NMP on the basis of water uptake. For this reason, DMAc was selected as the casting solvent for membrane preparation on which further studies will be carried out. Water uptake of DMAc membrane, prepared without sonication, was comapred with those reported in previuos literature. Table 2 illustrates the comparison between SPEEK-PAMPS and other relevant membranes previuosly reported in literature. In addition, this table serve to provide insight into the possible potentials of the composites. From this table, it was observed that the water uptake for SPEEK-PAMPS is within the range of values of other membranes on the table. Its water uptake is less than that of Nafion 112 which is a commercial membrane. The differences observed between the water uptake values of SPEEK – PAMPS and other membranes from the table can be explained based on three important factors - SO_3H group, the degree of sulfonation and experimental temperature (Gil et al. 2004; Shan et al. 2006; Wu et al. 2006). Water uptake of SPEEK – PAMPS is higher due to the presence of SO_3H group in the PAMPS molecules (Figure 2). SO_3H group have been reported to enhance membrane acidity and hydrophilic properties (Shan et al. 2006). Infact, the higher the SO_3H group the larger the water uptake. The degree of sulfonation has been reported to affect the water uptake of membranes. The higher the degree of sulfonation, the larger the value of water uptake (Gil et al. 2004). Thus, water uptake of SPEEK – PAMPS is lower than SPEEK -3 due their different degree of sulfonation, 53.3 and 72% respectively and the experimental temperature 24 and 60°C, respectively. The experimental temperature of SPEEK – 2 is close to that of the composite. However, the results cannot be compared in this case since the DS of SPEEK – 2 was not known. Above explanation is also true for the values of water uptake for SPEEK – 1.

Table 2 Comparison of water uptake values obtained with those from Literature (Xing et al. 2005)

Membrane	Water Uptake (wt%)	Experiment Temperature (°C)	**D S (%)	Reference
Nafion 112	22.85	N/A	-	(Xing et al. 2005)
SPEEK – 1*	16.00	30	45	(Ren et al. 2005)
SPEEK – 2*	44.90	25	N/A	(Zhong et al. 2008)
SPEEK – 3*	85.00	60	72	(Wu et al. 2006)
SPEEK– 4*	15.00	N/A	N/A	(Handayani and Dewi 2011)
SPEEK/PAMPS (3wt%)	21.21			This work

*The numbering have been included for explanation purposes.

**DS means Degree of Sulfonation

Conclusion and Recommendations

Blend of SPEEK and PAMPS was prepared by solvent casting method and its performance was evaluated with respect to its morphology and water uptake. Effect of sonication time (10 – 90minutes) and solvent types (DMAc and NMP) were also investigated. Performance of membrane prepared using DMAc was observed to be better than that of NMP. There was no correlation between the sonication time and membrane properties, but it was observed that the sonication deteriorate the membrane morphology. Membrane prepared without sonication shows better surface morphology, whereas the best water uptake was observed between 10 and 20 minutes. Thus, other properties need to be investigated in order decide the best sonication time. Therefore, further investigation such as thermal stability, methanol permeability, ion exchange capacity and gas (O_2 and H_2) permeability of the membrane are recommended for a complete evaluation of the membrane performance for direct methanol fuel cell (DMFC) application. All these investigations are in progress in our laboratory.

Acknowledgement

Authors would like to thank the National Science, Technology and Innovation Program (NSTIP) of KACST for funding this research under project number 10-ENE1374-04. Also, authors extend their appreciation to Center of Petroleum & Minearls, Center of Excellence in Renewable Energy, Mr. Dan Jerry D.Cortes of the Nanotechnology Laboratory, Chemical Engineering Department, King Fahd University of Petroeleum & Minerals, Saudi Arabis for their supports.

REFERENCE

Florjańczyk, Z., et al. (2003). "Proton conducting electrolytes based on poly(2-acrylamido-2-methyl-1-propanesulfonic acid)." Electrochimica Acta **48**(14–16): 2201-2206.

Gil, M., et al. (2004). "Direct synthesis of sulfonated aromatic poly(ether ether ketone) proton exchange membranes for fuel cell applications." Journal of Membrane Science **234**(1–2): 75-81.

Handayani, S. and E. L. Dewi (2011). "Influence of Silica/Sulfonated Polyether-Ether Ketone as Polymer Electrolyte Membrane for Hydrogen Fueled Proton Exchange Membrane Fuel Cells." Internat. J. of Sci. and Eng. **2**(2): 26-30.

Iqbal, M., et al. (2008). "Solvent effect on morphology and CO2/CH4 separation performance of asymmetric polycarbonate membranes." J. Membr. Sci. **318**(1–2): 167-175.

Joly, C., et al. (1999). "Residual solvent effect on the permeation properties of fluorinated polyimide films." Separation and Purification Technology **16**(1): 47-54.

Lam, C.-k., et al. (2005). "Effect of ultrasound sonication in nanoclay clusters of nanoclay/epoxy composites." Materials Letters **59**(11): 1369-1372.

Mohr, J. M., et al. (1991). "Gas transport properties of a series of poly(ether ketone) polymers." Polymer **32, 13**: 2387-2394

Recio, R., et al. (2007). "Gas separation of 6FDA–6FpDA membranes: Effect of the solvent on polymer surfaces and permselectivity." J. Membr. Sci. **293**(1–2): 22-28.

Ren, S., et al. (2005). "Surface modification of sulfonated poly(ether ether ketone) membranes using Nafion solution for direct methanol fuel cells." Journal of Membrane Science **247**(1–2): 59-63.

Rich, D. C., et al. (1997). "Effect of cure conditions on Probimide 32 polyamide-imide." Journal of Applied Polymer Science **63**(9): 1113-1126.

Santos, H. M., et al. (2009). The Power of Ultrasound. Ultrasound in Chemistry: Analytical Applications. J.-L. Capelo-Martinez. Weinheim, Wiley-VCH.

Shan, J., et al. (2006). "Sulfonated polyether ether ketone (PEEK-WC)/phosphotungstic acid composite: Preparation and characterization of the fuel cell membranes." Pure Appl. Chem **78**(9): 1781 - 1791.

Swier, S., et al. (2005). "Sulfonated poly(ether ketone ketone) ionomers as proton exchange membranes." Polymer Engineering & Science **45**(8): 1081-1091.

Wu, H.-L., et al. (2006). "Preparation and characterization of poly(ether sulfone)/sulfonated poly(ether ether ketone) blend membranes." European Polymer Journal **42**(7): 1688-1695.

Xing, D. M., et al. (2005). "Characterization of Sulfonated Poly(Ether Ether Ketone)/Polytetrafluoroethylene Composite Membranes for Fuel Cell Applications." Fuel Cells **5**(3): 406-411.

Xing, P., et al. (2004). "Synthesis and characterization of sulfonated poly(ether ether ketone) for proton exchange membranes." Journal of Membrane Science **229**(1–2): 95-106.

Ye, Y.-S., et al. (2012). "Water Soluble Polymers as Proton Exchange Membranes for Fuel Cells." Polymers **4**(2): 913-963.

Zhong, S., et al. (2008). "Modification of sulfonated poly(ether ether ketone) proton exchange membrane for reducing methanol crossover." Journal of Power Sources **180**(1): 23-28.

Fixed Plant Facility Siting – A Mega Project Lifecycle Framework

M.F. Haider
mech_haider@yahoo.com
mhaider@suncor.com

Prepared for Presentation at
American Institute of Chemical Engineers
2014 Spring Meeting
10th Global Congress on Process Safety
New Orleans, LA
March 30 – April 2, 2014

Fixed Plant Facility Siting – A Mega Project Lifecycle Framework

M.F. Haider
mhaider@suncor.com
mech_haider@yahoo.com

Keywords: Facility Siting, Project Management, Mega Project, Consequence Analysis (CA), Quantitative Risk Assessment (QRA), Adaptive, Change driven, Agile methodology

Abstract

Facility Siting based on tools such as Consequence Analysis (CA) and Quantitative Risk Assessment (QRA) is often times successfully applied for Brownfield operating unit projects, projects with relatively smaller footprints and for projects lasting relatively short time-periods. Challenge remains to apply and realize the full potential and value of Facility Siting for Mega-Projects. This is mostly due to project complexity and length – this paper will introduce a Management Framework to successfully apply CA and QRA as part of Facility Siting for Mega-Projects. The paper will also outline key strategies to ensure successful (i.e. value adding) implementation of facility siting, suggest a list of activities, associated sequence and pitfalls to avoid. Project Management team (e.g. Project Sponsor, Project Engineers, Managers) of both owner/operator and contractor/vendor may find this paper more useful as it provides a broad overview (i.e. when to initiate, how to plan, execute and close such assignment to maximize value) including real life examples of facility siting process over the project lifecycle.

1. Introduction

The driving force for process safety has been primarily based on catastrophic events [1]. With an increasing number of incidents, the process industry and governments started to take measures such as Facility Siting initiatives with hope to minimize loss of life and property. In the 1970s and 1980s, some of the world's most tragic industrial accidents took place. The Flixborough explosion in 1974 was reinforced by that of the Seveso tragedy in 1976 [2]. The Macondo blowout in the Gulf of Mexico on April 20, 2010, killed 11 employees and led to an uncontrolled oil spill lasting 87 days. This blowout was the most significant offshore incident in the US to-date. Finally, the Fukushima Daiichi nuclear plant incident in Japan in March 2011 drew the attention of the global process and power industries and encouraged them to incorporate natural disaster risks in a hazard analysis study [3]. As part of inherently safer design, many companies now-a-days apply CA and QRA as part of facility siting for their new projects. Due to the scarcity of information at the beginning, the studies in many cases start later in the project lifecycle for them to be effective and value adding. This paper will focus more on how to manage such initiatives for new Mega-Projects over the technical intricacies of the CA/QRA processes. The intention of the paper is to enhance the Body of Knowledge publicly available for management team in as much non technical languaga as possible to make their greenfiled Mega-Project initatives a success i.e. safer for people, environment and property.

So, what is different for industrial Mega-Projects?

- Industrial Megaprojects, as the name implies, are very large. To provide a simple and simply applied definition, we are defining a megaproject as any project with a total capital cost of more than $1 billion (U.S. dollars) as measured on January 1, 2003 [4].
- Most often engineering and construction activities run in parallel and not in sequence
- Multiple sub-projects under the 'Mega-Project umbrella' run in parallel
- Complex – design is almost 'never' complete in early stages – goes through many iterations
- Progressive elaboration is evident
- Scarcity of critical workforce with mega-project experience
- Scarcity of information – requires project team to make assumptions for CA/QRA
- Timeline: 2-5 years

2. Key Strategies

2.1 Facility Siting Charter

It is not only important or good to do but critical that the stakeholders need to identify, understand and agree on the purpose and value to conduct facility siting for their projects. The Author's personal experience over the years has been in order to maximize value of Facility Siting the project team has to start 'early' to develop project Charter and start with the end in mind. As can be seen from the following generalized Figure 1 that as the company starts to appraise opportunities, it is

Figure 1: Stages and Milestones to apply CA/QRA in a Mega-Project

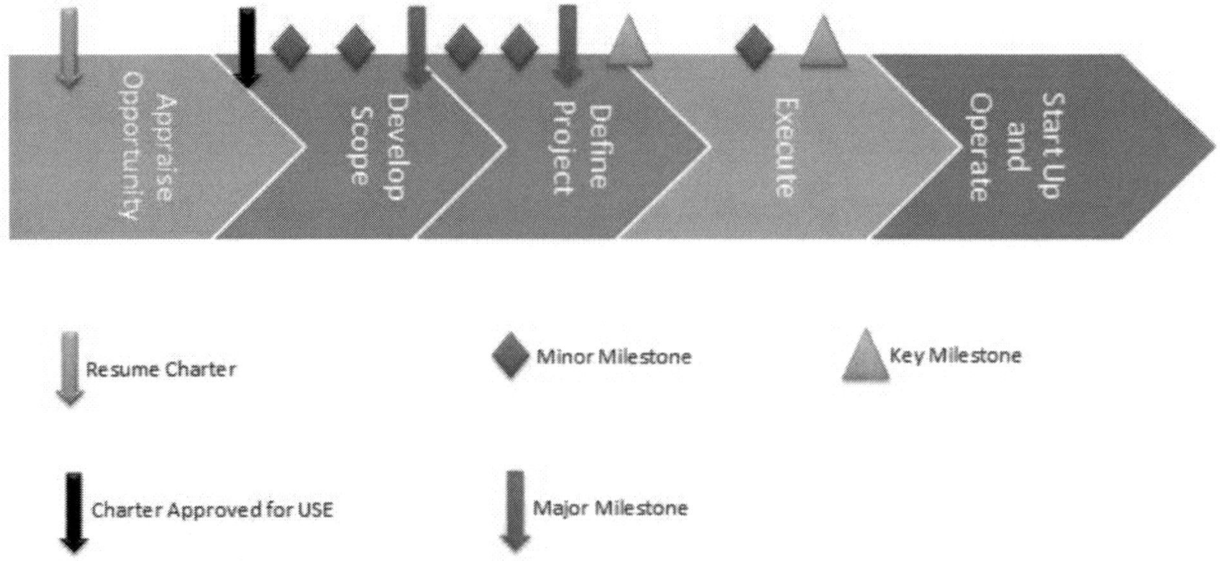

prudent to resume developing the Facility Siting Charter by bringing in the stakeholders together. Figure 1 shows a generalized project lifecycle for a typical Petro-Chemical Mega-Project. The

Figure also shows that the Key Deliverable is developed through a successive series of repeated cycles over the project life-cycle. The Milestones will be defined later in the paper. Part of starting with the end in mind is to clearly outlining ultimate goal/s, e.g. vision is to design a safer facility and utilize Mitigated QRA Report to formally document and communicate that facility is within the Risk Threshold or justification for accepting risk profile as is by Senior Management. The Vision in the Charter also provides opportunity to strike a balance between business goals and climbing the Risk Mountain – at each stage of increasing safety (decreasing risk). As part of striking a balance, the Project Engineer has to Identify options up-front if and/or when it becomes apparent that aggregate Residual Risk will not be within the Risk Threshold or acceptable so that team can put more effort in developing justification for either accepting or rejecting the risk profile by Management

Figure 2: Strike a balance between Business goal and Climbing the *expensive* Risk Mountain

2.2 Management Strategy

In order to successfully manage Facility Siting, CA and QRA for a Mega-Project, the stakeholders have to align on a management strategy. There are two primary management approaches for the purpose of managing such work:

1. Predictive or Plan driven
2. Adaptive or Change driven

The Author's past experience has been if the organization undertook very 'similar' projects in the near past (<5 years), critical experienced workforce is retained, Lessons Learned is readily available, and most importantly, the past project is still considered to be a success, it is advisable to use Plan driven approach. On the other hand, if the organization is introducing new technology, has limited past experience with similar project, involves relatively high number of stakeholders, capital cost > $1B, project life spans >2 years, it is recommended to utilize Change driven approach to maximize value. It is to be noted that it is not essential to meet all the suggested criteria mentioned above to pick one management strategy over the other.

2.2.1 Benefits of Adaptive or Change Driven Approach

In Adaptive/Change Driven Planning, project team intentionally repeats one or more project activities as the project team's understanding of the product increases ……….Also known as …agile methods …..intended to respond to high levels of change and ongoing stakeholder involvement [5]. The heart of Facility Siting involves making timely but informed decision. Some of the characteristics of this framework:

1. To add small incremental improvements that will deliver value to stakeholders
2. Milestones develop the product (e.g. Mitigated QRA Report) through a series of repeated cycles, while increments successively add to the functionality of the product
3. For each Milestone, a deliverable (read 'model') or set of deliverables will be completed. Future iterations may enhance those deliverables or create new ones. Each iteration incrementally builds the deliverables until the exit criteria for the phase are met, allowing the project team to incorporate feedback.

The biggest challenge with Mega-Project is scarcity of key information needed as inputs to develop scenarios and models. Some of the critical benefits, a project may gain by applying Change Driven approach:

1. The approach acts as Leading Indicator on Process Safety for Plant Layout - allows for multiple opportunities to make critical decisions on plant layout over the project life-cycle capitalizing on what is known at a given time. This lets the project team to use "what they know", to ensure "what they don't know" falls into the right place
2. Opportunity to make decisions over the project lifecycle not only works to save life but also helps with striking a balance between business needs and climbing the risk mountain
3. Aides design-development with progressive elaboration while producing final CA/QRA deliverable for the project

2.3 Execution Strategy

The main idea of Change Driven process starts with decomposing the known preliminary layout and process information into a set of requirements, work to be performed and their priority sequence, sometimes referred to as a *Product Backlog*.

It is advisable that the owner/operator forms a multi-use agreement (contract) with the chosen vendor. This way the plant areas under the Mega-Project umbrella can get their specific work done while the CA/QRA Project Team provides overall oversight and guidance.

At the beginning of an iteration, the team will work to determine how many of the highest priority items on the backlog list can be delivered within the next iteration. The time-length of each iteration should be less than 4 weeks. As shown in Figure 1, during "Develop Scope" and "Define Project", the length of each iteration could be even less than 2 weeks. This iterations (otherwise known as "Minor Milestone" as shown on Figure 1) will develop scenarios, assumptions and preliminary CA/QRA models. Assumptions are required due to lack of available quality information in the early stages of a project.

At the end of each iteration, the Minor Milestone/deliverable should be ready for review by the customers/stakeholders. This does not mean that the customer is required to accept delivery, just that the product should not include unfinished and/or incomplete features.

The sponsor and customer representatives should be continuously engaged with the other project stakeholders to provide feedback to Contractor on deliverables as they are created and to ensure that the *Product Backlog* reflects Sponsor/stakeholders' current needs. Below, you will find two examples using two different management strategies. For confidentiality purpose, the specific name/location is kept generic and few items were modified to suit.

2.3.1 Project X Facility Siting Using Change Driven Approach

A CA/QRA team was given the assignment to support design development during "Develop Scope" phase (Figure 1). The Sponsors already selected the site for the project. But the fixed plants were yet to be located on the site. The Vision for the initiative was for the QRA result to come under the organization Risk Threshold. Multiple EPC's were given to engineer different plant areas. The CA/QRA team selected a vendor to support them with defining the *Product Backlog* based on the preliminary Process Design Basis, intended product the facility is expected to sell, the feedstock necessary etc. The CA/QRA team had prioritized the Product Backlog based on hazard level of feedstock, preliminary PHA schedule and project schedule. This also involved obtaining feedback from different Process Engineering and Operations teams to collect necessary inputs to put together the assumptions and scenarios to be modeled within the 1st iteration.

Out of 25+ scenarios, only 7 perceived to be the most hazardous were modeled for CA that gave ample input (e.g. explosion, toxic and fire contours) for the CA/QRA team to work with the stakeholders to recommend location for specific process units on the given site. Note that, CA/QRA team integrated the potential consequence to understand local incident and global consequence events. At this time QRA was not performed. This was used as input for minor modification of preliminary plant layout. It is to be noted that the site plan was still preliminary after the first iteration but assisted different EPC accountable for their specific area to recognize where to locate the specific plants i.e. finalized key plans for the overall site plan.

With further design-development, more refined information became available from each EPC. This new information led to 130+ scenarios and was used for further CA/QRA iterations that created either new model and/or embellished the past ones. Each EPC was accountable for facility siting within their boundary while the CA/QRA team provided technical oversight and integration support for the overall project. The Sponsor/stakeholders were engaged on a regular basis to ensure that the Product Backlog reflects future modeling requirements to properly site facilities e.g. Fire Water Pump house, Tank Farms, site Medical Centre etc. Early engagement of CA/QRA team and higher management willingness to incorporate their feedback with the progressive elaboration of the project led to QRA result meeting the organization risk threshold and ultimately safer facility. Changes suggested based on CA/QRA work in earlier stage (i.e. Develop Scope, Define Project) were easier to incorporate into the design.

2.3.2 Project Y Facility Siting Using Plan Driven Approach

The sponsors for project Y selected the site for the project and decided to hold on to not perform CA/QRA work with the progressive elaboration of the project activities. This was mostly decided based on the fact that not all information was available to properly conduct the 'studies'. Fundamentally, the sponsors looked CA/QRA as deliverables requiring a more 'complete' plan. Once the Develop Scope (as referred in Figure 1) is almost complete, the sponsors decided to engage CA/QRA team and a vendor. By this time, the project had significantly high number of stakeholders. Many of the stakeholders were not very familiar with CA/QRA and the value it could bring to them. There was not much time to schedule and train the stakeholders as well to bring them up to speed. As a result, the stakeholders were unable to align on a Vision. The Vendor was asked to provide explosion, toxic and fire contour maps for specific scenarios and locations that were almost 'fixed' on the site plan at that point in time. The resultant contour maps showed significant consequence on nearby facility with over 700+ people. It was also noted that the plant area consuming the most hazardous substance would impact the nearby facilities significantly from consequence perspective. It should be noted that the result was only based on a selected few scenarios. By the time the project reached the "Execution" phase, more scenarios were added as part of the scope. But it was already too late to incorporate most of the recommended changes resulting from CA/QRA work. The project went through several area by area risk reduction workshops only to accept (by the company executives per organization risk matrix) the aggregate residual risk level which was also found to be not within the organization risk threshold.

2.3.3 Importance of Starting Early

Figure 3: Impact of variable based on project time [5]

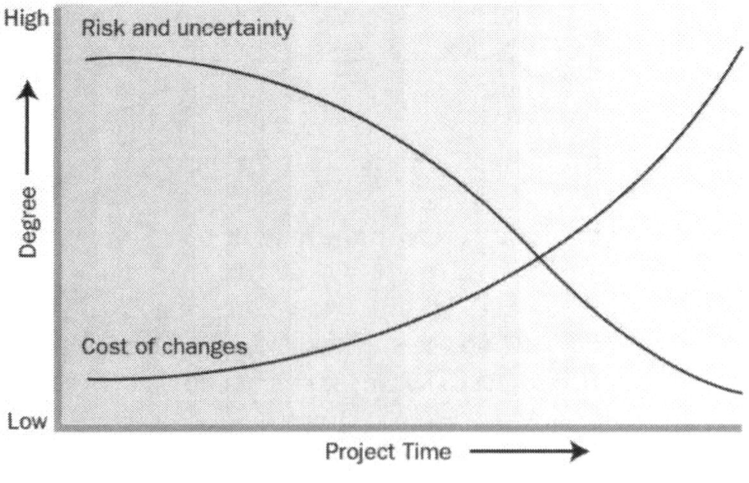

The above two examples highlight the importance of starting CA/QRA work early enough to realize and maximize the value CA/QRA brings to a Mega-Project. The high-level view provides common frame of reference for comparing projects for CA/QRA initiative, even if the projects were dissimilar. Change driven methodology allows for applying the known information at early stage in a way that supports laying out plant facilities to reduce risk and thus maximize potential value. While these characteristics remain present to some extent in almost all project life cycles, they are not always present to the same degree [5]. Change Driven framework, in particular, keeps stakeholder influences higher and the costs of changes lower especially in the early stages of the life cycle than in predictive/plan life cycles. The Figure 3 highlights a well-known fact of project management that even though risk and uncertainty remains high in the early stages of a project, cost of incorporating changes remains much lower. This translates into a key opportunity for project stakeholders to align on laying out the key areas on a site plan intelligently to reduce risk and maximize value by applying tools such as CA and QRA.

2.4 Key Decision Points

Facility Siting, from a management point of view, is all about bringing the stakeholders closer and find the 'right' spot for the facilities. Finding the 'right' spots requires making timely decisions. The challenge has always been to make timely decisions as the stakeholders more often find themselves forced to make those decisions based on information that are frequently of poor quality and at times uncertain, especially in the early stages of a project. At times, due to the scarcity of information, stakeholders rely on their past experience to make those decisions on plant layout. As a result, depending on the individual experience, two different teams for a project area may come up with two very different plant layout. CA/QRA work, if started early enough, will also depend on input that most often requires embellishment through assumptions to create needed scenarios and models. Even with this limitation, CA/QRA provides a key 'input' in allocating real estate for key facilities on a plant site at early stage of a project lifecycle. With each iteration and deliverable 'Change driven' methodology allows for multiple opportunities to make critical decisions on plant layout over the project life-cycle. It allows the stakeholders to decide when to 'sharpen the pencil' (e.g. PHAST to Computational Fluid Dynamics model), relocate facility or accept the consequence and/or risk, 'pull the plug' on further modeling/refinement and start developing justification, if required, to recommend management to accept/reject the aggregate risk profile.

2.5 Alignment with Process Hazard Analysis (PHA) Planning

CA/QRA results are a useful input to PHA (e.g. HAZOP, LOPA) activities. This is particularly true as projects move forward after establishing the plant layout on a high level and starts to conduct "Define Project" (Figure 1). PHA activities may also result in recommendations (i.e. new scenarios) that goes into the *Product Backlog* and requires project team to develop new models or refine old ones. Mega-projects have multiple sub-projects running in parallel most of the time, having an Integrated Schedule for all the reviews is a good strategy so that the right stakeholders are booked for the sessions. The integrated PHA schedule also supports prioritization of *Product Backlog* for CA/QRA activities over the lifecycle of the project.

2.6 Integration Strategy

It is critical for a Mega-Project involving multiple plant areas/silos to understand the BIG picture so that site risk can be dealt with on an aggregate basis. There is strong possibility of one area to impact other plants in the vicinity. It is recommended to review integration aspect and aggregated results on a regular basis. The following Figure 4 shows that the origin of an incident may reside in one particular plant area, while it may impact other areas in the vicinity.

Figure 4: Consequence of an incident in one area impacting multiple areas

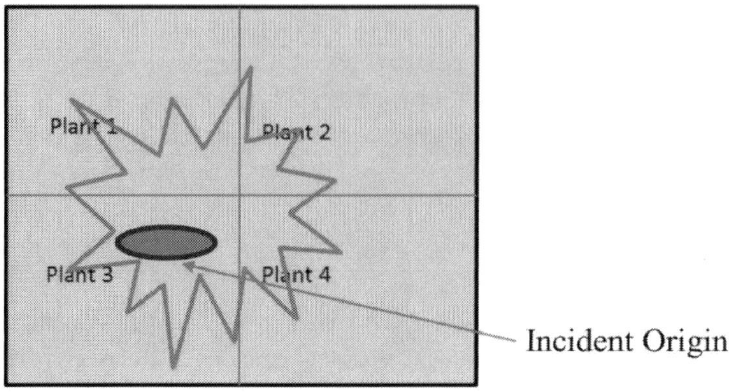

Recommended Strategy:

- Setup Strong Center - Task Force Single Point of Contact (SPOC)
- Collaborate – SPOC will promote Collaboration among Plants/Areas

3. List of Activities (suggested)

3.1 Overall Sequence

The CA/QRA work typically should start with the approved Charter as shown on Figure 1. A project Task Force i.e. SPOC is formed comprising of Project Engineer, Operations rep, Process rep and CA/QRA Technical Expert. The SPOC is ultimately accountable to produce the CA/QRA deliverable studies but also acts to support the design development and site plan finalization efforts in the early stages of the project.

The team collects 'input' such as Preliminary site Layout and design information, works with area process and operations team to identify scenarios, develop assumptions for known-unknowns and unknown-unknowns, develop modeling 'Product backlog', prioritize backlog based on PHA schedule and produce modeling output (i.e. deliverable) for stakeholder use. The stakeholder usually has the option of accepting or rejecting consequence. If the result of an iteration is acceptable within area causing the incident – communicate to outside stakeholders to ensure there is no cross boundary impact. If the result of the iteration is NOT acceptable to outside of the area – use the result as input to PHA for the areas to mitigate. Collaboration recommended between origin area and impacted area/s to identify appropriate mitigations. There is also a possibility that the consequence resulting from the iteration is not acceptable to the area causing the incident, in that case the area will initiate a PHA to mitigate the consequence. As part of developing mitigations, consider the following:

- Develop QRA if appropriate
- Perform Building Assessments for Critical Buildings
- Apply Cost-Benefit analysis to optimize risks/opportunities

- Determine Safe Haven locations and their characteristics

If Mitigation is not possible from Business perspective – perform As Low As Reasonably Practicable (ALARP) justification for Senior Management to formally accept the posed risk

- To apply mitigations on a site wide basis, review Milestone deliverables by taking on a BIG picture perspective
- Future iterations may enhance those deliverables or create new ones as new information is available. Each iteration incrementally builds the deliverables to produce the Mitigated CA/QRA report to support Facility Siting for the stage

The overall effort is expected to produce a maximum of two single reports for the overall site:

1. Mitigated version of QRA Report
2. and Consolidated CA Report

It is recommended to develop as-built, pre-start up CA/QRA reports prior to start-up. Table 1 and 2 outlines typical activities performed by SPOC and plant areas respectively once key process areas are established/finalized on Site Plan. Please note that the activities are project dependent and may warrant modifications to suit specific situation.

Table 1. SPOC Sequence of Activities to conduct Facility Siting

Initiative	Key Steps
Setup Facility Siting Team	Sponsor to support to setup a team (i.e. Project Engineer, Technical Expert, Operation, Process representative) to act as SPOC
Integration	Provide functional (technical and management) guidance to Areas
	Assess cross boundary impact/opportunity
	Communicate incremental outputs after each Area PHA/Risk Reduction
IFA versions (prior to "Define Project" stage)	Consolidated CA Report
	Mitigated and Integrated QRA Report
IFD CA/QRA (prior to "Execution" stage)	Re-validate IFA CA/QRA post PHA and vendor PHA
As-built versions	Post-oil version

3.2 Typical Area Activities

Table 2. Area (Typical) Sequence of Activities to conduct Facility Siting

Initiative	Key Steps
Plant 1 (Typical)	Perform Modeling and Calculations
	Conduct Risk Reduction Workshop (if appropriate) – Apply mitigations
	Incremental output for use based on reduced risk level
	Assess Buildings/facilities
	Identify Safe Haven/shelter Locations (if required)
	Estimate Area Project Scope impact
	Perform area scope updates/changes e.g. passive fire protection requirement
	Carry out QRA (if appropriate)
	Update Product Backlog
	Develop/update contract to support area specific activities

4. Pitfalls to Avoid

The following observations were noted by the Author during his time to manage Facility Siting scope:

1. Facility Siting does not replace good and common sense engineering practice e.g. locating a plant that uses high volume of hazardous material will significantly increase risks to other plants/areas (e.g. camp) within the vicinity
2. Evaluate effectiveness of 'Cost Reduction' philosophies as opposed to added risk e.g. different modularization techniques may decrease the plant footprint by reducing pipe lengths but makes the plant more compact and likely to increase severity of explosion consequence
3. Avoid developing models with hope to realize risk reduction
4. Sharpening the pencil for models most likely may not reduce risk
5. Avoid jumping into CFD or other similar methods without evaluating cost-benefit for undertaking such effort
6. Utilization of QRA results is much more controversial than the methodology. Watson [6] has suggested that QRA should be considered as an argument, rather than a declaration of truth. In this view – focus on accuracy over precision
7. CA/QRA Task Force should be part of Management of Change (MOC) process
8. Evaluate carefully before locating facilities e.g. tank farms close to main entrance/egress as part of Facility Siting

9. Keep the focus on VISION for the overall effort to avoid scope creep
10. Note the difference between Facility Siting as a Deliverable and tool for design-development activities

5. Conclusion

This paper presented Adaptive/Change driven framework to manage initiatives like facility siting/CA/QRA as part of Mega-Projects. Change driven framework does not mean "lack of planning or no planning". As demonstrated in the paper, this methodology starts with a vision and then develops the ultimate product both iteratively and incrementally when it is possible to define small incremental improvements that will deliver value to stakeholders. This allows making key decisions on laying out the facilities in early stages of a project rather than waiting for design to develop to a significant extent. Often times, project team wait too long for design to mature enough before applying CA/QRA to validate consequence impact and/or risk level. Most often, especially for Mega Projects, this leads to delays in executing the CA/QRA work properly at the right time to influence site layout. This delay in execution leads to scenarios when scope changes cannot be incorporated anymore. This translates into value erosion as project was unable to reap the full benefit of Facility Siting. The management framework recommended in this paper i.e. Change driven/adaptive framework allows for the stakeholders to start early in the project lifecycle to apply CA/QRA in laying out the plant facilities. The key is to start EARLY but START with the END in mind. Alignment among stakeholder on a VISION up-front is also important as this is expected to save life while striking a balance between business needs and climbing the Risk Mountain. Vision also allows avoiding scope creep while applying Adaptive/Change driven approach. It is highly recommended that stakeholders focus more on accuracy over precision. In the early stages of project lifecycle, quality information is scarce. Change driven/Adaptive framework gives you as the project stakeholder the opportunity to use "what you know" intelligently to ensure "what you don't know" falls into the right place.

6. References

[1] M. Mannan, A. Chowdhury and O. Reyes-Valdes, "A portrait of process safety: From its start to present day," College Station, 2012.

[2] M. Mannan, Lees' Loss Prevention in the Process Industries, Volumes 1–3 (3rd Edition), , 2005, Elsevier, 2005.

[3] K. McAndrews, "Consequences of Macondo: A Summary of Recently Proposed and Enacted Changes to US Offshore Drilling Safety and Environmental Regulation," in *Society of Petroleum Engineers, Americas E&P Health, Safety, Security and Environmental Conference*, 2012.

[4] E. W. Merrow, Industrial Megaprojects Concepts, Strategies and Practices for Success, New Jersey.: John Wiley & Sons, Inc., 2011.

[5] A Guide to the Project Management Body of Knowledge (PMBOK® Guide) – Fifth Edition, Newtown Square: Project Management Institute, Inc., 2013.

[6] S. Watson, "The Meaning of Probability in Probabilistic Risk Assessment," *Reliability Engineering and System Safety,* vol. 45, pp. 261-269, 1994.

348480 Global Energy and Transportation and Mobil Oil

Abstract:

Biofuels, Computational Methods, Engineering Economics, Polymers, Thermodynamics. Polymers-plastic, Thermodynamics- oil, Engineering Economics – the money it takes to finish the method of Global Engineering Research, Computational Methods – numerical input of analytical data, Biofuels – the transport fluids that utilized in more than one method from industrial usage to everyday living which enters the atmosphere. Exxon has completed a $600 million dollar deal for production of algae biofuel, amongst Mobil Oil has performed major research Global Energy and the effects of oil. Reading a most recent article on Mobil website.

348748 Characterization of Iron Phthalocyanine As the Active Material for Lithium Batteries

Abstract:

Results on the determination of the lithium-ion diffusion coefficient in the iron phthalocyanine (FePc) active material as a function of intercalation level, using a modified electrochemical impedance spectroscopy technique, will be presented. Also, developed thermodynamic functions such as the Gibbs free energy, entropy, and enthalpy changes for the lithium insertion reaction with FePc, derived from cell open-circuit voltage measurements as a function of intercalation level, are presented. Also presented are estimates of the maximum amount of lithium that can be intercalated into the FePc based on constant volume and temperature conditions.

349086 Design of a Free-Fall Reactor for Fast Pyrolysis of Waste Plastics

Abstract:

Thermal pyrolysis has earned significant reputation in the recovery of valuable chemicals, especially petrochemical based fuels from various materials, including biomass and plastics. Of many available technologies, *fast* pyrolysis was preferred in his study owing to its minimal heat and mass transfer limitations and better control on process conditions. The focus of this study is towards efficient conversion of waste plastics into gaseous products consisting primarily of alkanes, alkenes and aromatics. Lab scale experiments were performed using a micro reactor to determine favorable conditions, including reaction temperature and particle size that would result in maximum yield of ethylene. Using these results, a simple industrial size radiatively heated free-fall reactor was designed for the purpose of pyrolyzing waste plastics. The goal is to design a *fast* pyrolysis reactor that improves upon inefficiencies in traditional reactors while resulting in high yield of ethylene.

349732 Biosorptive Dehydration of Ethanol/Water Azeotropes Using Compound Starch-Based Adsorbent

Abstract:

Abstract: Recently, the interest in the production of bioethanol has greatly increased as a result of the increasing fossil fuel prices and associated environmental issues. Bioethanol is made from either grains or biomass and the broth contains 6 to 12 percent ethanol. In addition, water removal represents a serious problem as ethanol form azeotrope with water at atmospheric pressure. Distillation to water-free ethanol consumes 50 to 80 percent of energy used in the fermentation ethanol manufacturing process. Various techniques have been developed to break the azeotropic point, such as salting-out method, azeotropic distillation, extractive distillation, reactive distillation, adsorptive distillation and pervaporation. Among the methods, the adsorptive option is the most attractive, especially the biosorption, which has distinct advantages over conventional methods as it is environmentally friendly, highly selective, easily available, easy to operate, cost effective and reusable in repeated cycles in the treatment of aqueous organics.

In this work, a specially formulated compound starch-based adsorbent (CSA), which consists of corn, sweet potato and foaming agent, was developed. As an exploratory study, this research was focused on the investigation of CSA for biosorptive dehydration of aqueous ethanol and optimization of conditions for its maximum adsorption under various conditions of vapor feed flow rate, bed temperature and bed height through an orthogonal design method. Furthermore, to investigate the adsorption mechanism, dynamic saturated adsorbance of CSA was studied and characterization of the adsorbent before and after adsorption, as well as after regeneration, was performed.

In the first step, the net retention time and separation factor of ethanol and water were obtained using inverse gas chromatography (IGC), the results indicated that the net retention time of water on CSA is significantly longer than that of ethanol which attested the feasibility of this adsorptive separation purpose and low temperature could be propitious to the process which was consistent with the traditional theory of adsorption.

Then, through a batch adsorption experiment, the optimum adsorption condition was determined under vapor feed flow rate, bed temperature and bed height by orthogonal design method, which indicated that the optimization condition for ethanol was at vapor feed flow rate of 1.1g/min, bed temperature of 83 and bed height of 25cm. The packing density was 445kg/m^3. The residence time is determined by vapor feed flow rate and bed height. The residence time is more influenced than bed temperature. The longer the residence time, the more the output is. But it cannot be ignored that the adsorption selectivity may decreases. Low bed temperature is benefited to the adsorption process which agrees with the result of IGC. Five cycles experiments of adsorption-regeneration were carried out at the optimum adsorption and regeneration condition, the results showed that there is practically no change in the regeneration efficiency of CSA after the cyclic process adsorption–regeneration. It is due to the properly chosen regeneration conditions assuring complete desorption of the water adsorbed without changes in the CSA structure. This unchanged adsorption capacity after a number of cycles demonstrates reusability of the adsorbent.

To study the adsorption mechanism, experiment was conducted to study the dynamic saturated adsorbance of CSA. The water adsorption capacity of CSA has been found to be similar to some molecular sieves, which is about 0.15g/g. Besides, field emission scanning electron microscopy (FESEM) and mercury porosimetry (MP) were applied to investigate the change of surface morphology and microstructure of CSA before and after the adsorption, as well as after regeneration, as shown in Fig.1 and Fig.2. The FESEM images showed that compared with the particles before the adsorption, they became plump and the void spaces became smaller after adsorption, which indicated that the CSA was saturated with water. However, the CSA particles after regeneration became wizened like the nature particles, which demonstrated that almost all the water had been desorbed from CSA. The pore size distribution (PSD) indicated that compared with the PSD curve before adsorption, the peak at 6000nm disappeared and the peak at 17000nm became remarkably smaller after adsorption, which manifested that CSA is mainly composed of macropores which play a major role in the adsorption. However, the peaks at 6000 and 17000nm reappeared after regeneration, and the curve looked exactly the same with that before adsorption, which further demonstrated that CSA after regeneration can be totally regenerated and recovered to the original structure.

These results indicated that the starch-based adsorbent can be considered as an effective and low-cost bio-sorbent to dehydrate alcohol.

349781 Purification of 2-Amino-1-Phenylethanol Enantiomers By a New Technique Combining Distillation and Crystallization

Abstract:

Stripping crystallization (SC) is a new separation technique which combines distillation and crystallization. Basically, SC is operated at a triple-point condition, in which the liquid is simultaneously vaporized and crystallized due to the three-phase equilibrium. In essence, the process is continued until the liquid phase is completely eliminated and only the pure solid crystals remain in the feed. Thus, crystal washing is not required since only pure solid crystals remain in the feed and no impurities are adhered on the crystal surfaces at the end of SC. In the previous research, SC has been successfully applied to separate p-xylene from the mixed xylenes by Shiau et al. (2005, 2008).

Phenylethanolamine, also known as 2-amino-1-phenylethanol (2-APE), is an important basic skeleton in many adrenergic drugs and is often used as ingredients of cold medicines and anorectics. As the biological activity and pharmacokinetics of the enantiomers of a chiral drug can be different, the objective of this research is to apply SC to purify R-2-APE from the chiral 2-APE mixture. The experiments show that no crystallization occurs during SC without seeding due to subcooling of the liquid enantiomers. On the other hand, seeding can eliminate subcooling during SC, leading to the R-2-APE crystals gradually formed preferentially on the seeds. Thus, seeding with magnetic stirring can be successfully employed to purify R-2-APE from the 2-APE enantiomers.

1. Shiau, L. D., Wen, C. C. and Lin, B. S. Separation and purification of p-xylene from the mixture of m-xylene and p-xylene by distillative freezing, *Ind. Eng. Chem. Res*, **44(7)**, 2258-2265, 2005.

2. Shiau, L. D., Wen, C. C. and Lin, B. S. Separation of p-xylene from the multicomponent xylene system by stripping crystallization, *AIChE Journal*, **54(1)**, 337-342, 2008.

350841 Fabrication of Low Cost Insulating Material from Kaolin Clay for Construction Purposes

Abstract:

A new low cost, thermal insulator, construction mineral polymeric material was synthesized from local Saudi Arabia resources using Inorganic Mineral Polymerization technique (MiP). In this technique, the mineral polymer is produced from the following components: clay mineral rich in koalinite, sand, sodium hydroxide solution, and water. Inorganic polymerization experiments are conducted by mixing known amounts of clay minerals and sand in a variable speed stainless-steel mixer for some period of times until a homogenous solid mixture is obtained. Sodium hydroxide solution is added to the mixture and the polymerization reaction takes place at different mixing speeds and times. The produced mixture is divided into a number of samples of known weights. Each sample is moulded in a stainless-steel cylinder at a medium pressure. Then the moulded samples are cured in an oven for 24 hours. The curing temperature was 80°C. The mineral polymerization conditions such as the ratio of clay to sand filler, particle size of the clay, particle size of the sand, concentration and amount of sodium hydroxide, speed and time of mixing were optimized to obtain stable and hard mineral polymer. Compressive strength for different specimens of fabricated mineral polymers was tested under dry and wet conditions. Values of compressive strength with values ranging from 11.5 MPa to 27.9 MPa were obtained.

351141 Thermodynamic Study of Binary PAH (Anthracene + Phenanthrene) Solid Mixtures

Abstract:

Polycyclic aromatic hydrocarbons (PAHs) are common components of many materials, such as petroleum and various types of tars. They are generally present in mixtures, occurring both naturally and as byproducts of fuel processing operations. It is important to understand the thermodynamic properties of such mixtures in order to better understand and predict their behavior (i.e., fate and transport) in the environment and in industrial operations. The Knudsen effusion technique was combined with gas chromatography and differential scanning calorimetry to measure the vapor pressure, composition, and phase behavior of binary anthracene and phenanthrene mixtures at near ambient temperatures. Mixtures of anthracene and phenanthrene exhibited non-ideal mixture behavior: In mixtures with less than 80% phenanthrene, this more volatile component preferentially sublimed and the vapor pressures decreased as the mixtures became enriched in anthracene. The vapor pressure data do not follow Raoult's law (that is, ideal mixture) predictions. Mixtures with high phenanthrene content (> 80%) showed azeotrope like behavior, in that anthracene and phenanthrene sublimed at nearly constant vapor pressure and molecular ratio. These results confirm that the behavior of few-component PAH mixtures does not follow ideal mixture models. Hence, as tars and petroleum (and potentially other organic compound-containing mixtures) weather to few component mixtures over time, complicated, non-ideal phase behavior might be observed.

351160 How Confident Are You That Your Major Accident Risks Are Under Control?

Abstract:

Major accidents in the process industries including those at Texas City , Buncefield UK and in the Gulf of Mexico have highlighted the critical role played by senior managers and executives. Effective leadership is essential if companies are to develop a positive safety culture that remains constantly vigilant towards the risk of a major process safety incident. In the UK the HSE has made leadership a key priority for the high hazard industries and there is a cross-industry initiative to develop and deliver training focussed towards board members responsible for process safety performance.

This paper explores the leadership failings that contributed to recent major accidents, and the cost to both the company reputation and the individuals involved. Key principles are discussed that can help improve leadership in the major hazard industries. These include; attitudes and decisions of senior manager affecting the investment strategy and the safety culture of the organisation, reinforcing the importance of safety by personal example, thorough understanding of major accident hazards and key risk control systems, investigating process safety incident and near misses to find the underlying causes, developing world class safety management systems, and identifying weaknesses in these systems using targeted performance indicators.

The author bases this paper on the development of a training standard for 'Process Safety Leadership' involving a cross-industry 'expert panel'. Training of senior executives in compliance with this standard has been carried out from late 2011 and feedback from these courses and the key requirements for effective leadership will be discussed.

351236 Sil Determination: Shortcomings with the Use of LOPA

Abstract:

Proposed Topic Area: Managing Risks

For a number of years, industry has been trying to apply the principles outlined in IEC 61508. In the process sector, this has resulted in a focus on the requirements of IEC 61511 (2003) and ANSI/ISA 84.00.01 (2004). These are risk based standards. They involve the identification of potential hazardous events, the causes that could lead to each event and the necessary risk reduction for each event. Where the risk reduction involves an instrumented function (e.g. a trip or an interlock), the required performance of the instrumented function can described in terms of the average probability of failure on demand. In the standards, the performance for such instrumented functions is assigned to one of four Safety Integrity Levels. The assessment of what performance level is needed in respect to a specific hazardous event is referred to as SIL Determination. In the field of SIL Determination, IEC 61511 provides examples of a number of different approaches – these are shown in IEC 61511 Part 3. These include Event Trees, Risk Graphs, Safety Matrices, Fault Tree Analysis and Layer of Protection Analysis (LOPA).

In the first few years, following the publication of IEC 61508 Edition 1 in 2000, industry generally embraced the use of Risk Graphs as the method of choice. However, over the subsequent few years, many companies using this method found that it was (a) giving higher safety integrity level targets than really needed and (b) was less flexible than other methods and consequently more difficult to use in many real situations. This has resulted in a general move to other methods and in particular to the use of Layer of Protection Analysis. Indeed, LOPA may now be the predominant "Method of Choice" for SIL Determination. As a method, LOPA gives the impression of simplicity and appears quite straightforward. It almost invites anyone with a minimum of understanding of the method to be a practitioner.
However, there are a number of shortcomings in the use of the LOPA method that have become evident through its use across the globe.

This paper will review the application of LOPA and provide insights into what may be seen as the proper use of the method and its core principles. It will also highlight various shortcomings in use that have been found during expert reviews carried out of a number of existing studies. The paper will be of benefit both to those who are new to the LOPA method, but also of value to those who have been using the method for some time.

Shale Oil and Gas – Wellhead to Fuel

Dr. Lawrence Kremer - Baker Hughes
AIChE - Hilton New Orleans Riverside
April 1, 2014

14 AIChE Spring Meeting, New Orleans, LA

BAKER HUGHES

Outline Unconventional Shale Oil

- Properties of shale oils
- Problems
  - Transportation
  - Refining

14 AIChE Spring Meeting, New Orleans, LA

BAKER HUGHES

General Characteristics of Shale Oils

- API gravity typically varies 30-55+
- Very paraffinic
- Low sulfur
- Can have hydrogen sulfide and mercaptans
- Low TAN
- Very low asphaltenes
- Light crude with very little residual material
- Many samples show great quality variation

Factor	Bakken	WCS
API	39.7	21.6
S%	0.25	3.44
Light ends	2.6%	
LSR	3.5%	3%
Light Nap	12.4	3.0
Heavy Nap	9.7	5.0
Jet	10.5	10.0
LFO	27.6	12.0
Gasoil	27.2	37.0
VacTar	6.7	29.8

14 AIChE Spring Meeting, New Orleans, LA

BAKER HUGHES

Shale Oil Property Comparison

Parameter	Eagle Ford	Bakken
API	52	40.8
TAN (g KOH/g)	< 0.05	0.09
Sulfur (wt %)	<0.2	0.304
Asphaltene (wt %)	0.1	0.41
Resin (wt %)	1.6	4.95
Filterable Solids (PTB)	225	76

14 AIChE Spring Meeting, New Orleans, LA

BAKER HUGHES

Utica Shale Crude - Three Phase

14 AIChE®
Spring Meeting, New Orleans, LA

BAKER HUGHES

Eagle Ford Crude Samples

14 AIChE®
Spring Meeting, New Orleans, LA

BAKER HUGHES

Analytical Breakdown

Value	Yellow	Red	Black
API	55.0	44.6	52.3
TAN	<0.05	0.07	<0.05
Sulfur	<0.20%	<0.2%	<0.2%
Na	1.0 ppm	1.6 ppm	1.6 ppm
K	0.3 ppm	0.4 ppm	0.5 ppm
Mg	3.4 ppm	2.9 ppm	3.0 ppm
Ca	2.6 ppm	2.8 ppm	3.8 ppm
Asphaltene	0%	0%	0.1%
Resin	0.5%	3.2%	1.6%
Filterable solids	176 PTB	295 PTB	225 PTB
P	5.3 ppm	4.6 ppm	5.0 ppm

14 AIChE
Spring Meeting, New Orleans, LA

BAKER HUGHES

Micrograph of Sludge

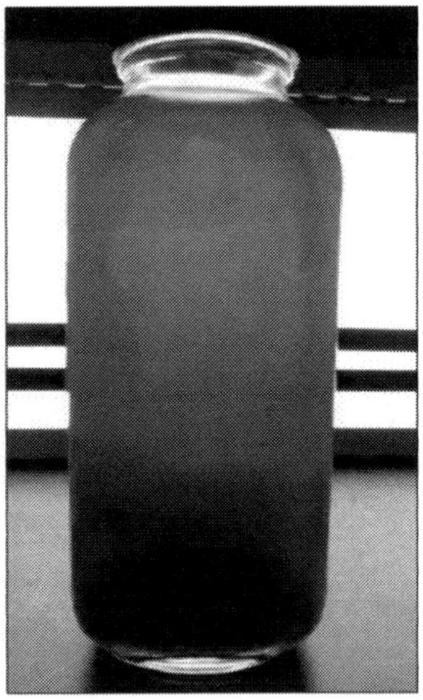

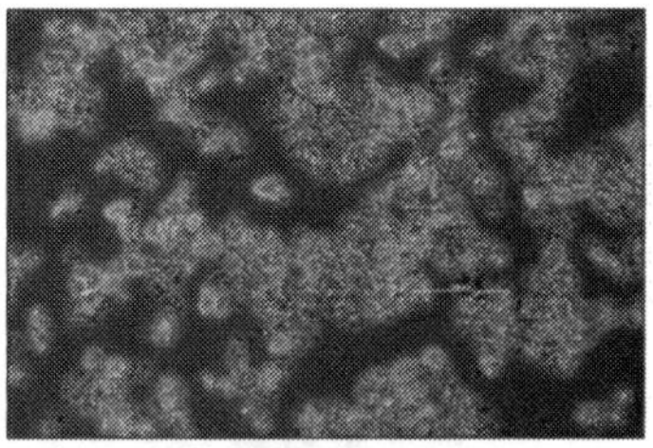

100 X

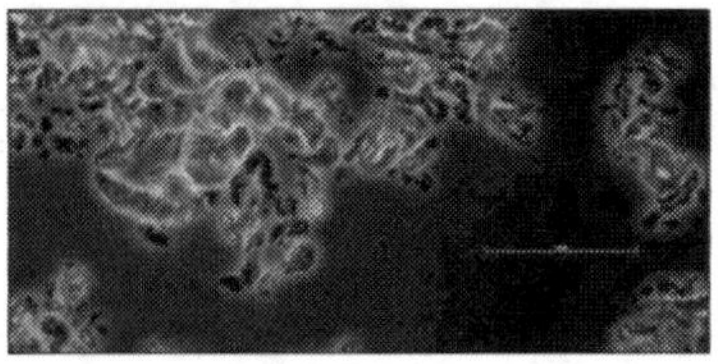

400 X

14 AIChE
Spring Meeting, New Orleans, LA

BAKER HUGHES

Paraffin Analysis of Sample

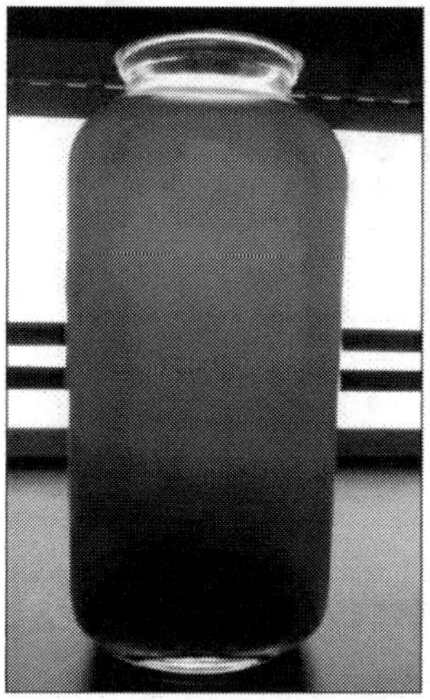

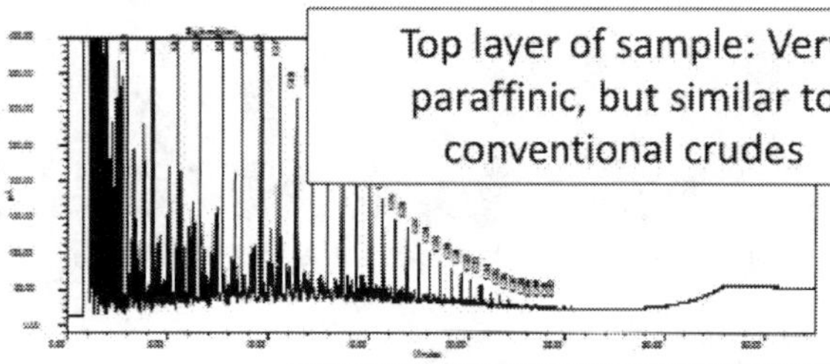

Top layer of sample: Very paraffinic, but similar to conventional crudes

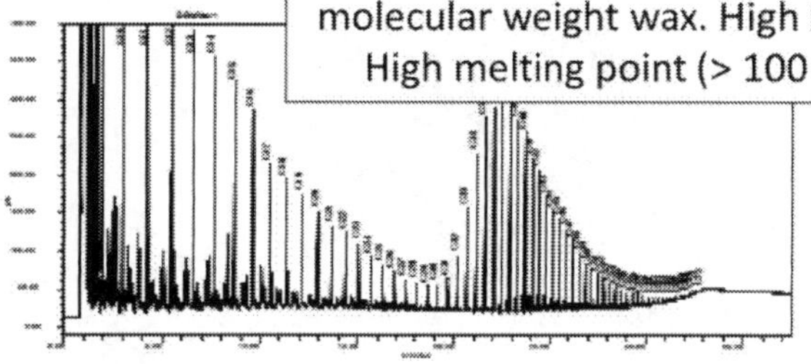

Bottom layer: Considerable high molecular weight wax. High MW = High melting point (> 100 °C)

14 AIChE® Spring Meeting, New Orleans, LA

BAKER HUGHES

Transport of Shale Production

- Infrastructure does not exist to meet needs
  - Most shale oil moved by rail or trucks
  - Limited pipelines expanding to meet transportation needs
- Unit trains to East Coast, West Coast, and Gulf Coast
- Problems:
  - Tankers and barges with heavy wax deposits ROB
  - Tight emulsions that look like oil or white condensate but high BS&W
  - Cars not uniformly treated for H_2S
  - Corrosion in rail cars

14 AIChE® Spring Meeting, New Orleans, LA

BAKER HUGHES

Viscosity of ROB Before Treatment

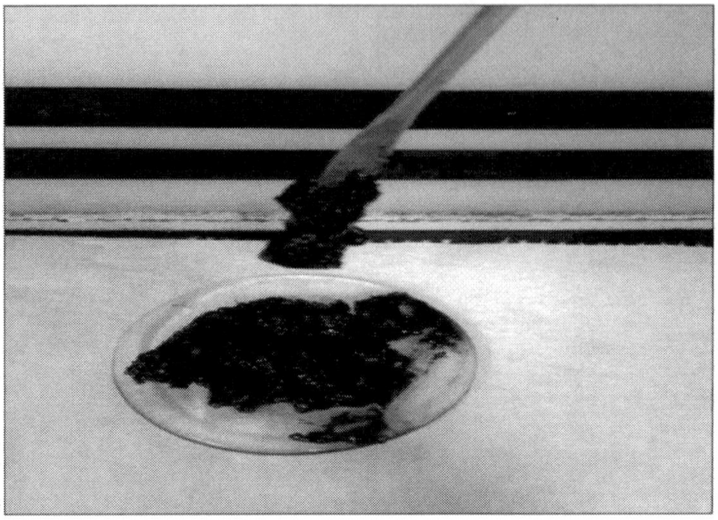

ROB from untreated compartment

14 AIChE Spring Meeting, New Orleans, LA

BAKER HUGHES

Eagle Ford Shale Oil Wax Dispersion Testing

Several products were effective in dispersing the paraffin wax in an Eagle Ford shale oil sample at ambient temperature for over a month.

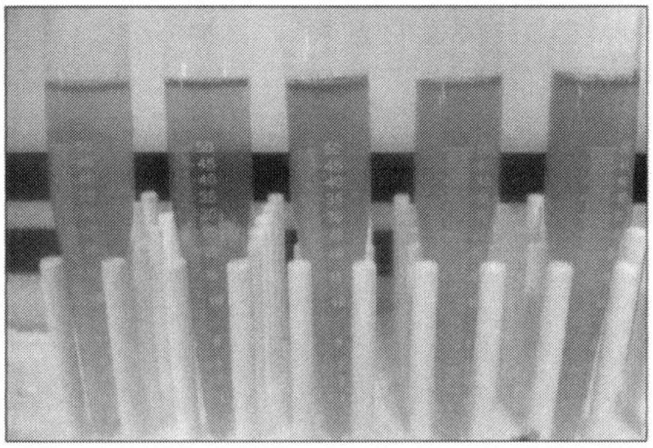

14 AIChE Spring Meeting, New Orleans, LA

BAKER HUGHES

Product 'A' Effect on Wax Deposition (Utica)

Blank **200 ppm** **400 ppm**

14 AIChE
Spring Meeting, New Orleans, LA

BAKER HUGHES

Refinery Issues

- Crude Blending
- Fouling
- Corrosion
- Product quality

14 AIChE
Spring Meeting, New Orleans, LA

BAKER HUGHES

Crude Unit Fouling

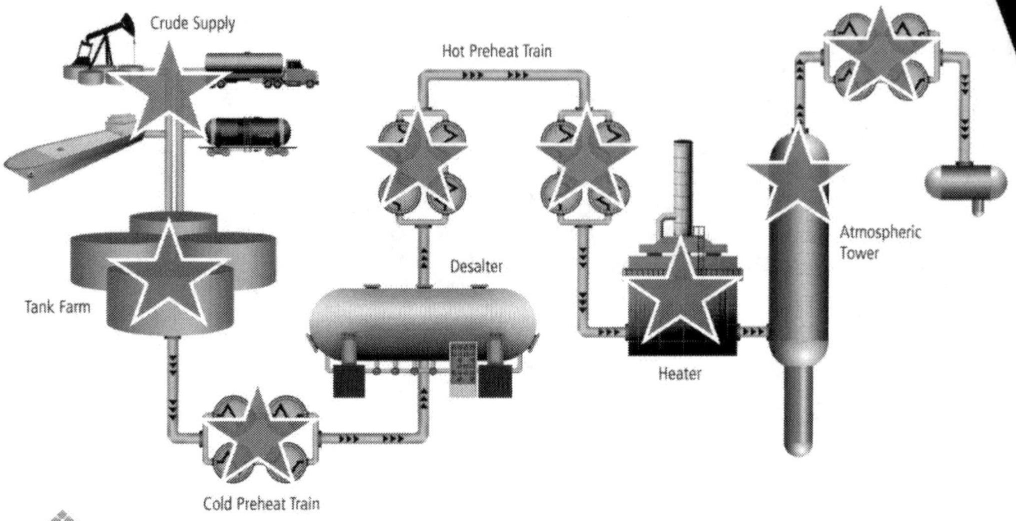

★ = Susceptible to Fouling

14 AIChE Spring Meeting, New Orleans, LA

BAKER HUGHES

Asphaltene Precipitation

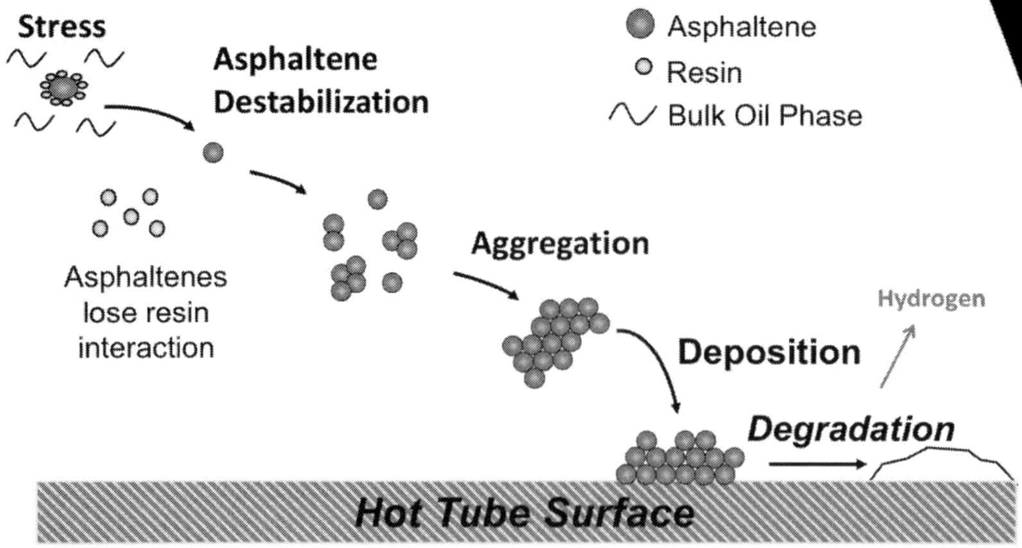

14 AIChE Spring Meeting, New Orleans, LA

BAKER HUGHES

Asphaltene Stability Index Test

Field ASIT services™ technology measures the onset of the flocculation of the asphaltenes with high accuracy by inducing the asphaltene precipitation via titration with a paraffinic solvent.

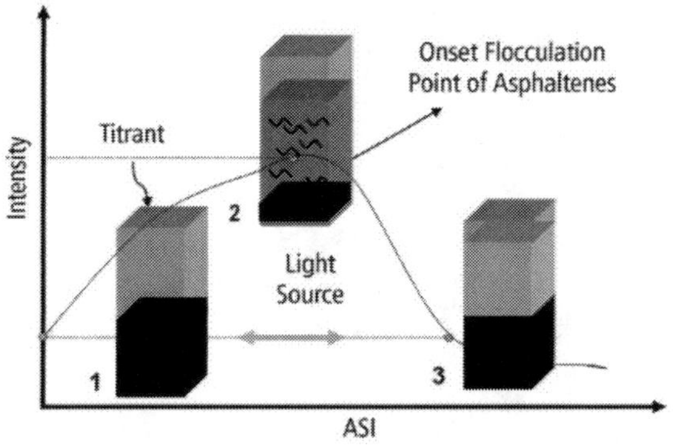

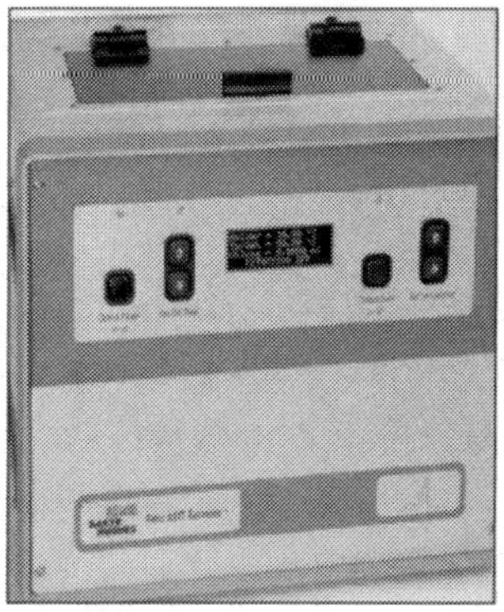

14AIChE®
Spring Meeting, New Orleans, LA

Field ASIT services is a trademark of Baker Hughes Incorporated

BAKER HUGHES

ASI Shows Eagle Ford Can Destabilize Crude

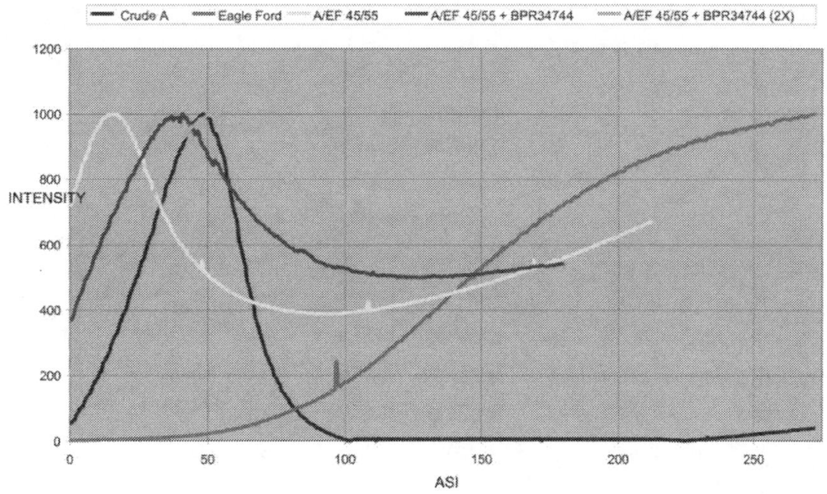

14AIChE®
Spring Meeting, New Orleans, LA

BAKER HUGHES

Tower Tray Fouling - Phosphorus

- Seen by refineries running Bakken, Eagle Ford, and WTI
- Operationally manifested as increasing tower dP
  - Typically in the kero/jet zone
- Tower inspection shows deposits on trays in kero/jet zone
  - Some deposition may also show a few tray levels above and/or below kero/jet zone
  - Deposit scale typically tightly bonded to tray surface restricting flow and valve function

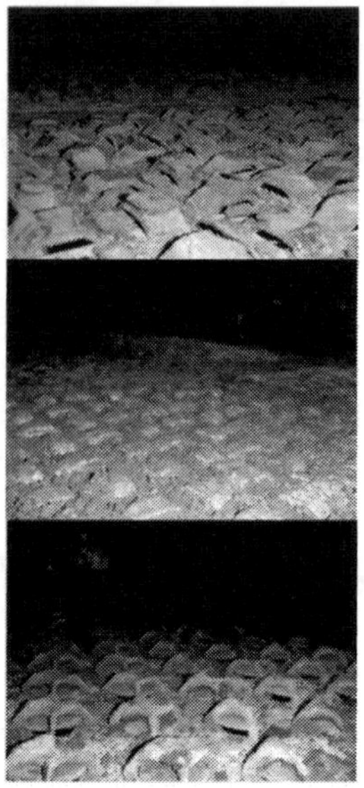

Above kero/jet zone

In kero/jet zone

Below kero/jet zone

14 AIChE Spring Meeting, New Orleans, LA

BAKER HUGHES

Shale Oil Impact – MEA Contamination

Corrosion and Fouling

- Shale oils transported by truck, rail, barge.
- Many shipments are treated with H_2S scavengers:

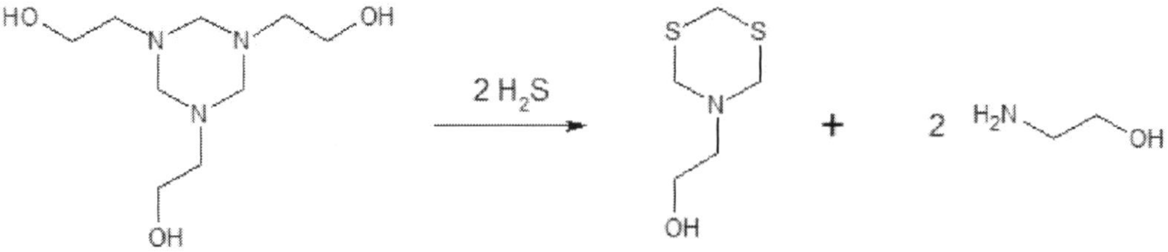

- Potential Impacts:
  - H_2S scavengers leave byproduct amine in crude oil (MEA, MA).
  - Tramp amines distill to OH system, create amine-HCl salts:
  - Corrosive to system metallurgy.

14 AIChE Spring Meeting, New Orleans, LA

BAKER HUGHES

Impact of Tramp MEA on OH Water

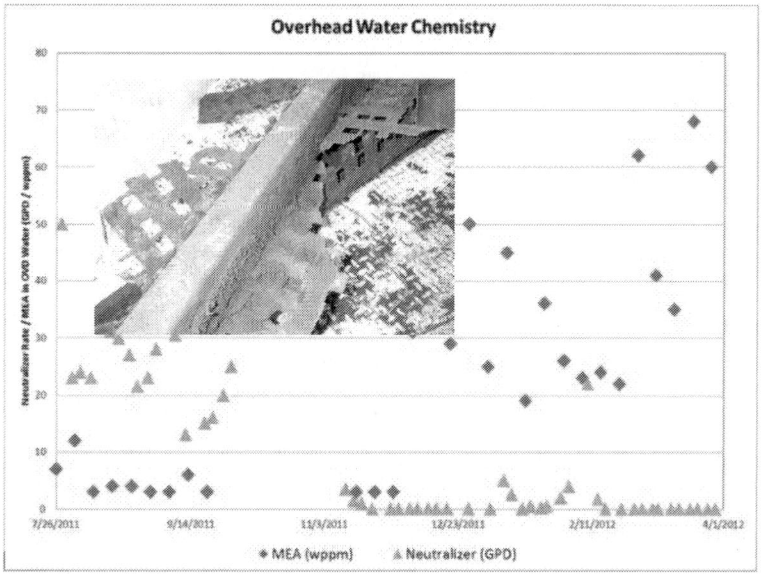

14 AIChE Spring Meeting, New Orleans, LA

BAKER HUGHES

Salt Risk w/MEA Concentration

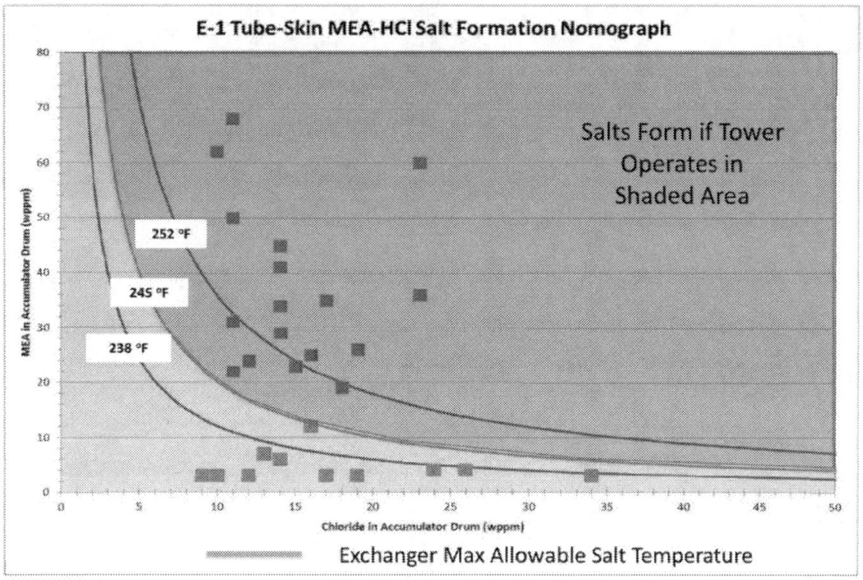

14 AIChE Spring Meeting, New Orleans, LA

BAKER HUGHES

Shale Oil – Refinery Process Concerns and Monitoring

BAKER HUGHES

CHARACTERISTIC OF SHALE OIL	CONCERN IN REFINERY	PROCESS DATA	SAMPLES AND ANALYSES Samples	 Analysis	OUTPUT DATA	FREQUENCY	BHI PROGRAMS
Crystalline Wax	Tank bottom sludge	• Tank Sludge Depth	• Tank bottom	• Visual sludge • GC analysis	• Tank Sludge Depth • Crystal Wax Presence	• Monthly	• XERIC™ Pretreatment
	Cold Preheat Fouling	• Desalter Temperature			• Temp vs. Time	• Update 2x/mo.	• LIFESPAN™ Wax Dispersants
	Desalter emulsions	• Emulsion layer growth	• Triline emulsions	• Visual wax • GC analysis	• Crystal Wax Presence	• After Shale Oil begins • When emulsion grows	• XERIC Demulsifiers
Asphaltene Stability in Blends	Desalter emulsions	• Emulsion layer growth	• Triline emulsions	• Precipitated Asphaltenes	• Precipitated Asphaltenes	• After Shale Oil beings • When emulsion grows	• XERIC Asphaltene Stabilizers • Field ASIT services™
	ColdPreheat Fouling	• Desalter temperature	• Crude blend	• ASIT	• Temp vs. Time • ASIT #	• Pre-Shale Oil • Update 2x/mo.	• LIFESPAN Antifoulants • Field ASIT services
	Hot Preheat Fouling	• Furnace inlet temp • Furnace tube temps • Furnace fuel usage • Preheat DP	• Crude blend	• ASIT	• Temp vs. Time • Fuel use vs. Time • ASIT #	• Pre-Shale Oil • Update 2x/mo.	• LIFESPAN Antifoulants • MILESTONE™ Furnace Fouling Control • Field ASIT services
Paraffinic	Furnace Coking	• Furnace tube temps • Furnace fuel usage	• Crude blend	• HLPS	• Coking tendency	• Pre-Shale Oil • W/Shale Oil	• MILESTONE Furnace Fouling Control
Tramp MEA	Tower/OH corrosion (Early Warning)	• OH Neutralizer rate • EXCAL usage rate	• OH water • Desalter brine	• OH pH • Brine pH	• Early Sign of Amine • Early sign of Amine	• Pre-Shale Oil • W/Shale Oil	• TOPGUARD™ Overhead Corrosion Control • EXCALIBUR™ Contaminant Removal Technology
	Tower/OH corrosion (Quantitative)	• Inputs for CRM	• OH water • Desalter ww • Desalter brine	• IC Amines	• Confirm MEA in Crude • OH MEA ppm • Salt Risk Tower • Salt Risk OH System	• Pre-Shale Oil • W/Shale Oil • Update 2x/mo.	• TOPGUARD Corrosion Control • EXCALIBUR Technology • TOPGUARD Corrosion Risk Monitor (CRM)
Phosphorus	Tower P Fouling	• Tower DP			• Tower DP increase	• 2x/mo.	• LIFESPAN Antifoulants Dispersants
			• Crude	• ICP for P	• P in Crude	• 2x/mo.	
			• Kerosene	• Filt Solids • XRF Filter P • ICP for P	• Incr. Kero Solids • P in Solids • P in Kerosene	• 2x/mo.	
			• Kerosene • Diesel	• ASTM D86	• Incr. in Diesel IPB – Kero EP overlap	• 2x/mo.	

352529 A Study on Exergy-Based Thermodynamic Analysis and Process Synthesis of Mixed-Refrigerant Systems for Ethylene Plants

Abstract:

Refrigeration system holds an important role in most chemical/petrochemical processes. A refrigerant system transfers heat from low-temperature sources to high-temperature sinks through vapor-compression cycles at the expense of mechanical work, magnetism, laser or other means. In many industrial processes, the refrigeration systems are employed for chilling and freezing the final liquid/solid products or liquefying the gas intermediates/products for the downstream cryogenic separation. Optimal synthesis of these systems is attractive not only because of their economic benefits but also because they incorporate most of energy transfer operations in widespread use in the chemical industry. The traditional cascade refrigeration system used in ethylene plants includes multiple refrigerants and multiple temperature and pressure levels. Recently, mixed-refrigerant system (MRS) has shown some advantages comparing to the single-component one. A MRS uses a mixture as the refrigerant instead of several pure components as in conventional cascade refrigeration systems. This system operates within smaller temperature differences at the lower limit; therefore, it leads to a smaller increase in entropy and consequently a smaller loss in energy. Besides, an MRS also requires a simpler unit configuration and fewer maintenance problems.

In this study, MRS used in ethylene plants is studied based on in-depth exergy analysis, a powerful method to explore system energy loss within a large temperature range crossing the ambient temperature due to various compositions of the employed mixed refrigerant. An exergy-based mixed-integer linear programming (MINLP) model is developed to maximize the energy efficiency of each system, where multiple refrigerant combinations and their recycling loops are simultaneously addressed and all of cooling demands are fulfilled as inputs for a heat exchanger network synthesis. The efficacy of the developed methodology is demonstrated a case study of a mixed-refrigerant system used in a typical ethylene plant.

352748 Plant-Wide Simulation of Ethanol Oxidation Process for Acetic Acid Production

Abstract:

Acetic acid is an important chemicals that is widely used in manufacturing for many products, such as medicine, photographic film, synthetic fiber, food additives, and paint. The annual global demand of acetic acid is about 6.5 million tons. There are three major processes for acetic acid productions: i) low-pressure carbonylation of methanol synthesis process, which employs methanol and carbon monoxide reaction to produce acetic. It is currently most popular method for acetic acid production. ii) Ethylene oxidation process, where ethylene is oxidized to acetaldehyde first, and then further oxidized to acetic acid. iii) Ethanol oxidation process, where acetaldehyde is prepared by ethanol oxidation, and then acetaldehyde is oxidized to acetic.

In this study, a plant-wide model for ethanol oxidation process to produce acetic acid is developed by using Aspen Hysys. The modeled process is divide into two main sections: the ethanol oxidization to acetaldehyde process and the acetaldehyde oxidization to acetic acid process. In the first section, the raw feeds of ethanol and water is mixed, evaporated, and directed to a reactor for oxidation. Besides the major product of acetaldehyde, there are some byproducts of CO,CO_2, methane, tr-Crotonal, small quantity of acetic acid as well as unreacted materials. After the reaction, the outflow goes to an absorber system to separate the insoluble gases from the liquid. Next, the liquid effluent enter two distillation columns to separate unreacted ethanol for recycle and the byproduct of formaldehyde, as well as obtain the high-purity acetaldehyde. In second manufacturing section, the enriched acetaldehyde, air, and the catalyst of manganese acetate are sent into a distillation column for oxidation reaction and separation. then the bottom outflow goes to a distillation column to get the product of acetic acid, while the light component goes to another two distillations to separate oxidation byproduct of lipid and recycle unreacted acetaldehyde. The recycled acetaldehyde goes back to the oxidation column. We simulate the entire plant with these both process in steady-state model first, and obtain the simulation results agreeing well with real plant data. Then some critical sections are further examined by dynamic models to examine the process control ability. Some valuable process and control knowledge have been disclosed by our simulations, which can provide more technical support for plant future manufacturing.

351249 Dynamic Modeling and Optimization of Ring Opening Polymerization: Modeling Challenges and Computation Improvements

Abstract:

Recent work has focused on modeling and optimizing semi-batch reactors for ring opening polymerization. Dynamic models have been developed, using first principles (mass balance, heat balance, reaction kinetics, etc.) one based on population balances and another using the method of moments. These models have been validated with historical plant data, matching reactor pressure profiles. The moment model has the added benefit of increased computational efficiency. When using these models to optimize the reactor temperature and monomer feed rate, reductions in polymerization time are seen. Because of the promise from initial product recipes, these simulations and optimizations should be applied to many similar polymerizations.

A reformulation procedure has already been applied to the above system, to separate the fast dynamic model from the differential equation, modeling the fast kinetics algebraically in their quasi-steady state form. However, further challenges with a large, nonlinear system make it difficult for simulations and optimizations to converge. The most realistic system inputs must be used, and the best initial conditions for the optimization are critical. From the chemistry of polymerization, heuristics were developed to determine these initial conditions. Here we present rules of thumb, and general assumptions that enable good initial conditions to be calculated, given a realistic polymerization system. With these informed initial guesses, improved computation time is seen for these optimizations.

POLYMERIZATION IN A SPRAY DRYER – DESIGNING A PRE-REACTION BEFORE ATOMIZATION TO BOOST THE REACTION IN THE SPRAY

Magnus Tewes, Technische Universität Bergakademie Freiberg, Germany
Professor Dr.- Ing. Urs A. Peuker, Technische Universität Bergakademie Freiberg, Germany

Introduction

Powder polymers with well defined size and a spherical shape are commonly generated by the use of spraying devices, e. g. spray dryer. Here polymers in solution or suspensions of polymers are atomized to small droplets and dried to particles. These particles are mostly used in a wide field of industrial applications like pharmaceutical, cosmetics and chemical industries. Especially water soluble or water absorbing polymers are present as additives in drugs, hair styling products or cleaning agents. Polymers as powders instead of gels or solutions are in advantage concerning transport and storage or further processing because of their flowing characteristics and stability. But the production of these powders is complex. After the polymerization in different reaction steps for dissipating the high amount of exothermic heat formation, processes are required like drying, milling and agglomeration. The idea to combine all the single steps in a so called one step process is obvious and theoretical described in U.S. patent from 1993 [1]. Krüger adopted this new process concept, designed a plant and called the process spray polymerization. The plant is a vertical heated pipe reactor with an ultra sonic droplet generator that creates droplets in a row in the sizes of 90–220 μm. He investigated the generation of superabsorbent polymer particles [2]. Previous works show that efficiency and yield of the spray polymerization is a key problem. Most of the educts evaporate during the drying and aren't available for the reaction consequently. The modification of the spray polymerization process and a development of an economical concept is the motivation of this research.

The new concept is based on a common spray dryer, the main aggregate, where the liquid is atomized by a high velocity gas and dries by a hot gas in a co-current stream. The dryer itself combines three different sub-processes, atomization of the liquid to small droplets, drying and solid formation of the dispersed droplets. Finally the dried particles are separated from the gas stream. The advantage of a spray dryer is the fast process time (< 10 s) because of the small droplets and the high convective heat transfer of the drying gas. It is used to generate solid particles from a liquid rapidly but gently. Even a high drying temperature causes low surface temperature of the droplets, so glass transition temperature of the polymer will not be reached. A new process will be created if additional to these three operations a reactive system is sprayed. In this case polymer reactions are applied, so the idea is a one step process from a liquid monomer solution to a ready powder polymer.

The reaction in a droplet is compared to a small batch reactor, which is heated by the drying gas in the spray dryer. If the monomer is volatile, it will evaporate during the heating and the concentration of monomer in solution changes. With the heating the reaction accelerates and the concentration of monomer decreases. Polymer molecules are formed and change the viscosity of the liquid in the droplet. In addition to that, the polymer will tend to build a skin on the surface of the droplet and change the heat and mass transfer from the inner droplet to the outside. If the monomer isn't volatile, only the solvent evaporates and the concentration increases first, but then the reaction decreases the concentration again. It becomes clear now that the process becomes more complex by the competition

of all four sub-processes. The advantage of the reaction in the droplets is that no safety problems will occur because of the small volume. A runaway of the reaction, as the worst case, will only cause unintended particle morphologies. But the advantage of the small droplets and the fast drying times of a spray dryer is also a bottle neck of the spray polymerization.
Due to the low retention time in the spray dryer the process requires a very intensive reaction with fast polymerization kinetics. Therefore the polymerization is a radical reaction of acrylic acid and its acrylates. This system is also used by Krüger because of the fast growing rates [2]. The acrylates are nonvolatile and nonflammable compared to the acid and therefore, easier to use without taking account of explosion protection.

Although the monomer system is very fast and has been tested for a polymerization in a droplet, first considerations about the reaction kinetics show that the retention time is still too low. A pre-treatment of the reactive liquid has to be done to start and properly progress the reaction before it gets atomized. The focus of this research is on the pre-reaction before the atomization in order to guarantee an effective one step process and production of well defined particles in size and shape. The kinetics of the pre-reaction is also important to figure out the impact of the four different sub-processes in the droplet during the spray drying.
The start of the polymerization is mainly depended on the decomposition kinetics of the starter or of the monomers. The decomposition rate is a function of temperature and concentration of the educts. Radicals activate monomers and macro molecules are formed. These oligomers or molecules with higher polymerization degree effect the atomization. In earlier works the influence of molecular mass of a polymer solution has been investigated. It has a big influence on the atomization if molecular mass is high enough to cause a complex rheology with non-Newtonian viscoelastic behavior of the polymer solution. Then filaments instead of droplets are formed [3].

The pre-reaction is necessary to overcome the problem of the short retention time. If the start of the reaction is before the atomization, the properties of the liquid will change depending on the amount of pre-treatment. Therefore a special nozzle is designed to investigate the effect of the retention time of the reactive liquid within the nozzle. This nozzle is modular designed to change the retention time in the nozzle and to heat the nozzle inside. The next chapter is dealing with the design.

Because of the fact that the rheology has a big influence on the atomization, a new approach has been developed to measure the rheokinetics or function of viscosity over the time during the polymerization. Here results are presented and used to demonstrate the progress of reaction within the nozzle. A big challenge is to handle the experimental data, but a more ambitious work is to generate kinetic data that are able to describe the chemistry of polymerization in the nozzle and in the droplets.

Designing the pre-reaction

The reason for developing a pre-reaction is to enlarge the process time the reactive liquid has to polymerize. This pre-treatment will be done before the atomization in a special nozzle. In order to have a compact process and keep the idea of a one-step-process a first idea was to design a new nozzle, a so called 3-fluid nozzle. Here the mixing of the monomers with a starter in solution was intended to take place within the nozzle. Due to mixing problems within the nozzle this concept is frozen by now. The alternative nozzle is therefore a 2-fluid nozzle with special properties.

One special design feature is the double jacket that is passed through by a heating liquid to heat up the feeding monomer/starter solution. The radical starter is a thermal initiator and therefore the temperature has a high impact on the decomposition rate. Because of safety aspects the

monomer/starter solution is fed at a low temperature of 5 to 15 °C. Consequently a rapid heating to the reaction temperature is necessary.
The nozzle has a modifiable length. Variation of the length influences the retention time of the reactive liquid in the nozzle at a constant flow rate. A constant flow rate is set in order to have no effect on the atomization later. The variation of the pre-reaction time is available by changing the length. An optimum of the geometry will be needed where, on the one hand the temperature for reaction is reached and, on the other hand the progress of the reaction is sufficient. The more the reaction is progressing the better the efficiency of the spray polymerization, but the higher the degree of polymerization at the nozzle tip. With the length of the polymer chain the viscosity raises. This can hinder the droplet formation or block the nozzle.

For atomization the second fluid, nitrogen, is used at a high pressure. The gas is fed through the double jacket and concentrical to the tip of the nozzle. Through thin channels and with a tangential direction to the liquid's exit the gas is swirled and expands outside the nozzle. High velocity gas atomizes the reactive liquid compared to standard twin fluid nozzles. After the atomization the reaction progresses in the droplet. This process is comparable with a small batch reactor, whereby the nozzle is more a laminar flow reactor.

Another advantage of the nozzle is that its design is like a lance. So every inlet is on the bottom of the nozzle including the outlet of the double jacket. The outlet of the spraying fluids is at the tip of the nozzle. The application of this special nozzle in a common spray dryer is without any problems. The constructed nozzle uses a nozzle tip from Buchi® Corporation with a liquid diameter of 0.7 mm and a gas diameter of the annular gap of 1.5 mm.

Measuring the effect of the rheokinetics

A free radical polymerization is mostly an exothermic reaction, because the combination of small monomer molecules to a polymer chain means losing of degree of freedom. The amount of reaction heat becoming free is a safety problem, especially if very fast reactions are processed. These fast reactions are used for polymerizations in a spray dryer.

The kinetics of the polymerization are generally described in three different steps; initiation, propagation and termination. Monomers are inactive under certain conditions so an initiator is used. The initiator is decomposed and forms radicals that activate monomers if an initiator radical is connected with a monomer. The polymerization starts and the radical monomers form a polymer chain by connecting more and more monomers. This chain growing step is called propagation. If the growing is interrupted and the chain becomes inactive the reaction is at its last step, the termination. Often the combination or disproportionation of two active polymer chains ends the reaction.

The kinetic of the reaction's beginning is normally described by the following equation (see eq. 1). The rate of the propagation reaction rate is described as the change in concentration in the time interval by the use of some simplifications [4].

$$r = \frac{dc_M}{dt} = \sqrt{\frac{k_i}{k_t}} \cdot k_p \cdot \sqrt{c_i} \cdot c_M \tag{1}$$

The indices *i*, *p* and *t* shows the influence of the three different steps. All constants *k* are depended on temperature and the material and describes by the Arrhenius law.
In view of the complete transformation of monomers to polymers the radical polymerization is more complex. Regarding this, the reaction is described by the help of statistic values, like the distribution or mean value of the molecular weight. The molecular weight will be measured with special

chromatography. This analysing technique often uses light scattering or viscosity measurement to determine the volume of the polymer coils. The intrinsic viscosity is measured in dilute solution and is a reciprocal concentration which can determine the mean molecular weight. Equation 2 shows the relation of mean molecular weight $\overline{M}$, intrinsic viscosity η and concentration c and was developed by several scientists; Staudinger, Fikentscher, Houwink, Huggins, Hess and Sakurada [5-9].

$$\eta = K \cdot c^b \cdot \overline{M}^a \qquad (2)$$

The idea of measuring the viscosity over time during a polymerization and therefore be able to describe the kinetics is first described in detail by Malkin and Kulichikhin and called rheokinetics. The aspiring target of this method is to generate a semi empirical description. For the initial stage of the radical polymerization of acrylamide in water they find equation 3. This function is developed with the use of equation 1 and 2 and describes the viscosity η over the time t [10].

$$\eta(t) = K' \cdot c_{M,0}^{a} \cdot c_{I,0}^{b-a} \left(k_p \sqrt{\frac{k_i}{k_t}}\right)^{a+b} \cdot t^b \qquad (3)$$

The parameter *K*' takes into account the dependency of the viscosity with change of temperature. It is like the kinetic parameters k calculated with the Arrhenius law.

In addition the knowledge of the function of viscosity over the reaction time is of great benefit in terms of the atomization. The results of former investigations on the atomization of polymer solutions are helpful optimizing the retention time of the reactive solution in the nozzle. To measure the rheokinetics a modification of a rheometer is necessary, as it is seen in figure 1.

The standard rheometer is a HR-1 by TA Instruments and has a cone-plate geometry. The 2° cone with 40 mm diameter is connected to a motor. The plate is coupled with a Peltier temperature control. The Peltier element controls a temperature range between -40 and 200 °C with a controlled heating rate of up to 10 °C/min.

The modification is done by covering the sample with two semi caps. They are connected to a nitrogen supply and at a certain step the sample gets purged with nitrogen during the measurement. Here the nitrogen is when turned on, starting the reaction, while oxygen that inhibits the polymerization is displaced rapidly. The geometry of the caps fits with no contact to the moveable cone. Through small gaps the gas is forced from the caps over the holes and sample between caps and cone to the outside. Several holes are drilled in an additional plate and arranged in a circle around the upper cone. This modified plate has two functions. It fits with the Peltier plate and the two are assembled together. The holes in it are filled with water, so during the measurement the nitrogen flows over the water and is saturated. Thus, evaporation of the sample is suppressed because of a saturated atmosphere.

With heating up the sample to a temperature of 75 °C the procedure to get the time dependence of the viscosity starts. The sample is a solution of monomer, water and initiator and its volume is 525 μL. A damage of the rheometer because of the high reaction enthalpy is therefore excluded. Then nitrogen flow is turned on (0.1 L/min) and the lack of an inhibition of the radicals will start the polymerization immediately. Normally an inhibitor is present in monomers to avoid a self-ignition, but it works in combination with oxygen only [11]. At a constant shear rate of 300 1/s and a sampling rate of 0.5 Hz the function of viscosity during the polymerization is plotted.

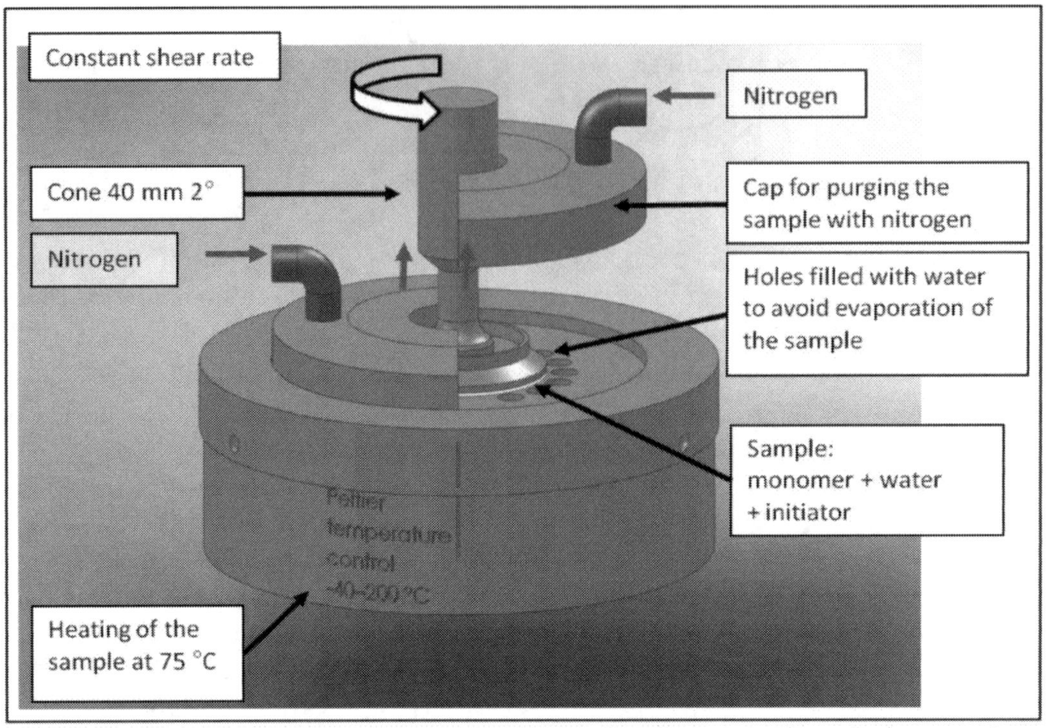

Figure 1. Modification of a rheometer to measure the viscosity over time during a polymerization

Results of the polymerization in the rheometer

The reactive system used is acrylic acid and its acrylates. A work of Krüger shows that a polymerization with this material in a droplet during a free fall is successful [2]. The monomer is applied by BASF SE and contains an inhibitor. With a vapor pressure of 4.3 hPa the acid is less volatile than water. But acrylic acid is flammable and explosion protection has to be guaranteed if used in spray dryers. That is why the salts of the acrylic acid, acrylates, are less harmful and to be preferred. They are produced by neutralization of the acid e.g. with sodium hydroxide in a solution.

The results of the rheokinetics have two benefits. On the one hand the viscosity function is measured and an optimal nozzle design is possible. On the other hand it is an ambitious target to use the method of Kulichikhin and Malkin to get information about the kinetic of the reaction as it is mentioned in equation 3. But this equation has a big disadvantage because of the dimensions. For the initial boundary, $t = 0$, the viscosity would also zero if b is not equal to zero and is consequently not logical.

All following results are done with the same testing procedure as described before and with a minimum of three repetitions. In order to explain the analyzing, four different measurements of acrylic acid (AA) and sodium acrylate (SA) with two initiator ratios, 0.8 mmol/mol (dotted lines) and 1.2 mmol/mol (plain lines) related to the monomer's molar amount, and two different concentrations, 2.8 and 3.5 mol/L, are used as an example. In figure 2, left, the raw data of the measurement are shown. The shear viscosity is a function of time, which, after a plateau, is getting a high gradient after a characteristic time, called zero time or time of the viscosity´s increase t_0. It can be directly captured by the raw data for acrylic acid. Sodium acrylate does not show such a rapid increase and an overall threshold has to be defined.

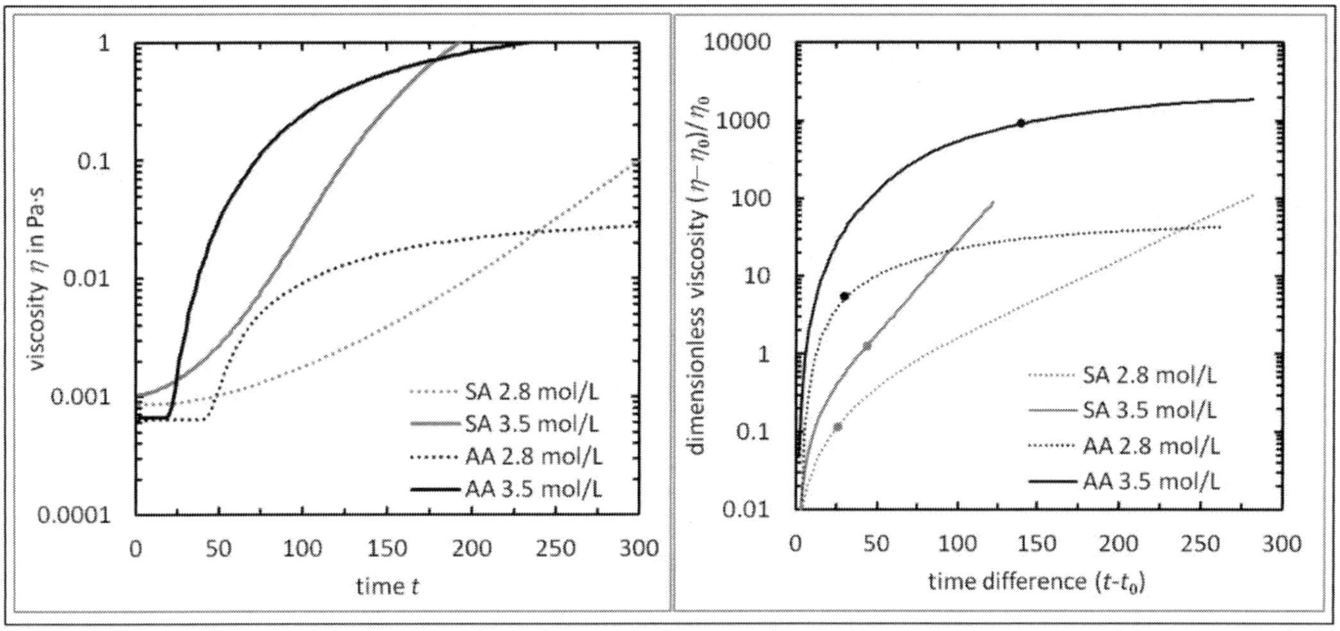

Figure 2. Working with rheokinetics – l. data by rheometer, r. dimensionsless view for the analysis

The next information that is necessary to describe the function $\eta(t)$ is the gradient after passing the zero time. In order to work with the data a transformation to dimensionless quantities is necessary. Therefore the viscosity is calculated by the quotient of the difference between start and current time and the starting point. To focus the gradient easily the Savitzki and Golay smoothing is an efficient tool to smooth the data first and calculate the gradient with the first derivation of the smoothing [12, 13]. Then a fitting to a power function and an exponential function is done mathematically with a linear regression. The breaking point between the power and exponential function is marked in figure 2 with a dot. This intersection gives a good basis to discuss the differences between the polymerization of acrylic acid and the acrylates. Although sodium acrylates show an earlier rise in the viscosity the gradient is lower. The polymerization kinetic is therefore different and the only reason results from material properties. It seems that the pH value, which differs from 1 (acrylic acid) to 7 (acrylates) is the main influence, because it affects the solubility and morphology of a polymer in the solution. Another difference is the exponent of the exponential fitting. Acrylic acid has a negative exponent that decreases the gradient of the viscosity but the acrylates' gradient is still increasing and positive.

The result of the zero time is presented in figure 3. Here acrylic acid and first results of sodium and lithium acrylate are plotted with view of the concentration of the initiator in the liquid. Different analyses have shown that only the amount of starter is determining the resulting zero time. And it is obvious, that the more initiator is present the more radicals are getting free during the decomposition and consequently the zero time is decreasing. But the comparison between acrylic acid and the acrylate is not significant by now. It seems that the zero time of the acrylate is less than for acrylic acid. For every concentration the zero time can be calculated by using an exponential regression. A first simple rule of optimizing the nozzle could be, that the retention time has to be smaller or equal to the zero time. Then the viscosity of the liquid is low enough (compared to water) for excluding problems concerning the atomization.

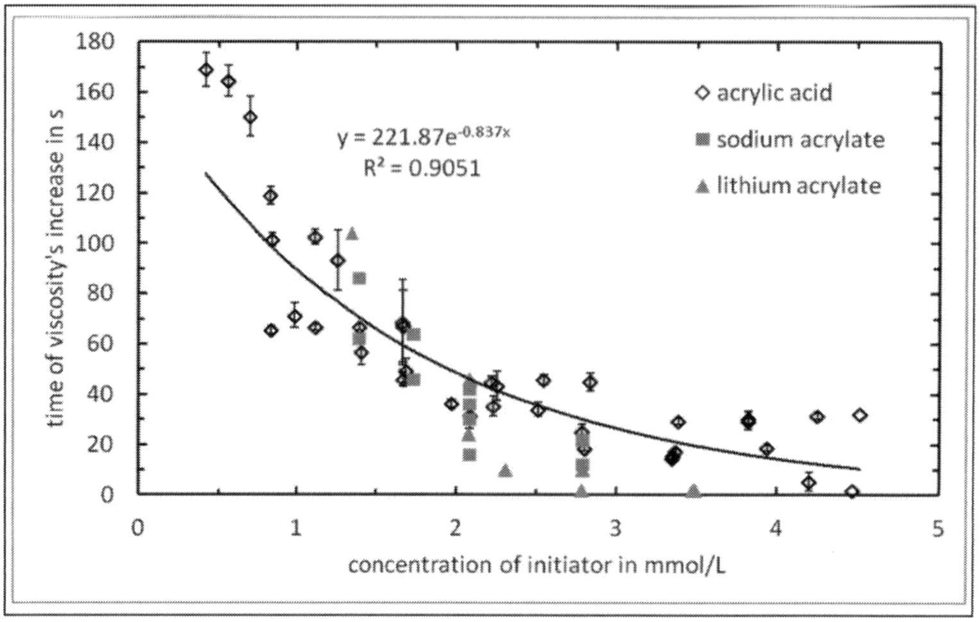

Figure 3. Time of the viscosity increase for acrylic acid and acrylates

Figure 4 is presenting the missing information to describe the function of viscosity over the whole time when only the power function describes the increase. Taking into account the work of Kulichikhin and Malkin the equation 3 can be simplified by combining all constants and get a simple equation. This is done in equation 4. Now the viscosity is only a function of time and two constants, a factor β and an exponent b.

$$\frac{\Delta\eta(t)}{\eta_0} = \beta \cdot (\Delta t)^b \quad (4)$$

As it is seen in figure 4 left, the factor β is a function of the molar fraction of the monomer and of the initiator at the beginning when $t=t_0$. This dependence is given by a product of two power functions and a factor. The result shows that the influence of monomer molar fraction is very high compared to the amount of the initiator. In addition the range of values is very broad, so it has to be noticed that although the square of the correlation coefficient is near to one the variance of the power function is high. There is a good agreement with the power function of equation 3.

But on the right of figure 4 another mathematical correlation is in no agreement with the results of Kulichikhin and Malkin. Here a similar power function is correlated with other coefficients. The exponent b is therefore also depended on the monomer and initiator molar fraction at the reaction's start. Here the influence of the monomer is only the power of 3 higher than the influence of the initiator. Thus, this result is in contrast to the rheokinetic of equation 3, because no influence of the initiator concentration is mentioned. The meaning is that during the polymerization the influence of the monomer and initiator start concentration is present at all time. The influence of the monomer is, at the start of the viscosity´s increase, higher than afterwards.

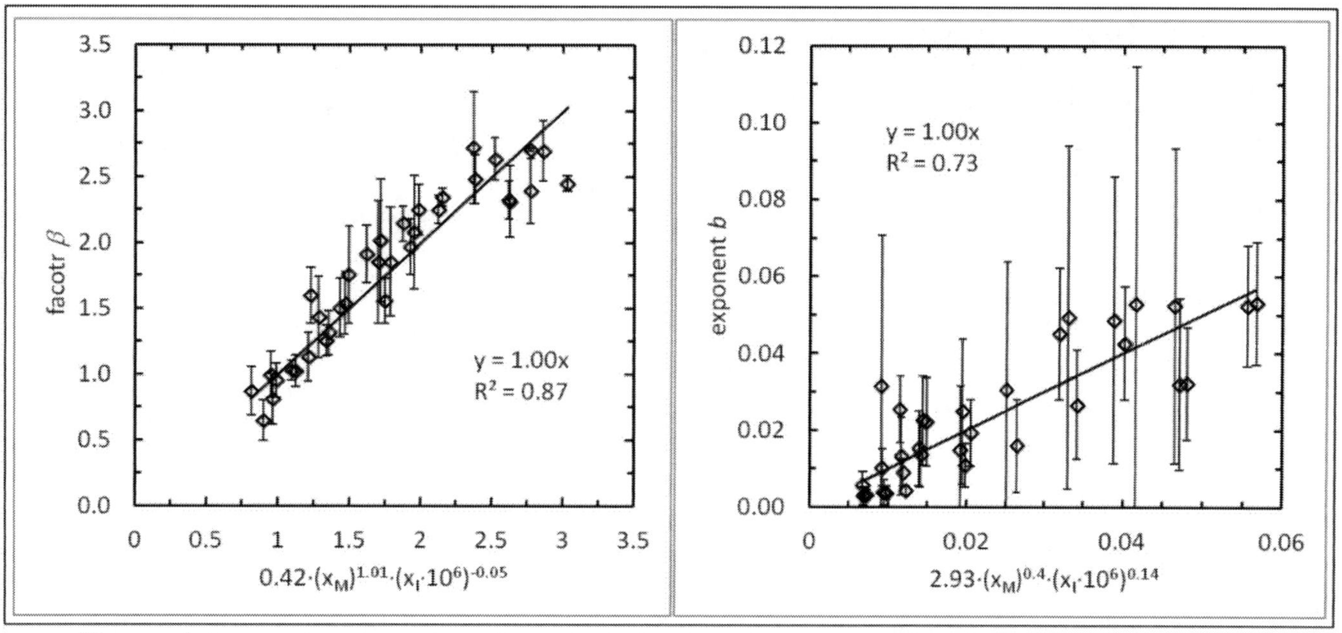

Figure 4. Presenting the values for calculating the viscosity gradient in the nozzle over time

In figure 5 a good agreement to the power function is shown. The biggest different between the calculated and measured values is due to the small differences in the zero time. The table in figure 5 demonstrates the good agreement of the power function and a relative error of 7 % between the exponents.

The presented results are able to extend the following theoretical model of the progress of the polymerization within nozzle. That model is based on a simplified heat and mass transfer for a plug flow reactor. Now the function of viscosity with time will be integrated. The extended modeling will generate base points for the optimization. For validation of the theoretical model plenty of tests with the nozzle have to be arranged. In an intermediate device, where the nozzle is processed in a dripping mode and no drying is applied, results for temperature distribution are generated. Now the focus is on the theoretical model.

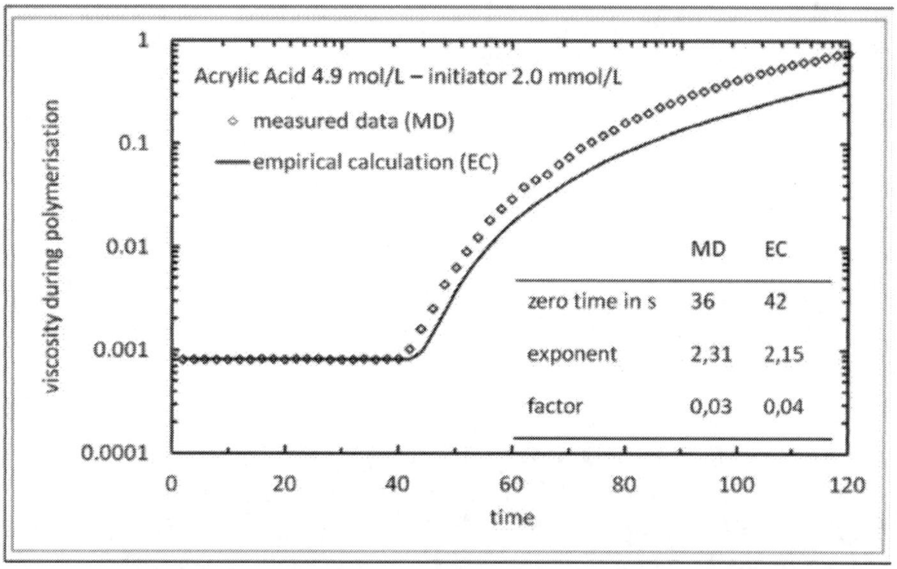

Figure 5. An example of the calculation of the viscosity function during polymerization

Modeling the progress of the polymerization within the nozzle

Before testing the reactive system, theoretical considerations are necessary. The first reason for the consideration is that with a good model the number of tests can be reduced. But the more important reason is the fact that the tests deal with a fast reactive monomer, which will probably cause safety problems. A runaway of the 3-fluid nozzle is possible, because the high reaction enthalpy will accelerate the reaction. The pressure within the nozzle will rapidly increase, if the nozzle gets blocked because of the resulting viscosity´s increase.

The modeling is based on the heat and mass transfer equations. These differential equations are presented in equation 5 and 6. As it is seen a transformation to a dimensionless form is made. On the one hand the mass transfer is the quotient of the conversion and the change of the dimensionless length *dz/L*. It is calculated with the use of the Damköhler number. The rate *r* describes the reaction kinetic and was introduced in equation 1. The heat transfer is given as the quotient of the change of the dimensionless temperature and the dimensionless length. The heat transfer is depended on the mass transfer, making a simultaneous solution of both equations necessary.

$$\frac{\mathrm{d}U}{\mathrm{d}z/L} = \frac{\tau \cdot r_0}{c_{10}}\left(\frac{r}{r_0}\right) = Da_0\left(\frac{r}{r_0}\right) \quad (5)$$

$$\frac{\mathrm{d}\vartheta}{\mathrm{d}z/L} = Da_0\left(\frac{r}{r_0}\right)\frac{\Delta T_{ad}}{T_0} - \frac{K_W A_W}{\dot{V}\rho\tilde{c}_p}(\vartheta_W - \vartheta) \quad (6)$$

To solve the problem of the simultaneous solution easily the differential quotients are approximated as difference quotients. By choosing a small increment for the control variable *z*, length of the pipe, the solving is done numerical by several iterations.

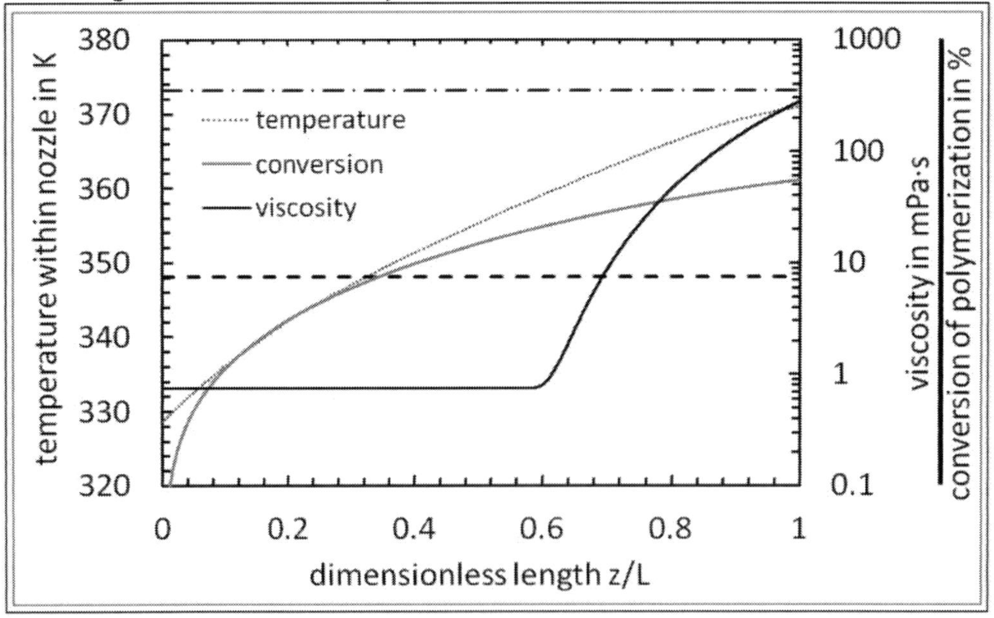

Figure 6. Calculating the temperature, conversion and viscosity in the nozzle for 5.6 mol/L acrylic acid in water and an initiator concentration of 2.3 mmol/L during polymerization

A result of the calculation is shown in figure 6. Here the temperature and conversion is plotted over the dimensionless length. It is an example for a high conversion within the nozzle when using a 5.6 mol/L acrylic acid solution in water with an initiator ratio of 0.6 $\mathrm{mmol}_{\mathrm{initiator}}$ / $\mathrm{mol}_{\mathrm{monomer}}$. The dashed line marks the reaction temperature of 75 °C. As an upper limit the dash-dot line marks the boiling temperature of water at 1 bar. At this point the solution overcomes the boundaries of constant pressure and one-phase state. It is shown that the temperature and the conversion increase until the end of the nozzle at 400 mm. That is obvious because of the enthalpy of the reaction. After a high rise of the conversion its gradient becomes lower because the polymerization consumes the monomers; the maximum of 100 % is the theoretical limit. The lower gradient of the conversion at the end damps the temperature gradient because of the linking of heat and mass transfer. If only the temperature and the conversion would be considered, the safety of the pre-reaction seems to be guaranteed, because the temperature is below the boiling point of water. The high conversion should increase the efficiency of the process very much as no reactive mass is evaporated.

But the result of the viscosity development is overcritical. The function of viscosity is calculated with the results presented before. The start of the function is when the temperature crosses the first mark at 75 °C. After the plateau ending with the zero time the viscosity function is calculated by the use of the power function. If the rheokinetics can be transferred to the behavior within the nozzle and a critical viscosity for the atomization is known, the model is able to find optimal parameters for a safe reaction. In the example above a viscosity of 324 mPas would cause the blocking of the nozzle sooner or later. Because it is just a simple model, a testing of the forecast is absolutely necessary.

In the intermediate device, described before, tests with different temperatures are done and the heating up within the nozzle are measured with thermo elements and finally validate the model concerning the energy balance.

Conclusions

The presented work shows the development that has been done to design a spray polymerizing process. The focus is on the pre-reaction before the atomization in order to improve the efficiency of the process as only low retention times are present. First a special nozzle has been mentioned and explained in detail. In order to do first practical tests it was said that there is still a big lack of information concerning the right use of a high reactive radical polymerization. Therefore a new measuring technique is presented to study the influence of the polymerization, especially the influence on the viscosity. The effect is called rheokinetics. Results of acrylic acid and acrylate are presented. A semi empirical function of the starting point of the viscosity increase is only depended on the initiator concentration. The function of the viscosity increase of acrylic acid is determined and it is demonstrated that here the influence of the concentration of monomer is higher than of the initiator.

In addition a modeling of the heat and mass transfer during the pre-reaction in the nozzle is presented. A model calculates the temperature and conversion distribution along the axis of the nozzle. With the combination of the viscosity function the model is able to find basic points to plan further experiments with a reactive solution and avoid a runaway of the reaction and consequently risks.

Finally the outlook work is to test the nozzle with a reactive solution in the intermediate device. Then the step to work in a spray dryer with atomization is not very far away. Another research is the linking of the function of viscosity to kinetics of the polymerization to have a better rheokinetic approach for acrylic acid and its acrylates.

Acknowledgements

We would like to thank the German research foundation for the financial support and the BASF SE for donating the materials. A special thank is to all the co-workers in the SPP 1423 and to our team at our institute, especially to my students.

Nomenclature

Latin

a	exponent of power function
A	area of heat transfer
b	exponent of power function
c	concentration in mol/L
$\tilde{c}_p$	heat capacity at constant pressure in J/(kg·K)
Da	Damköhler number
k	reaction constants in L/(mol·s) or 1/s
K	empirical constant
K	overall heat transfer coefficient in W/(m²·K)
l	length in mm
L	length in mm
M	molecular mass in g/mol
r	rate of polymerization in mol/(L·s)
t	time in s
T	temperature in K
U	conversion
x	molar fraction
z	length in mm

Greek

β	factor of power function
η	viscosity in Pa·s
ρ	density in g/L
ϑ	dimensionless temperature
τ	retention time in s

Indices

0	initial condition
ad	adiabatic
i	initiation
I	initiator
ideal	ideal
M	monomer
p	propagation
t	termination
W	wall

References

1. Levendis, Y. A.; Panagiotou, T. and Flagan, R. (1993), Patent U.S. 5,269,980.
2. Krüger, M. (2003) "Sprühpolymerisation," PhD thesis, Hamburg.
3. Tewes, M. and Peuker, U. A. (2012), "Polymerization in sprays – one step from a monomer to a powder polymer," *12th Triennial ICLASS*, Germany, September.
4. Stickler, M. (1979), *Makromol. Chem*, 180, pp. 2615.
5. Fikentscher, H. and Mark, H. (1929), *Kolloid Z.*, 49, pp. 135.
6. Staudinger, H. (1930), *Kolloid Z.*, 51, pp. 71.
7. Hess, K. and Sakurada, I. (1931), *Ber. Dtsch. Chem. Ges.*, 64, pp. 1183.
8. Huggins, M. (1938), *J. Phys. Chem.*, 42, pp. 911 (1938)
9. Houwink, R. (1940), *J. Prakt. Chem.*, 157, pp. 15.
10. Kulichikhin, S. G. and Malkin, A. Y. (1997), *Polym. Eng. Sci.*, 37, pp. 1331.
11. Becker, H. (2003), PhD thesis, Darmstadt.
12. Savitzky, A. and Golay, M. J. E. (1964), "Smoothing and differentiation of data by simplified least squares procedures," *Analytical Chemistry*, 36, pp. 1627.
13. Gorry, P. A. (1990), "General least-square smoothing and differentiation by the convolution (Savitzky-Golay) Method," *Analytical Chemistry*, 62, pp. 570

UTILIZATION AND BENEFITS OF A NANOPARTICLE EXTRACTION PROCESS IN A DROP COLUMN FOR PRODUCTION OF HIGH QUALITY ORGANOSOLS

Jacqueline V. Erler[1]*; Tom Leistner*[2]*; Urs A. Peuker*[1]
[1]*TU Bergakademie Freiberg, Freiberg, Germany*
[2]*Helmholtz-Institute Freiberg of Resource Technology, Freiberg, Germany*

Abstract

The emphasis of this research is the development of a continuous particle extraction process at the interface of two liquids to produce stable colloidal and functionalized particles. To demonstrate the phase transfer process we designed a miniplant using a drop column as transfer device. For process development, the investigation of the miniplant in various operating methods is required. The necessary experimental setup, materials and relevant parameters for a stable process behavior are presented. The process behavior within the drop column is characterized by a particle-free phase transfer, whereby the measured total organic carbon (TOC) is only dependent on the used surfactant. Based on the modeling of the system the transfer behavior can be characterized, if the transfer fluxes of the surfactant are evaluated. Furthermore, the mass flows and the yield of transferred magnetite are determined by ICP-OES measurements. The additional investigation of interfacial tensions using the pendant-drop method provides conclusions regarding to the impact of interaction of the elementary processes and procedures at the phase boundary. Moreover, the transfer kinetics is specified and compared with calculated theoretical values resulting from a kinetic approach. As a consequence of this, the colloidal stability of the resulting organosols is estimated and compared with the results of previous investigations using a laboratory centrifuge as transfer device. In both devices different forces affect the phase transfer so different organic solvents are applied. The differences in colloidal stability are characterized by a dispersion analyzer as well as theoretically with the use of solubility distances by Hansen.

Introduction

The establishing of high quality organosols requires the production of stable colloidal and functionalized particles with a minimum amount of stabilizing surfactant in a liquid organic medium. The magnetite nanoparticles used have special magnetic properties and therefore, they have a great technological application potential. Thus, they are very interesting for a broad range of research areas for example in the field of functional materials and magnetic fluids applied in low friction dynamic systems, vibration dampers and tweeters (1, 2). Furthermore, possible areas of application are coatings and composite materials (3, 4) used in chemical reactors as magnetically separable catalyst material (5) as well as in the biomedical sector (6, 7, 8). Particularly in combination with polymers it is necessary, that the magnetite nanoparticles are available as stabilized colloids in the organic phase (9, 10, 11). But on an industrial scale the nanoparticles are synthesized mainly in the aqueous phase. The transfer of the particles into the organic phase is conventionally done by strategies based on filtration with subsequent drying and redispersion steps. But the properties of magnetite nanoparticles are limited, because they tend to oxidation and agglomeration due to their increased surface area/volume ratio (12).

Hence, a new type of particle extraction process is of great importance, which includes the transfer of magnetite particles from the aqueous into an immiscible organic phase directly through the liquid-liquid phase boundary. Further advantages are no exposure of the particles to a gas atmosphere and with it the avoidance of sintering by capillary forces and a high particle concentration in the receiving phase as well. The engineering concept is a miniplant, where the continuous phase transfer can be realized using a drop column. In 1968 the first liquid-liquid extraction experiments of ultrafine particles were conducted by Lai and Fuerstenau. For the separation of a mixture of alumina, water and oil they used a separatory funnel to put on top of each other the two phases (13). In our study in all operating methods a drop column is used in the simplest form without internals. The fundamental steps of the phase transfer process mechanism are at first the partly agglomeration of the synthesized magnetite nanoparticles caused by the instable conditions in the aqueous phase (14). After injecting the organic phase as drops through a distributor into the magnetite suspension in the drop column, the sedimentation is prevented by the stirring effect of the rising liquid drops. The organic phase represents the disperse phase of lower density and the magnetite suspension demonstrates the continuous aqueous phase, whereby collisions of particles and drops can be realized. However, the first step is necessary for the laboratory centrifuge as transfer device, because through sedimentation the agglomerates are transported into the organic phase. At the liquid-liquid interfaces the surfactant molecules from the organic phase interact with the hydrophilic particle surface, which leads to a chemical adsorption (15, 16). Thus, the formed partially nonpolar magnetite surface implicates the hydrophobization and functionalization of the particles as well as the wetting of the modified particle surface by the organic phase, which allows the phase transition of magnetite. Stable organic colloids are formed by deagglomeration of the particles as well as physicochemical dispersion due to the strong repulsive potential of the adsorbed surfactant molecules (17). The assumption of an efficient particle extraction process requires a high yield of transferred particles in the organic phase, low process times and a stable product. Therefore, a stable process in the drop column is necessary characterized by a stable drop formation and upward movement with a sufficient coalescence rate, which leads to a clearing off of the column.

Materials, Methods and Experimental Setup

By using a co-precipitation reaction the magnetite nanoparticles are synthesized. The procedure is widely applied and specified in (18) as well as the pH-conditioning of the aqueous phase. The used surfactants are ricinoleic acid and oleic acid, which are unsaturated fatty acids (FA) at the 9th C-atom with a carbon chain length of 18 C-atoms. In addition, ricinoleic acid has a hydroxyl group at the C-atom C12. Both surfactants are of technical grade with 90% purity from Sigma. The study presents the results with the parameters $x_{surf} = 1.4$ mass-% (mass fraction of surfactant in the solvent iso-octane as organic phase) and a specific amount of surfactant per magnetite $X_{S/M} = 0.2$ g/g. In (18) the methods are described in detail.

The drop column has a length of 700 mm in the partial recirculation operation. In the simple multi-level circuit as well as in the continuous circuit of both phases the length of the drop column is 920 mm caused by additional glass components for the implementation of the operating method. The internal diameter of the drop column of 25 mm is kept constant. Thus, a volume of 300 ml of the magnetite suspension is utilized at the first mentioned operation and 600 ml for the other one. The volume used corresponds to a mass of 6 g or rather 12 g magnetite nanoparticles. The dispersing system at the bottom of the transfer device is a single metal capillary with an inside diameter of 3.2 mm centrally mounted in a perforated plate. Both supply system and removal system consist of a peristaltic pump (Ismatec Reglo Analog) and a Tygon-tube with an inside diameter of 3.2 mm. In addition, the removal system has an automated sampling system with a pneumatic 3-way valve, manufactured by Swagelok, which is controllable by measurement software. The sample volume is 2.9 ml. To keep the volume of the disperse phase constant, the same volume is added simultaneously as organic phase in the receiver tank. The volume flow rate is 29 ml/min in the partial recirculation operation and in the simple multi-level circuit of the miniplant. To control the continuous circuit of both phases a volume flow rate of 14.5 ml/min is used.

Previous investigations using a laboratory centrifuge (Hettich Universal 30F) as transfer device are carried out with dichloromethane (DCM) as solvent, because the aqueous phase is placed on top of the organic phase as described in (14).

Results and Discussion

As mentioned for the development of a continuous particle extraction process the miniplant is investigated in different operating methods. The necessary parameters for a stable process behavior, which are the mass fraction of surfactant in the organic solvent x_{surf} as well as the specific surfactant amount to magnetite $X_{S/M}$, are determined using the miniplant in a partial recirculation operation (Figure 1a). This means, that the continuous phase is stationary and the disperse phase is in recirculation. Figure 1b represents the miniplant in continuous circuit of both phases, which is regarded later. Figure 1c outlines the procedures in the drop column during the extraction process. Thus, the disperse phase is pumped from the receiver tank through the capillary into the drop column, which is already filled with magnetite suspension, the formed drops rise through the column in the continuous phase, due to their lower density. This leads to collisions of particles and drops, whereby a phase separation in an organic phase, containing the extracted magnetite particles, and an almost particle-free aqueous phase takes place. Hence, the clearing off of the column and a black color of the organosol can be observed. Furthermore, a coalescence area is formed at the surface of the continuous phase (interface layer). The enrichment at the interface is a pickering emulsion consisting of droplets of the organic phase in water with only partial hydrophobized particles. If the coalescence is prevented, the particles can not be extracted in the organic phase resulting in a dispersion ribbon (19).

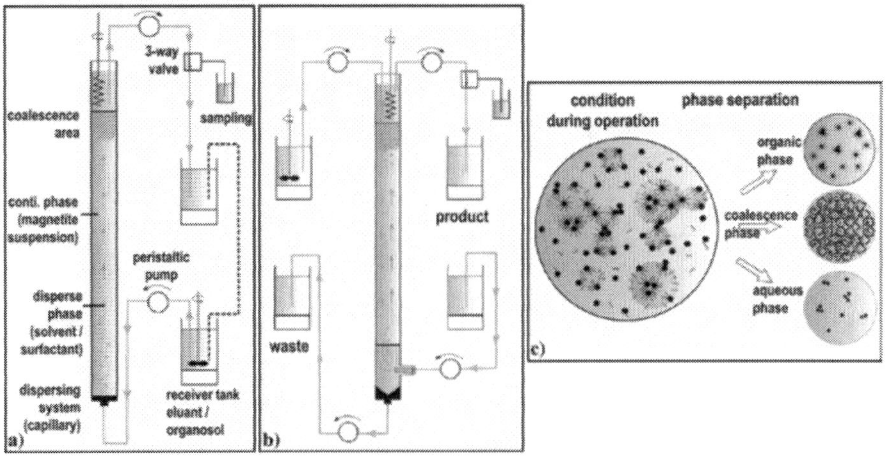

Figure 1. Experimental setup of the miniplant in a partial recirculation operation (a) as well as in continuous circuit of both phases (b) and illustration of the occurring procedures during the particle extraction process (c).

For the calculation of the corrected mass concentration of transferred magnetite particles in the receiver tank $\beta_{magn,rt}(t)$ a validation of the system is essential, due to the fact that after each removed sample volume the amount of magnetite is reduced. In consideration of the input and output to the receiver tank and the residence time distribution Equation (1) can be applied for the evaluation:

$$\beta_{magn,rt}(t) = \left(\beta_{magn,rt,0} + \frac{a_i * V_{rt}}{\dot{V}_{disp}} - b_i \right) * e^{\frac{V_{rt}}{\dot{V}_{disp}} t} + a_i t + b_i - \frac{a_i * V_{rt}}{\dot{V}_{disp}} \tag{1}$$

where $\beta_{magn,rt,0}$ is the initial magnetite mass concentration of the system, V_{rt} is the volume in the receiver tank, $\dot{V}_{disp}$ is the volume flow of the disperse phase and a_i as well as b_i are formed by linear interpolation between the removed samples.

For the characterization of the process behavior within the drop column in dependence from surfactants used the method of a particle-free operation is applied. Thus, it was possible to observe how the organic phase immediately disintegrates in drops directly above the opening of the metal capillary. With ricinoleic acid as surfactant a multitude of very small droplets as well as a turbidity in the column is formed. In contrast to oleic acid the noticed larger and significantly less drops without turbidity are similar like the mechanism of a single periodic drop formation (20). As a consequence of this, the available phase interface is smaller for the particle extraction process. The turbidity occurred is distributed in the whole column within a few minutes caused by the low buoyant force of the droplets. Owing to the additional hydroxyl group ricinoleic acid is also soluble in the salts and ammonia consisting aqueous phase, whereby the spontaneous formation of an emulsion is generated. As a result the small emulsion droplets with a median value of 5-10 µm measured by laser diffraction spectroscopy are iso-octane droplets, which are stabilized of ammonium ricinoleates existed at the interface by the dissolving of ricinoleic acid in the aqueous phase. This leads to additional available phase interface for the transfer process. This connection is also proved by the implementation of the experiments using a pH-indicator bromothymol blue in the continuous phase, whereby the formation of the surface active ammonium ricinoleates is observed resulting in color changes.

The measurements of the TOC content at defined time points of sampling in the aqueous phase (every 5 minutes till 45 minutes) confirmed these observations. By using ricinoleic acid as surfactant, a significant increase of the TOC content can be observed. Finally, after 5 minutes process time the aqueous phase is reached an average value of 2.0 g/l of dissolved ricinoleic acid. Whereas oleic acid is used, the poor ability of dissolving within the aqueous phase is expressed considerably, since the TOC content is on average 0.02 g/l. Thus, it can be recognized that in contrast to oleic acid approximately the 100-fold amount of ricinoleic acid dissolved in the aqueous phase irrespective of time.

Furthermore, the measurements of interfacial tensions on the particle system using the pendant-drop method (Figure 2) are demonstrated that the preferred dissolution of ricinoleic acid in the aqueous phase has a higher influence on the interfacial tension in contrast to the oleic acid. The interfacial tension between the aqueous and organic phase determines the drop size and the available phase interface for the particle extraction process. Thus, with ricinoleic acid as surfactant the interfacial tension is reduced from 31 mN/m to 8 mN/m and only for 2 minutes a stable drop can be formed. After this time the drop disintegrates caused by the dissolution into the aqueous phase resulting in a fast mass transport through the phase boundary. In the presence of oleic acid in the system an interfacial tension of 11 mN/m is reached and the time for a stable drop, which also means the equilibration at the interface amounts to 30-45 minutes. Due to the comparability a dimensionless representation of the time is chosen.

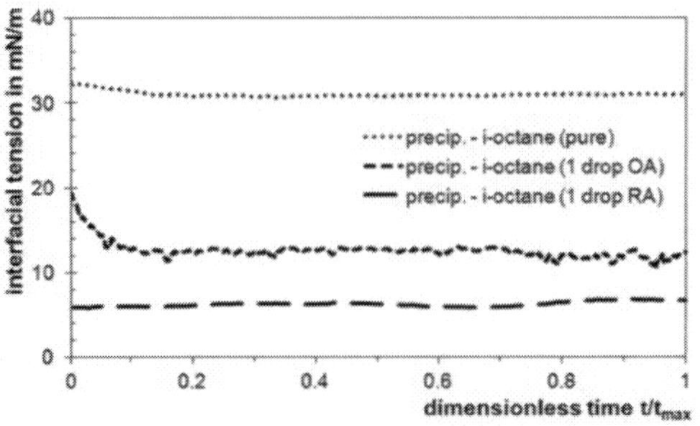

Figure 2. Demonstration of the interfacial tensions dependent on the surfactant used.

Consequently, the both surfactants used have a different process behavior with varying consequences to the particle extraction process. The determination of the yield of transferred magnetite in dependency of the process time indicates that the phase transfer with ricinoleic acid takes place rapidly. Because the increase of transferred magnetite to around 80% after 10 minutes is nearly twice the amount compared to oleic acid. The longer process times for the extraction of magnetite with oleic acid are caused by the lower dissolution in the aqueous phase resulting in a smaller available phase interface. This also can be recognized during long-time trials. Already after 15 minutes the yield of transferred magnetite with ricinoleic acid as surfactant amounts nearly 90% and is kept constant over the process time up to 165 minutes. In contrast with oleic acid after 270 minutes processing time approximately 86% of the magnetite particles have been extracted into the organic phase. Moreover, the clearing off of the column with ricinoleic acid is achieved after nearly 6 minutes, in contrast with oleic acid it can be observed after 160 minutes.

For the characterization of transfer kinetics, the reaction of 1st order is assumed. The experimental and theoretical calculated relative change in mass concentration conforms only for oleic acid, so that the particle extraction process can be estimated as a 1st order reaction. This means that the dissolution of oleic acid from the organic into aqueous phase does not influence the process kinetics, whereby the surfactant mass transfer is negligible. In contrast to ricinoleic acid another approximation represented by a sigmoid function in terms of a dose-response-curve has to be applied. Thus, the mass transport of the surfactant by dissolving in the aqueous phase strongly influences the kinetics of the particle extraction process within the column. This can be demonstrated as the dose for the process and is prevalent especially in the first minutes of the extraction process. This is the case, if the equilibrium of ricinoleic acid concentration in either phase is not reached. Furthermore, the response is represented by the functionalization of the particles resulting in the following extraction of them. Therefore, it can be assumed that the particle extraction process with ricinoleic acid as surfactant is controlled by process imbalance.

The colloidal stability of organosols is investigated using the calculation of the solubility distances $D_{FA-solvent}$ between the surfactant and the solvent (Equation 2) by the determination of the Hildebrandt parameter δ_t. This parameter can be calculated with the three Hansen solubility parameters in consideration of the dispersive δ_d, polar δ_p and hydrogen δ_h bonding components (Equation 3). Thus, the interaction between the used surfactants as pure substances and capped onto the magnetite surface can be characterized (17, 21). The calculated values are presented in Table 1 and it applies the following correlation: The smaller the value, the more soluble is the surfactant in the solvent. Ricinoleic acid has a lower solubility in iso-octane caused by the polar and hydrogen bonding parameters and therefore a better dissolution into the aqueous phase resulting in a more rapid particle extraction process. In contrast, oleic acid is more soluble in iso-octane, whereby more stable drops are formed during the phase transfer. Therefore, the colloidal stability of the organosol is increased. Compared with the produced organosols in the transfer device lab centrifuge using the solvent DCM the colloidal stability with ricinoleic acid is increased. Due to the very good solubility of ricinoleic acid in DCM, all surfactant molecules are very fast available at the phase interface to disintegrate the agglomerates in primary particles. Therefore, the result is a stable colloidal organosol.

$$D_{FA-solvent} = \sqrt{4*(\delta_{d,FA} - \delta_{d,solvent})^2 + (\delta_{p,FA} - \delta_{p,solvent})^2 + (\delta_{h,FA} - \delta_{h,solvent})^2} \quad (2)$$

$$\delta_t = \sqrt{\delta_d{}^2 + \delta_p{}^2 + \delta_h{}^2} \quad (3)$$

Table 1. Summary of the calculated Hansen solubility parameters and solubility distances.

substance –	δ_t in $MPa^{½}$	δ_d in $MPa^{½}$	δ_p in $MPa^{½}$	δ_h in $MPa^{½}$	$D_{FA-iOCT}$ in $MPa^{½}$	D_{FA-DCM} in $MPa^{½}$
iso-octane	14.30	14.30	0	0	–	–
DCM	20.20	18.20	6.30	6.10	–	–
RA	19.68	16.84	8.32	5.89	11.39	3.39
OA	18.31	17.30	5.73	1.78	8.49	4.71
RA@Fe_3O_4	19.40	17.27	6.89	5.55	10.66	2.03
OA@Fe_3O_4	17.90	17.68	2.81	0	7.32	7.11

For the analysis of the product stability a dispersion analyzer (Lumisizer) is applied, whereby the integral transmission as a function of time can be determined. By the increase of integral transmission, conclusions regarding to the colloidal stability can be drawn. These experimental observations are confirmed with the calculated solubility distances using the Hansen solubility parameters.

Due to the interesting process behavior and short processing times with ricinoleic acid in the partial recirculation operation of the miniplant as mentioned before, further operating methods are developed by experiments using ricinoleic acid as surfactant. Therefore, the simple multi-level circuit of the miniplant can be generated. This means that both phases are in circuit, but the removal and supply of the phases as well as the production of organosol are discontinuously. Due to stable processing control and yields of transferred magnetite about 80-90%, the continuous circuit is developed. Therefore, the organosol can be produced continuously including the simultaneous formation of waste and product as shown in Figure 1b. Yields of transferred magnetite are reached of about 90% (Figure 3a). Furthermore, the miniplant is utilized for a selective flotation process of magnetite and quartz particles in the partial recirculation operation. In Figure 3b the recovery of transferred magnetite dependent on the process time is demonstrated. Including the transferred particles of the interfacial layer (IL) a recovery of extracted magnetite particles of 76% is reached. This process engineering application presents, that the miniplant can be utilized for separation processes of two minerals for example.

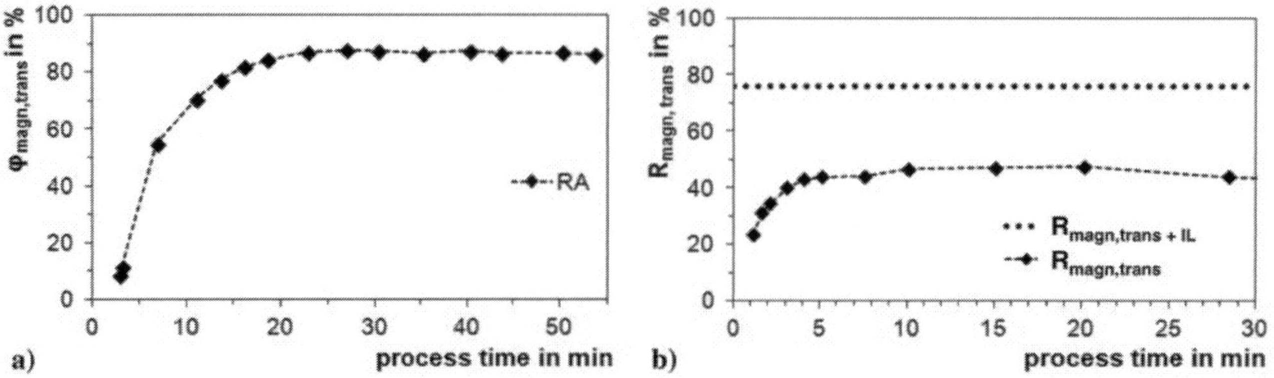

Figure 3. Demonstration of the process time dependent yield of transferred magnetite in continuous circuit of both phases using ricinoleic acid as surfactant (a) and the recovery of transferred magnetite by the implementation of a separation process in partial recirculation operation of the miniplant (b).

Conclusions

The study has been presented that the particle extraction process can be performed in different transfer devices. As a consequence of this, the interaction between the surfactants as well as the organic and aqueous phase used has to be considered, because this leads to different procedures at the phase boundary. A continuous particle extraction process with a drop column as transfer device was developed. Furthermore, the process behavior, transfer kinetics and colloidal stability were investigated and the process engineering application for a selective separation process was proved.

Acknowledgments

We give thanks to the Deutsche Forschungsgemeinschaft (DFG) for financial support (project no. PE1160/6-3). Additionally, we would like to thank Andre Rieger and the laboratory technicians from Institute of Thermal Process Engineering, Environmental and Natural Products Process Engineering of the TU Bergakademie Freiberg for the TOC measurements as well as Christian Strötges from Institute of Physical Chemistry of the TU Dortmund for the measurements of interfacial tensions.

References

1. Scherer, M. and Figueiredo Neto, A. M. (2005), “Ferrofluids: properties and applications”, *Brazilian Journal of Physics*, Vol. 35, No. 3A, pp. 718-727.
2. Hurlebaus, S. and Gaul, L. (2006), “Smart structure dynamics”, *Mechanical Systems and Signal Processing*, Vol. 20, No. 2, pp. 255-281.
3. Dallas, P., Georgakilas, V., Niarchos, D., Komninou, P., Kehagias, T. and Petridis, D. (2006), “Synthesis, characterization and thermal properties of polymer / magnetite nanocomposites”, *Nanotechnology*, Vol. 17, pp. 2046-2053.
4. Teja, A. S. and Koh, P.-Y. (2009), “Synthesis, properties and applications of magnetic iron oxide nanoparticles”, *Progress in Crystal Growth and Characterization of Materials*, Vol. 55, pp. 22-45.

5. Hickstein, B. and Peuker, U. A. (2009), “Modular process for the flexible synthesis of magnetic beads - process and product validation”, *Journal of Applied Polymer Science*, Vol. 112, pp. 2366-2373.
6. Banert, T. and Peuker, U. A. (2007), “Synthesis of magnetic beads for bio-separation using the solution method”, *Chemical Engineering Communications*, Vol. 194, No. 6, pp. 707-719.
7. Laurent, S., Forge, D., Port, M., Roch, A., Robic, C., Vander Elst, L. and Muller, R. N. (2008), “Magnetic iron oxide nanoparticles: synthesis, stabilization, vectorization, physicochemical characterization and biological application”, *Chemical Reviews*, Vol. 108, No. 6, pp. 2064-2110.
8. Mahmoudi, M., Sant, S., Wang, B., Laurent, S. and Sen, T. (2011), “Superparamagnetic iron oxide nanoparticles (SPIONs): development, surface modification and applications in chemotherapy”, *Advanced Drug Delivery Reviews*, Vol. 63, pp. 24-46.
9. Banert, T. and Peuker, U. A. (2006), “Preparation of highly filled super-paramagnetic PMMA-magnetite nano composites using the solution method”, *Journal of Material Science*, Vol. 41, pp. 3051-3056.
10. Kirchberg, S., Rudolph, M., Ziegmann, G. and Peuker, U. A. (2011), „Nanocomposites based on technical polymers and sterically functionalized soft magnetic magnetite nanoparticles: synthesis, processing and characterization”, *Journal of Nanomaterials*, Article ID 670531.
11. Rudolph, M. and Peuker, U. A. (2011), “Coagulation and stabilization of sterically functionalized magnetite nanoparticles in an organic solvent with different technical polymers”, *Journal of Colloid and Interface Science*, Vol. 357, pp. 292-299.
12. Machunsky, S. and Peuker, U. A. (2007), “Liquid-liquid interfacial transport of nanoparticles”, *Physical Separation in Science and Engineering*, Article ID 34832.
13. Lai, R. W. M. and Fuerstenau, D. W. (1968), “Liquid-liquid extraction of ultrafine particles”, *Transactions of the American Institute of Mining, Metallurgical*, and Petroleum Engineers, Vol. 241, pp. 549-556.
14. Erler, J., Machunsky, S., Grimm, P., Schmid, H.-J. and Peuker, U. A. (2013), “Liquid-liquid phase transfer of magnetite nanoparticles – evaluation of surfactants”, *Powder Technology*, Vol. 247, pp. 265-269.
15. Zhang, L., He, R. and Gu, H.-C. (2006), “Oleic acid coating on the monodisperse magnetite nanoparticles”, *Applied Surface Science*, Vol. 253, pp. 2611-2617.
16. Rudolph, M., Erler, J. and Peuker, U. A. (2012), “A TGA-FTIR perspective of fatty acids adsorbed on magnetite nanoparticles – decomposition steps and magnetite reduction”, *Colloid and Surfaces A: Physicochemical and Engineering Aspects*, Vol. 397, pp. 16-23.
17. Rudolph, M. and Peuker, U. A. (2012), “Phase transfer of agglomerated nanoparticles – deagglomeration by adsorbing grafted molecules and colloidal stability in polymer solutions”, *Journal of Nanoparticle Research*, Vol. 14, No. 7, Article 990.
18. Erler, J. V., Leistner, T. and Peuker, U. A. (2014), “Application of a particle extraction process at the interface of two liquids in a drop column – consideration of the process behavior and kinetic approach“, *Advances in Chemical Engineering and Science*, doi:10.4236/aces.2014, in press.
19. Blaß, E. (1988), “Bildung und Koaleszenz von Blasen und Tropfen”, *Chemie Ingenieur Technik*, Vol. 60, pp. 935-947.
20. Räbiger, N. and Schlüter, M. (2006), “Bildung und Bewegung von Tropfen und Blasen”, *VDI Wärmeatlas*, Hrsg. von VDI Gesellschaft, Springer, Lda1-Lda15.
21. Hansen, C. M. (2007), “Hansen solubility parameters“, A User´s Handbook, Second Edition CRC, Boca Raton.

Dimensional Analysis of Packed Bed Gas Absorbers

J. H. Worstell[1,2], M. Zhang[1], E. Doll[1], A. Aboalsaud[1], F. Kaleem[1], K. Parihar[1], S. Rich[1], and M. Doll[3]

[1] University of Houston
[2]Worstell and Worstell, Consultants and University of Houston
[3]Shell Global Solutions

Overview

Critique current method for designing and scaling absorption towers.

Propose a new method for designing and scaling absorption towers.

Show how to use this new method to design absorption towers.

Reaction-free Chemical Engineering

Chemical Engineering

involves more than chemical reaction.

Feed preparation

Product purification

Heat transfer

Heat recovery

Momentum transfer

Today's Subject

Feed preparation and/or product purification.

Specifically
Gas Absorption

Examples

Feed Purification
H_2S removal from process streams
Natural gas by alkaline absorption
Refinery off-gas by alkaline absorption

Production Purification
Coke-oven off-gas treatment
NH_3 by water absorption
Benzene/toluene by oil absorption

Gas Absorption

Moving molecules from gas-phase into liquid-phase

Moving molecules across a physical boundary

Thus
Larger the physical boundary surface area, the more absorption occurs

Equipment
Agitated Vessel
Spray Tower
Bubble-cap, Sieve-tray, Grid-tray Towers
Packed Bed Towers

Most common
Packed Bed Towers
Obtain highest possible physical boundary surface area

Current Design Method

Based on Equilibrium

1. Obtain equilibrium data --- assume tower operates at equilibrium;
2. Specify inlet ammonia partial pressure or liquid concentration;
3. Specify outlet ammonia partial pressure or liquid concentration;
4. Specify gas, water flow rates --- plot above information;
5. Ensure Operating Line is above Equilibrium Line (if below, stripping occurs, not absorption);
6. Calculate height of packing

$$G_M dy = k_{GSF} a_G P (y - y_{EQ}) dh$$

$$L_M dy = \frac{k_{LSF} a_L}{\rho_M} (x_{EQ} - x) dh$$

Current Design Method

7. Assume gas film or liquid film diffusion rate limited --- consider gas;
8. Integrate rate equation

$$h = \frac{G_M}{k_{GSF} a_G P} \int_{y_{Outlet}}^{y_{Inlet}} \frac{dy}{y - y_{EQ}}$$

9. Assume Equilibrium Line is linear --- Henry's Law applies

$$h = \frac{G_M}{k_{GSF} a_G P} \int_{y_{Outlet}}^{y_{Inlet}} \frac{dy}{y - He\,x}$$

10. Rearrange the Operating Line

$$x = \frac{y}{m} - \frac{b}{m}$$

Current Design Method

11. Substitute for x

$$h = \frac{G_M}{k_{GSF} a_G P} \int_{y_{Outlet}}^{y_{Inlet}} \frac{dy}{y - He\left(\frac{y}{m} - \frac{b}{m}\right)}$$

12. Rearrange and integrate

$$h = \left(\frac{G_M}{k_{GSF} a_G P}\right)\left(\frac{1}{1 - He/m}\right) \ln\left[\frac{\left(1 - He/m\right) y_{Outlet} + b\,He/m}{\left(1 - He/m\right) y_{Inlet} + b\,He/m}\right]$$

Current Design Method

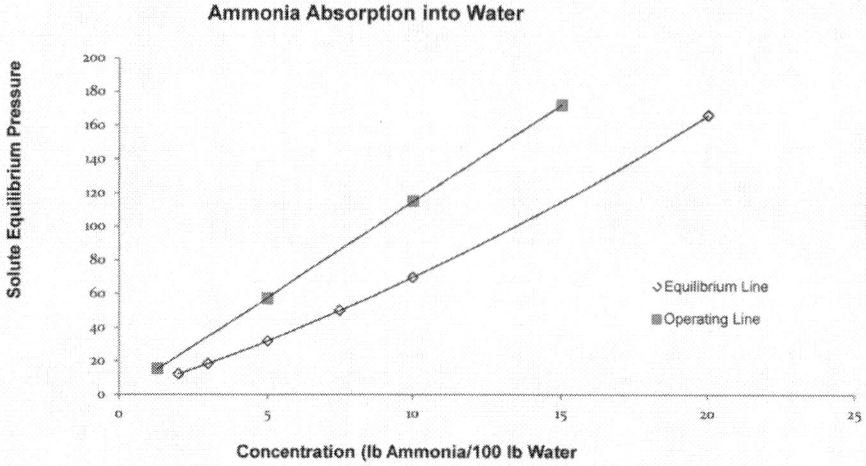

Issues

Assumes Tower achieves Equilibrium

Achieving Equilibrium requires infinite time --- commercially unviable
Achieve stable dynamic condition --- hopefully

Assumes gas film or liquid film diffusion rate limited

Estimates (measures ?) $k_{GSF}a_G$ or $k_{LSF}a_L$

Assumes Equilibrium Line linear during operation

In order to use Henry's Law assumption

Significant assumption that may not hold.

Is there a method with fewer assumptions ?

Consider Dimensional Analysis

Dimensional Analysis

A method for determining the relationships between process variables

Does not provide analytical solution to conservation equations

Forms the foundation for similitude and for the theory of models

Has been used by aeronautical, civil, and mechanical engineers
for past one hundred years
(processes use L,M, and T dimensions;
thus, dimensional analysis can be done by hand)

Not used much by chemical engineers
(processes use L, M, T, θ, and N;
thus, not solvable by hand)

Solvable by matrix algebra

Dimensional Analysis of Ammonia Absorption Tower

$f(dP/Z, D, L, c_{Gas,In}, c_{Gas,Out}, v_G, \rho_G, \mu_G, c_{Liq,In}, c_{Liq,Out}, v_L, \rho_L, \mu_L) = 0$

Design specifications: adiabatic operation and $\Delta T = 0$

Dimension Table

Dimension / Variable	$\Delta P/Z$	D	L	$C_{Gas,In}$	$C_{Gas,Out}$	v_G	ρ_G	μ_G	$C_{Liq,Out}$	$C_{Liq,In}$	v_L	ρ_L	μ_L
L	-2	1	1	-3	-3	1	-3	-1	-3	-3	1	-3	-1
M	1	0	0	0	0	0	1	1	0	0	0	1	1
T	-2	0	0	0	0	-1	0	-1	0	0	-1	0	-1
N	0	0	0	1	1	0	0	0	1	1	0	0	0

12

Definition of Characteristic Length

When we load a bed, what do we know?
The weight of packing put into the circular column.
Converting its $W_{packing}$ to volume, we obtain

$$W_{packing}/\rho_{LBD} = V_{Packing} = A_{CS} Z_{Packing}$$

The area available for fluid flow is εA

$$L_{Char} = \frac{\left(W_{packing}/\rho_{LBD}\right)}{\varepsilon A_{CS}} = \frac{A_{CS} Z_{Packing}}{\varepsilon A_{CS}} = \frac{Z_{Packing}}{\varepsilon} = L$$

If Z = 3 m and ε = 0.4, then L = 7.5 m; if Z = 3 m and ε = 0.97, then L = 3.1 m. Thus, L is an indication of tube length; i.e., flow path length through the packing.

13

Dimension Matrix and Rank Matrix

Dimension Matrix

$$\begin{bmatrix} -2 & 1 & 1 & -3 & -3 & 1 & -3 & -1 & -3 & -3 & 1 & -3 & -1 \\ 1 & 0 & 0 & 0 & 0 & 0 & 1 & 1 & 0 & 0 & 0 & 1 & 1 \\ -2 & 0 & 0 & 0 & 0 & -1 & 0 & -1 & 0 & 0 & -1 & 0 & -1 \\ 0 & 0 & 0 & 1 & 1 & 0 & 0 & 0 & 1 & 1 & 0 & 0 & 0 \end{bmatrix}$$

Rank Matrix

$$\begin{bmatrix} -3 & 1 & -3 & -1 \\ 0 & 0 & 1 & 1 \\ 0 & -1 & 0 & -1 \\ 1 & 0 & 0 & 0 \end{bmatrix}$$

Rank Calculation and Dimensionless Parameter Calculation

Rank Calculation

$$\begin{vmatrix} -3 & 1 & -3 & -1 \\ 0 & 0 & 1 & 1 \\ 0 & -1 & 0 & -1 \\ 1 & 0 & 0 & 0 \end{vmatrix} = 4$$

Rank = 4 because determinant is non-singular and it is a 4x4 matrix

$$N_P = N_V - R$$
$$N_P = 13 - 4 = 9$$

Will determine nine dimensionless parameters

15

Matrix Equation to Be Solved

$$\begin{bmatrix} k_1 \\ k_2 \\ k_3 \\ k_4 \\ k_5 \\ k_6 \\ k_7 \\ k_8 \\ k_9 \\ k_{10} \\ k_{11} \\ k_{12} \\ k_{13} \end{bmatrix} = \begin{bmatrix} I & 0 \\ -R^{-1}B & R^{-1} \end{bmatrix} \begin{bmatrix} k_1 \\ k_2 \\ k_3 \\ k_4 \\ k_5 \\ k_6 \\ k_7 \\ k_8 \\ k_9 \\ k_{10} \\ 0 \\ 0 \\ 0 \end{bmatrix}$$

Jonathan Worstell, *Dimensional Analysis*, Elsevier, Oxford, UK, May 2014.

T. Szirtes, Applied Dimensional Analysis and Modeling, Second Edition, Butterworth-Heinemann, Burlington, MA, 2007.

16

Inverse Rank Matrix and -R^{-1}*B Matrix Calculations

Inverse Rank Matrix

$$\begin{bmatrix} -3 & 1 & -3 & -1 \\ 0 & 0 & 1 & 1 \\ 0 & -1 & 0 & -1 \\ 1 & 0 & 0 & 0 \end{bmatrix}^{-1} = \begin{bmatrix} 0 & 0 & 0 & 1 \\ -1 & -3 & -2 & -3 \\ -1 & -2 & -1 & -3 \\ 1 & 3 & 1 & 3 \end{bmatrix}$$

-R^{-1}*B Matrix

$$-\begin{bmatrix} 0 & 0 & 0 & 1 \\ -1 & -3 & -2 & -3 \\ -1 & -2 & -1 & -3 \\ 1 & 3 & 1 & 3 \end{bmatrix} * \begin{bmatrix} -2 & 1 & 1 & -3 & -3 & 1 & -3 & -1 & -3 \\ 1 & 0 & 0 & 0 & 0 & 0 & 1 & 1 & 0 \\ -2 & 0 & 0 & 0 & 0 & -1 & 0 & -1 & 0 \\ 0 & 0 & 0 & 1 & 1 & 0 & 0 & 0 & 1 \end{bmatrix}$$

17

-R^{-1}*B Matrix

-R^{-1}*B Matrix

$$-R^{-1} * B = \begin{bmatrix} 0 & 0 & 0 & -1 & -1 & 0 & 0 & 0 & -1 \\ -3 & 1 & 1 & 0 & 0 & -1 & 0 & 0 & 0 \\ -2 & 1 & 1 & 0 & 0 & 0 & -1 & 0 & 0 \\ 1 & -1 & -1 & 0 & 0 & 0 & 0 & -1 & 0 \end{bmatrix}$$

18

Matrix Equation to Be Solved

$$
\begin{bmatrix} k_1 \\ k_2 \\ k_3 \\ k_4 \\ k_5 \\ k_6 \\ k_7 \\ k_8 \\ k_9 \\ k_{10} \\ k_{11} \\ k_{12} \\ k_{13} \end{bmatrix}
=
\begin{bmatrix}
1 & 0 & 0 & 0 & 0 & 0 & 0 & 0 & 0 & 0 & 0 & 0 & 0 \\
0 & 1 & 0 & 0 & 0 & 0 & 0 & 0 & 0 & 0 & 0 & 0 & 0 \\
0 & 0 & 1 & 0 & 0 & 0 & 0 & 0 & 0 & 0 & 0 & 0 & 0 \\
0 & 0 & 0 & 1 & 0 & 0 & 0 & 0 & 0 & 0 & 0 & 0 & 0 \\
0 & 0 & 0 & 0 & 1 & 0 & 0 & 0 & 0 & 0 & 0 & 0 & 0 \\
0 & 0 & 0 & 0 & 0 & 1 & 0 & 0 & 0 & 0 & 0 & 0 & 0 \\
0 & 0 & 0 & 0 & 0 & 0 & 1 & 0 & 0 & 0 & 0 & 0 & 0 \\
0 & 0 & 0 & 0 & 0 & 0 & 0 & 1 & 0 & 0 & 0 & 0 & 0 \\
0 & 0 & 0 & 0 & 0 & 0 & 0 & 0 & 1 & 0 & 0 & 0 & 0 \\
0 & 0 & 0 & -1 & -1 & 0 & 0 & 0 & -1 & 0 & 0 & 0 & 1 \\
-3 & 1 & 1 & 0 & 0 & -1 & 0 & 0 & 0 & -1 & -3 & -2 & -3 \\
-2 & 1 & 1 & 0 & 0 & 0 & -1 & 0 & 0 & -1 & -2 & -1 & -3 \\
1 & -1 & -1 & 0 & 0 & 0 & 0 & -1 & 0 & 1 & 3 & 1 & 3
\end{bmatrix}
\begin{bmatrix} k_1 \\ k_2 \\ k_3 \\ k_4 \\ k_5 \\ k_6 \\ k_7 \\ k_8 \\ k_9 \\ k_{10} \\ 0 \\ 0 \\ 0 \end{bmatrix}
$$

19

Columns of Interest

$$
\begin{bmatrix} k_1 \\ k_2 \\ k_3 \\ k_4 \\ k_5 \\ k_6 \\ k_7 \\ k_8 \\ k_9 \\ k_{10} \\ k_{11} \\ k_{12} \\ k_{13} \end{bmatrix}
=
\left[\begin{array}{ccccccccc cc}
1 & 0 & 0 & 0 & 0 & 0 & 0 & 0 & 0 & & \\
0 & 1 & 0 & 0 & 0 & 0 & 0 & 0 & 0 & & \\
0 & 0 & 1 & 0 & 0 & 0 & 0 & 0 & 0 & & \\
0 & 0 & 0 & 1 & 0 & 0 & 0 & 0 & 0 & \text{Zero} & \\
0 & 0 & 0 & 0 & 1 & 0 & 0 & 0 & 0 & & \text{Matrix} \\
0 & 0 & 0 & 0 & 0 & 1 & 0 & 0 & 0 & & \\
0 & 0 & 0 & 0 & 0 & 0 & 1 & 0 & 0 & & \\
0 & 0 & 0 & 0 & 0 & 0 & 0 & 1 & 0 & & \\
0 & 0 & 0 & 0 & 0 & 0 & 0 & 0 & 1 & & \\
0 & 0 & 0 & -1 & -1 & 0 & 0 & 0 & -1 & & \\
-3 & 1 & 1 & 0 & 0 & -1 & 0 & 0 & 0 & R^{-1} & \\
-2 & 1 & 1 & 0 & 0 & 0 & -1 & 0 & 0 & & \text{Matrix} \\
1 & -1 & -1 & 0 & 0 & 0 & 0 & -1 & 0 & &
\end{array}\right]
\begin{bmatrix} k_1 \\ k_2 \\ k_3 \\ k_4 \\ k_5 \\ k_6 \\ k_7 \\ k_8 \\ k_9 \\ k_{10} \\ 0 \\ 0 \\ 0 \end{bmatrix}
$$

20

Reading Dimensionless Parameters from Matrix

	Π_1	Π_2	Π_3	Π_4	Π_5	Π_6	Π_7	Π_8	Π_9
$\Delta P/Z$	1	0	0	0	0	0	0	0	0
D	0	1	0	0	0	0	0	0	0
L	0	0	1	0	0	0	0	0	0
$C_{Gas,In}$	0	0	0	1	0	0	0	0	0
$C_{Gas,Out}$	0	0	0	0	1	0	0	0	0
v_G	0	0	0	0	0	1	0	0	0
ρ_G	0	0	0	0	0	0	1	0	0
μ_G	0	0	0	0	0	0	0	1	0
$C_{Liq,Out}$	0	0	0	0	0	0	0	0	1
$C_{Liq,In}$	0	0	0	−1	−1	0	0	0	−1
v_L	−3	1	1	0	0	−1	0	0	0
ρ_L	−2	1	1	0	0	0	−1	0	0
μ_L	1	−1	−1	0	0	0	0	−1	0

21

Basic Dimensionless Parameters for Gas Absorption

$$\Pi_1 = \frac{(\Delta P/Z)\mu_L}{v_L^3 \rho_L^2}$$

$$\Pi_2 = \frac{\rho_L D v_L}{\mu_L}$$

$$\Pi_3 = \frac{\rho_L L v_L}{\mu_L}$$

$$\Pi_4 = \frac{C_{Gas,In}}{C_{Liq,In}}$$

$$\Pi_5 = \frac{C_{Gas,Out}}{C_{Liq,In}}$$

$$\Pi_6 = \frac{v_G}{v_L}$$

$$\Pi_7 = \frac{\rho_G}{\rho_L}$$

$$\Pi_8 = \frac{\mu_G}{\mu_L}$$

$$\Pi_9 = \frac{C_{Liq,Out}}{C_{Liq,In}}$$

22

Rearranging Dimensionless Parameters

$$\Pi_1 = \frac{\left(\Delta P/Z\right)\mu_L}{v_L^3\,\rho_L^2}$$

$$\Pi_2 = \frac{\rho_L\,D\,v_L}{\mu_L} = Re_{D,Liq}$$

$$\Pi_{367/8} = \frac{\rho_G\,L\,v_G}{\mu_G} = Re_{L,Gas}$$

$$\Pi_{3/2} = \frac{L}{D}$$

$$\Pi_{5/9} = \frac{C_{Gas,Out}}{C_{Liq,Out}}$$

$$\Pi_6 = \frac{v_G}{v_L}$$

$$\Pi_7 = \frac{\rho_G}{\rho_L}$$

$$\Pi_8 = \frac{\mu_G}{\mu_L}$$

$$\Pi_{9/4} = \frac{C_{Liq,Out}}{C_{Gas,In}}$$

23

Rearranging Dimensionless Parameters

$$\Pi_1 = \frac{\left(\Delta P/Z\right)\mu_L}{v_L^3\,\rho_L^2}$$

$$\Pi_2 = \frac{\rho_L\,D\,v_L}{\mu_L} = Re_{D,Liq}$$

$$\Pi_{367/8} = \frac{\rho_G\,L\,v_G}{\mu_G} = Re_{L,Gas}$$

$$\Pi_{3/2} = \frac{L}{D}$$

$$\Pi_{5/9} = \frac{C_{Gas,Out}}{C_{Liq,Out}} = 0$$

$$\Pi_6 = \frac{v_G}{v_L}$$

$$\Pi_7 = \frac{\rho_G}{\rho_L}$$

$$\Pi_8 = \frac{\mu_G}{\mu_L}$$

$$\Pi_{9/4} = \frac{C_{Liq,Out}}{C_{Gas,In}}$$

24

Experimental Apparatus

25

Packing Material Tested

26

Results

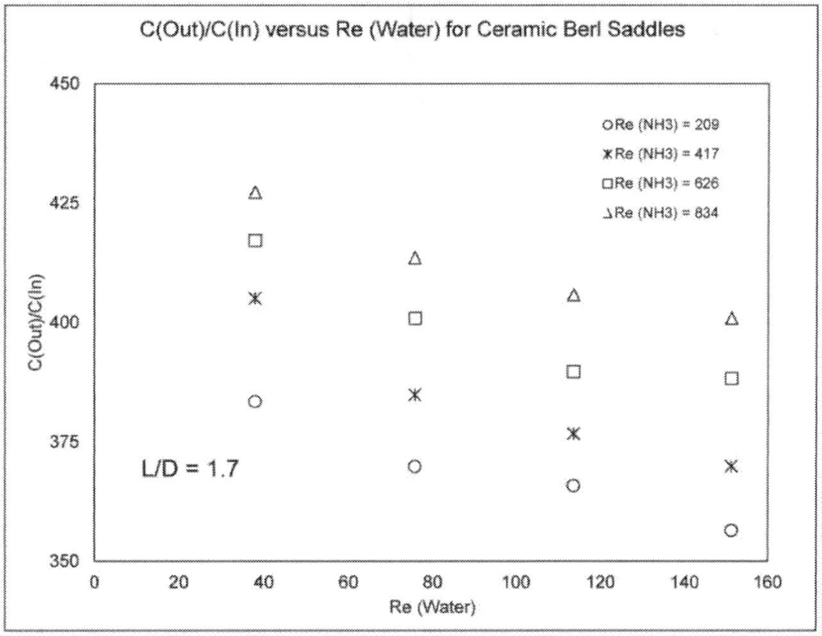

27

Results

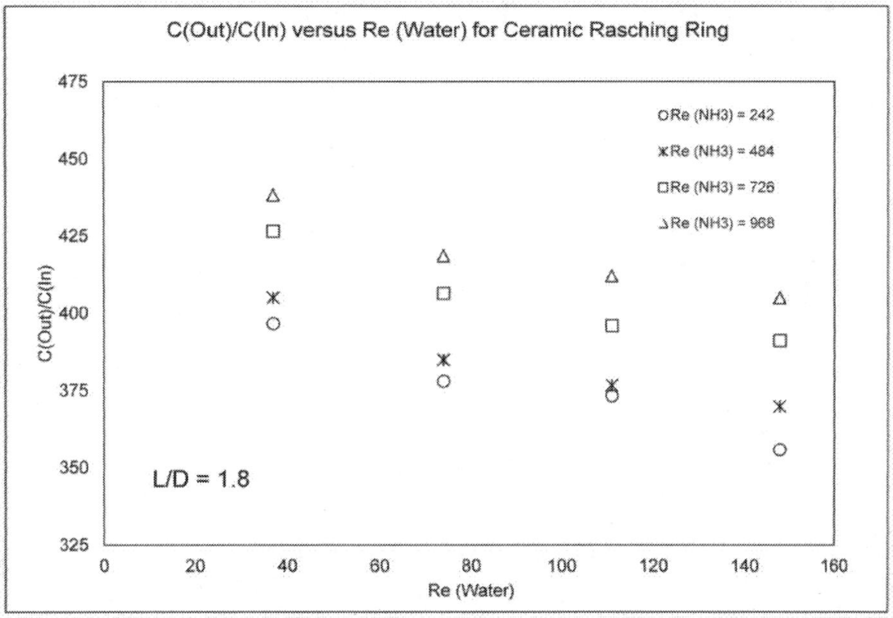

28

Results

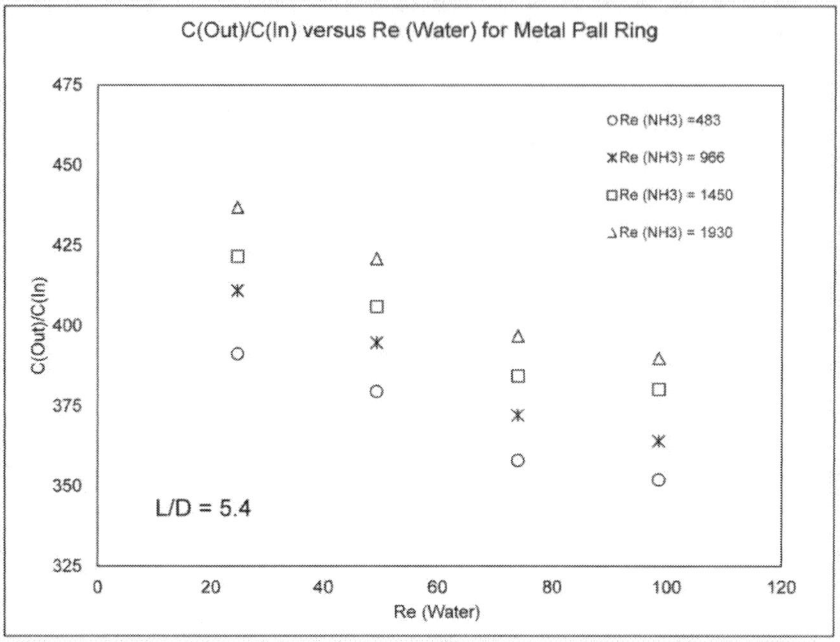

29

Results

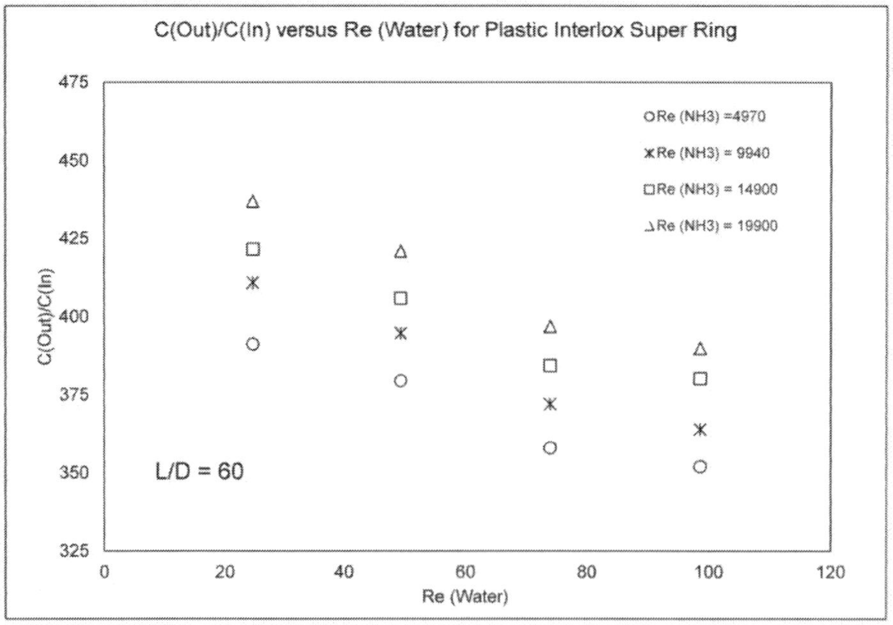

30

New Design Method

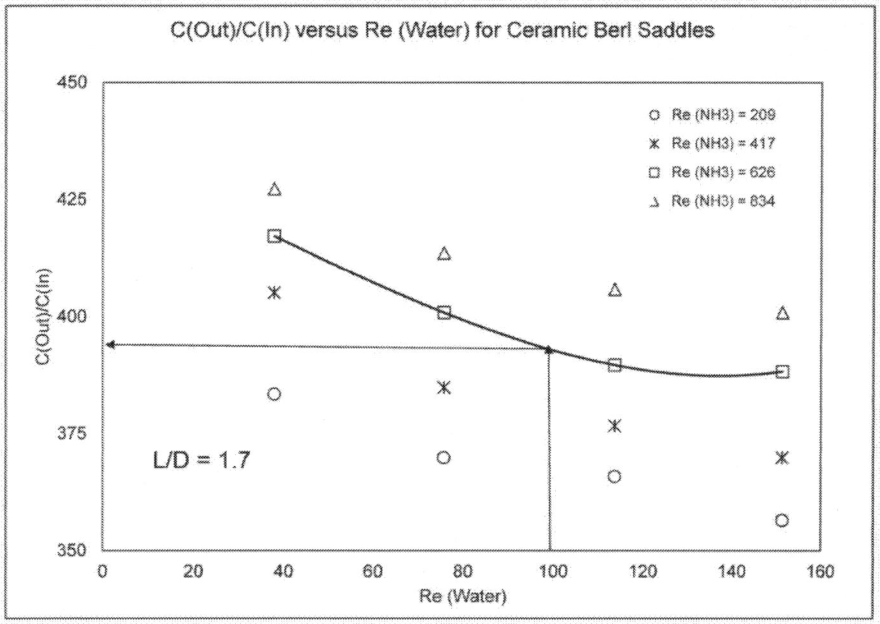

31

New Design Method

1. Choose Re(liq);
2. Choose Re(gas) to achieve desired C(Out)/C(In);
3. Choose diameter of plant absorption column;
4. Calculate L(plant)

$$\left(\frac{L}{D}\right)_{Expt} = \left(\frac{L}{D}\right)_{Plant}$$

$$L_{Expt}\left(\frac{D_{Plant}}{D_{Expt}}\right) = L_{Plant}$$

5. Calculate weight of plant packing

$$L_{Expt}\left(\frac{D_{Plant}}{D_{Expt}}\right) = L_{Plant} = \frac{\left(W_{Packing}/\rho_{LBD}\right)}{\varepsilon A_{CS}}$$

$$L_{Expt}\left(\frac{D_{Plant}}{D_{Expt}}\right)\left(\varepsilon A_{CS}\rho_{LBD}\right) = W_{Packing}$$

32

New Design Method

6. Calculate height of plant packing

$$\frac{Z_{Packing}}{\varepsilon} = L_{Plant}$$

$$Z_{Packing} = \varepsilon L_{Plant}$$

33

Advantage of New Design Method

No Assumptions.

34

Project "Multi-Phase" – a combined effort of industry and academia to investigate bubble column hydrodynamics

P. Rollbusch, M. Becker, M. Ludwig, M. Grünewald // 01.04.2014

SPONSORED BY THE

Federal Ministry of Education and Research

EVONIK INDUSTRIES

1. Motivation
2. Project Multi-Phase
3. Experimental activities
4. Simulation activities
5. Summary

1. Motivation - academia

EVONIK INDUSTRIES

Bubble columns in theory

- Easy design & construction
- Suitable for high pressures
- No moving internals (e.g. stirrers)
- Low specific power input for generation of mass transfer surface

around 7.500 publications*

- Mostly aqeous systems at atmospheric conditions
- Lab scale columns
- Results obtained at high gas throughputs
- Semibatch operating mode

* results obtained via SciFinder 2014-02-04

01.04.2014 Project "Multi-Phase" – a combined effort of industry and academia Page 3

1. Motivation - industry

EVONIK INDUSTRIES

Bubble columns in practice

- Large scale continuous, pressurized
- Equipped with internals for heat removal, suppression of backmixing etc.
- Usually organic, impure liquids are processed
- Complex interactions of hydrodynamics, transport processes and reaction kinetics

around 69 publications*

- Often dealing with single phenomena
- Still mostly aqueous systems
- Lab scale columns
- Controversial results

01.04.2014 Project "Multi-Phase" – a combined effort of industry and academia Page 4

1. Motivation – multiphase reactor engineering

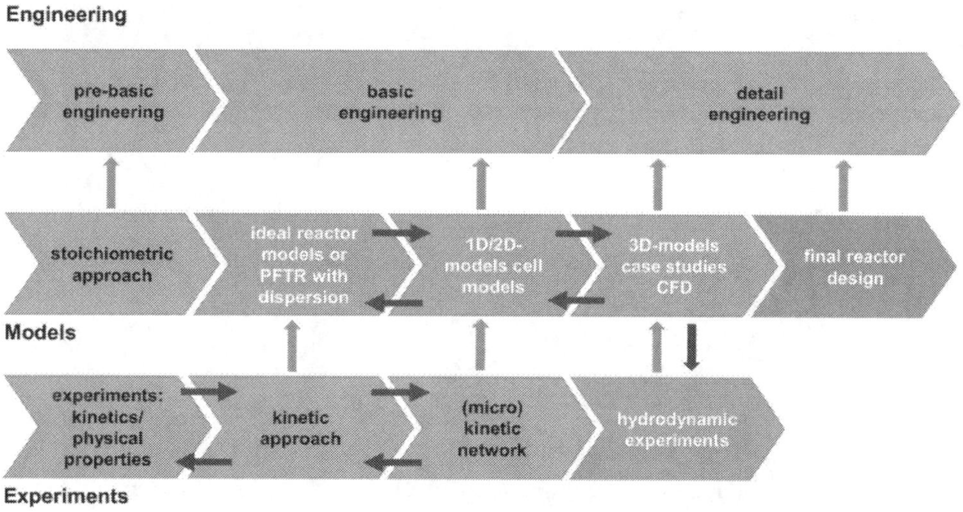

01.04.2014 Project "Multi-Phase" – a combined effort of industry and academia Page 5

2. Project Multi-Phase

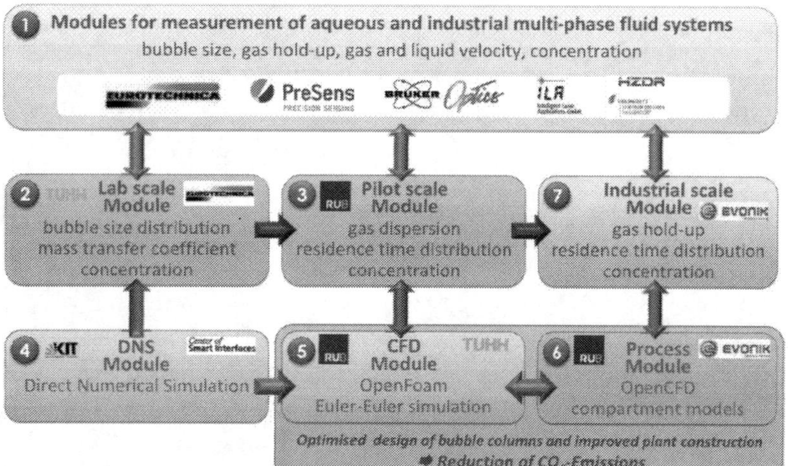

01.04.2014 Project "Multi-Phase" – a combined effort of industry and academia Page 6

3. Experimental activities – single bubble behaviour

EVONIK INDUSTRIES

Lab-scale bubble column, local measurements of:

- Bubble rise velocity and size distribution (DN100 and DN330 column)
- Liquid phase velocity, mass transfer coefficient, residence time
- EPIV measurements in cumene/N_2 and H_2O/N_2

TECHNICAL BIOCATALYSIS TUHH

TUHH Institut für ims Mehrphasenströmungen

ILA Intelligent Laser Applications GmbH

PreSens PRECISION SENSING

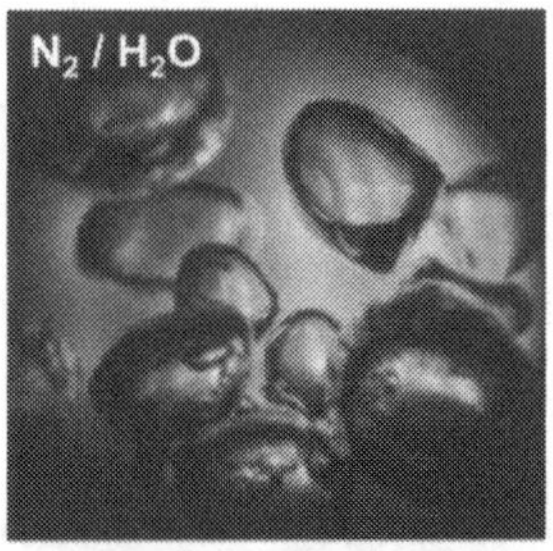

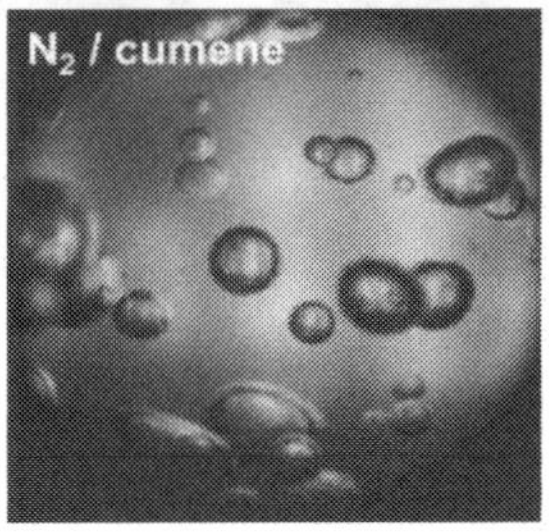

3. Experimental activities – wire mesh sensor

EVONIK INDUSTRIES

RUB

Plexiglas column

- Diameter: 0.3 m
- Height: 4 m (varying H/D ratios)

Wire-Mesh sensor

- 32x32 and 64x64 grid

Objectives

- Radial phase holdups
- Residence time distribution
- Interfacial area

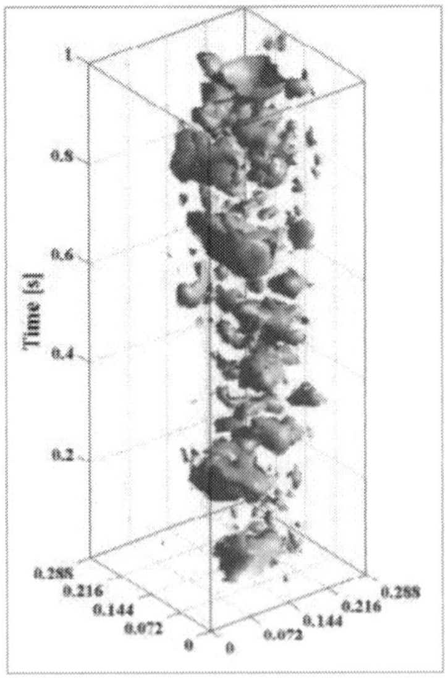

3. Experimental activities – pilot scale

EVONIK INDUSTRIES

Pressurized column

- Measurements in organic liquids (cumene) or water sparged with nitrogen
- Testing of developed measurement techniques (Gamma-CT, PIV, wire-mesh-sensor, ATR-probe)
- Objectives: phase and bubble size distributions, liquid phase residence time, radial holdup profiles

- Pressure up to 3.6 MPa
- Temperature up to 75 °C
- Diameter: 0.33 m, height: 4 m
- Gas/liquid in concurrent upflow
- Axial and radial holdup profiles

3. Experimental activities – Gamma - CT

EVONIK INDUSTRIES

HZDR HELMHOLTZ ZENTRUM DRESDEN ROSSENDORF

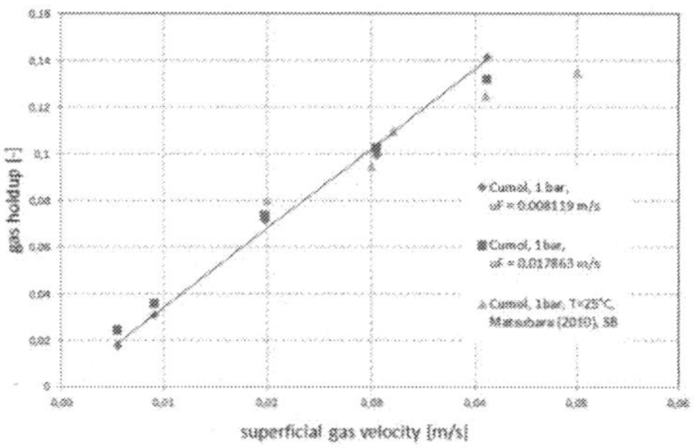

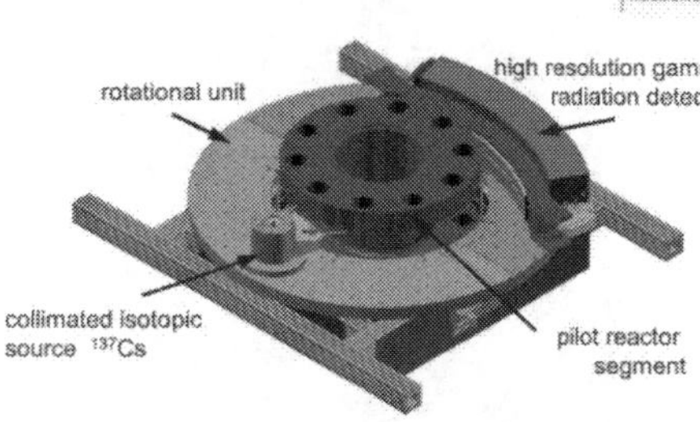

U. Hampel, et. al., Rev. Sci. Inst. Vol. 78 (2007) 103704.

- Radial holdup profiles of cumene and water at 0.1, 1.85 and 3.6 MPa
- Very good agreement with literature data and ΔP measurements

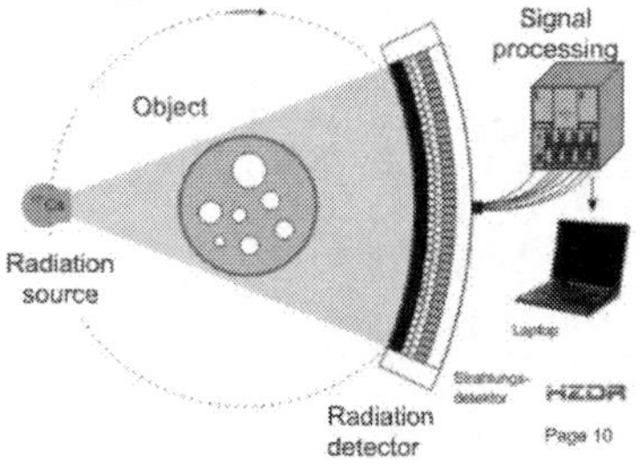

3. Experimental activities – influence of pressure

EVONIK INDUSTRIES

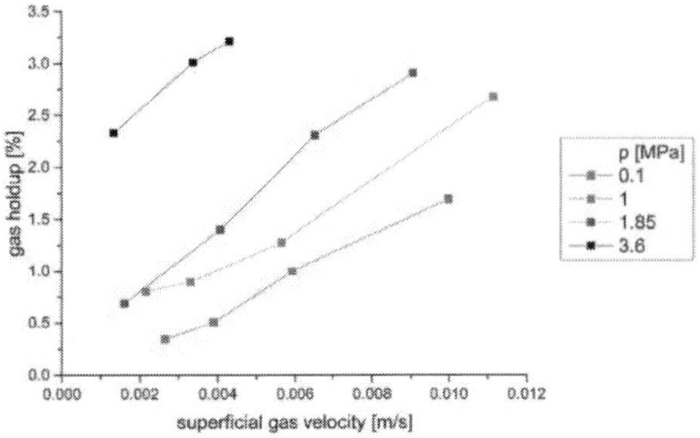

- N_2 holdups in water or cumene via Δp
- Radial holdups: Gamma-CT and WMS
- Time averaged measurements of 10 – 15 min (f = 1/s)

3. Experimental activities – pilot scale

EVONIK INDUSTRIES

Atmospheric glass columns

- Preliminary measurements in organic liquids (cumene/acetone), water and water-alcohol solutions sparged with nitrogen
- Objectives: axial gas phase distribution, liquid phase residence time, influence of scale and impurities on gas holdup

- Diameter: 0.16 m
- Height: 2 m

Scale-Up

- Diameter: 0.3 m
- Height: 4 m

3. Experimental activities – liquid properties

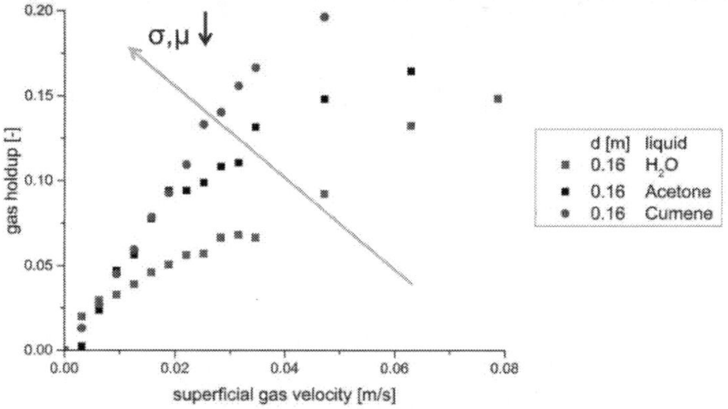

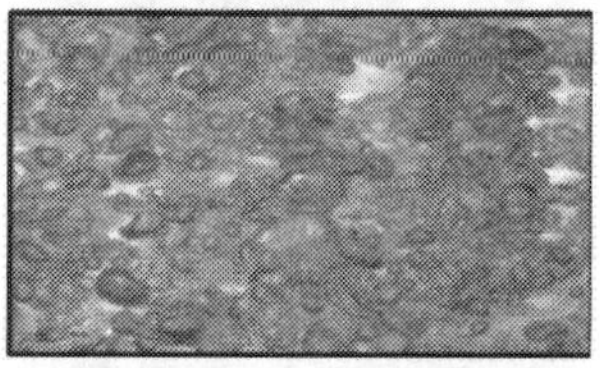

	ρ [kg/m³]	σ [mN/m]	μ [mPas]	Eo [-]	Mo [-]
H_2O	998	72.1	0.89	3.4	1.92e-10
Acetone	782	25.2	0.33	7.7	9.52e-12
Cumene	868	28.2	0.68	7.5	2.53e-11

3. Experimental activities – influence of scale

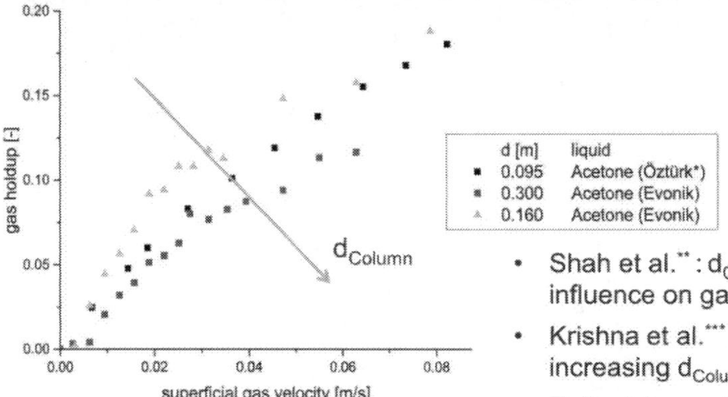

- Shah et al.**: d_{Column} > 0.15 m → no influence on gas holdup
- Krishna et al.***: gas holdup decreases with increasing d_{Column}
- Both statements valid for air/H_2O
  - ➢ Effect of different liquid properties?

* S. S. Öztürk, A. Schumpe, W. D. Deckwer, *AIChE Journal 1987, 33 (9), 1473-1480. DOI: 10.1002/aic.690330907*
** Y. T. Shah, B. G. Kelkar, S. P. Godbole, W. D. Deckwer, *AIChE Journal 1982, 28 (3), 353-379.*
*** R. Krishna, J. M. v. Baten, M. I. Urseanu, *Chemical Engineering & Technology 2001, 24 (5), 451-458. DOI: 10.1002/1521-4125(200105)24:5<451::aid-ceat451>3.0.co;2-8.*

4. Simulation activities – DNS & CFD

EVONIK INDUSTRIES

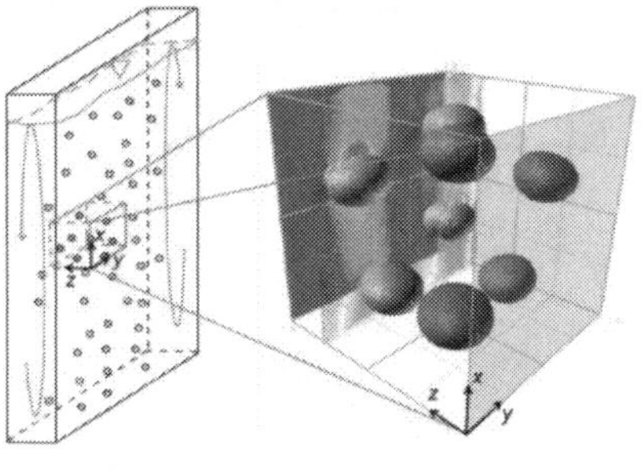

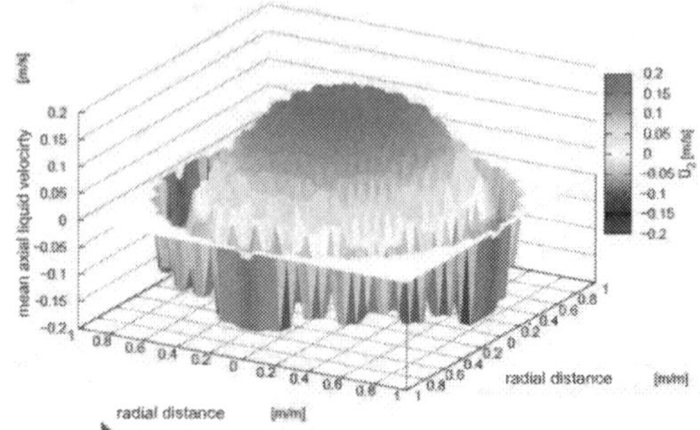

KIT Karlsruhe Institute of Technology — Bubble-bubble interaction → Multiscale approach → Flow field, Euler-Euler framework

- Simulation studies of:
- ➢ Single bubble
- ➢ bubble swarm behaviour

- ➢ Model validation with single bubble measurements

- Model validation of:
- ➢ Turbulence
- ➢ Mass transfer

- ➢ Model validation with pilot scale measurements

4. Simulation activities – Short-Cuts

EVONIK INDUSTRIES

RUB

- Bubble column partitioned into compartments
- Each compartment represents basic hydrodynamic models (CSTR, PFTR…)
- Reasonable computing time (and costs)

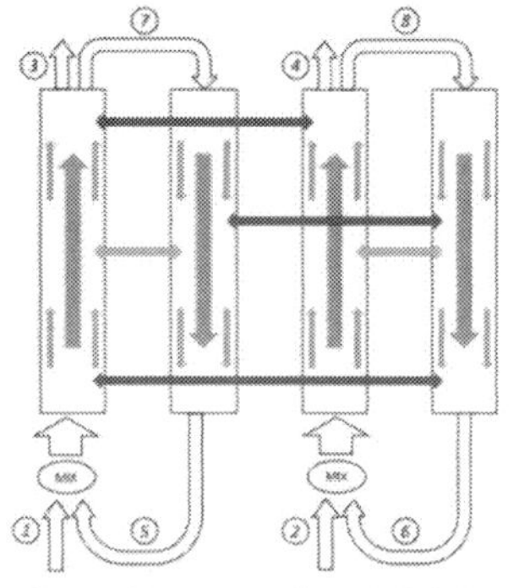

Abel et al., *Chemie Ingenieur Technik 2013, 85 (7), 1112-1117. DOI: 10.1002/cite.201300043.*

Benefit for reactor design?

- "simple" 1-D dispersion model
- Assess impact of uncertainties on economics and environment

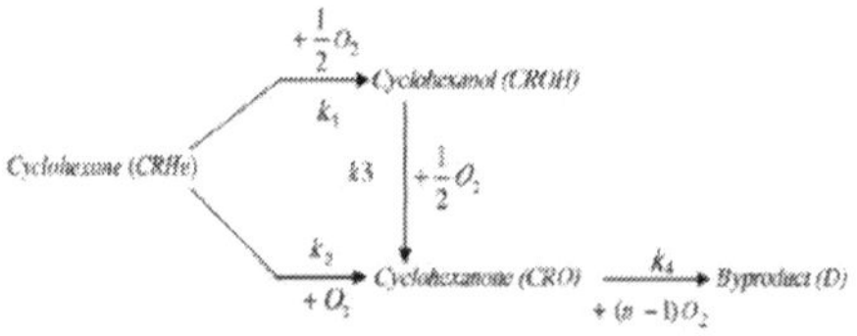

Rollbusch et al., *Chemie Ingenieur Technik 2013, 85 (9), 1425-1425. DOI: 10.1002/cite.201250725.*

5. Summary

EVONIK INDUSTRIES

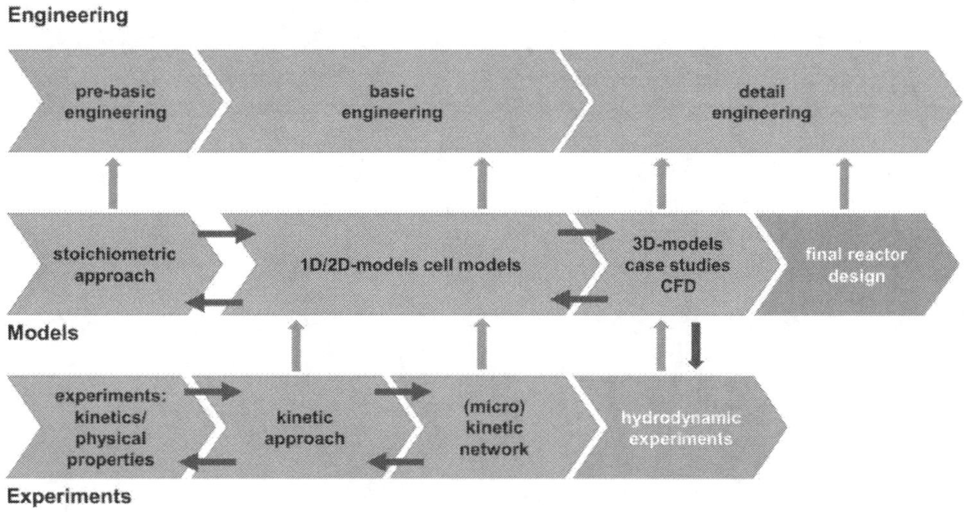

01.04.2014 Project "Multi-Phase" – a combined effort of industry and academia Page 17

5. Summary

EVONIK INDUSTRIES

- MultiPhase is:

...a scientific network comprising small to large sized companies and universities

...dedicated to optimize multiphase reactor design

- Within Multiphase:

...single bubble and pilot scale hydrodynamics are examined at processing conditions

...the overall objective of providing a reliable design methodology for bubble column reactors is pursued

01.04.2014 Project "Multi-Phase" – a combined effort of industry and academia Page 18

SPONSORED BY THE

Federal Ministry
of Education
and Research

The Multi-Phase project (full title: Increasing Energy Efficiency and Reducing Greenhouse Gas Emissions by Multiscale Modeling of Multiphase Reactors) is supported by the BMBF under support no. 033RC1102A.

351503 Scaling Solid Resuspension and Sorption for the Small Column Ion Exchange (SCIX) Processing Tank

Abstract:

The Small Column Ion Exchange (SCIX) process is being developed to remove cesium, strontium, and actinides from Savannah River Site (SRS) Liquid Waste using an existing 1.3 million gallon waste tank (i.e., Tank 41H) to house the process. Savannah River National Laboratory (SRNL) is conducting pilot-scale mixing tests to determine the pump requirements for suspending and resuspending Monosodium Titanate (MST), Crystalline Silicotitanate (CST), and simulated sludge. In addition, SRNL will also be conducting pilot-scale tests to determine the mixing requirements for the strontium and actinide sorption. As part of this task, the results from the pilot-scale tests must be scaled up to a full-scale waste tank. This presentation describes the scaling approach.

Project "Multi-Phase" – gas holdup in bubble columns in multidimensional space

P. Rollbusch, L. Schlusemann, M. Becker, M. Ludwig, M. Grünewald // 01.04.2014

RUB

EVONIK INDUSTRIES

1. Introduction
2. Influence of scale
3. Effect of pressure
4. Radial gas holdup distribution
5. Summary

1. Introduction

EVONIK INDUSTRIES

Production scale columns

- Diameters up to several meters
- Pressures above 0.1 MPa
- Continuous operation
- Liquids other than H_2O

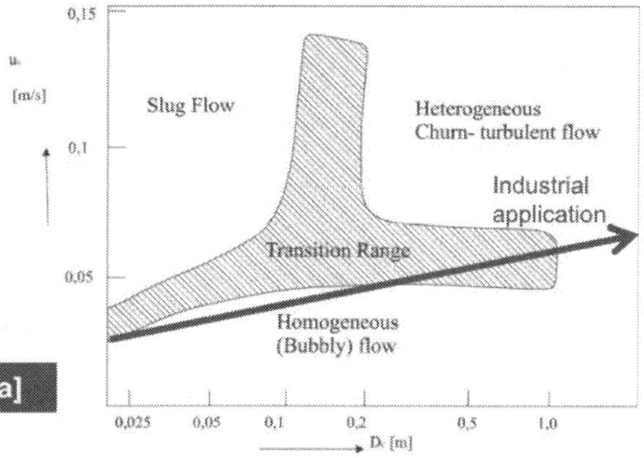

Process	H/D [-]	P [MPa]
Hock [2]	4.8	0.7
Coal liquefaction [3]	> 12.5	30
BIOHOCH[4]	< 2.5	-

[1] Y. T. Shah, B. G. Kelkar, S. P. Godbole, W. D. Deckwer, *AIChE Journal 1982, 28 (3), 353-379.*
[2] M. Weber, *Chemical Engineering and Technology 2002, 25 (5), 553-558.*
[3] A. Bakopoulos, *Chemical Engineering Science 2001, 56 (17), 5131-5145.*
[4] R. Steiner, *Chemical Engineering and Processing 1987, 21 (1), 1-8.*

1. Introduction

EVONIK INDUSTRIES

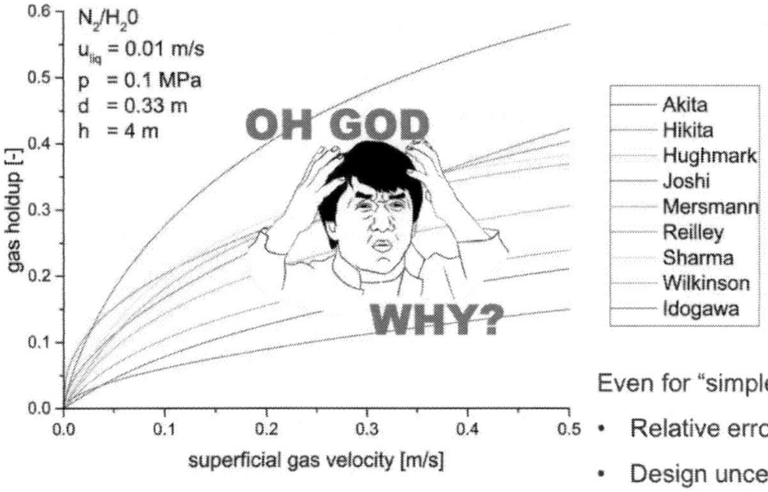

That`s why:

- H/D ratio
- sparger
- liquids
- surfactants

Even for "simple" systems:

- Relative error up to 420 %
- Design uncertainties
- Exp. validation needed

* references see Appendix A

2. Influence of scale

EVONIK INDUSTRIES

- only few correlations account for column diameter
- Examples: Akita/Yoshida[5], Zehner[6]
- Zehner correlation shows observable decrease of holdup with column diameter

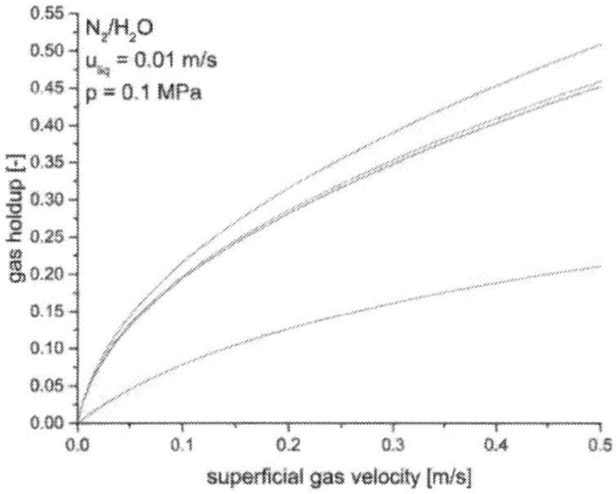

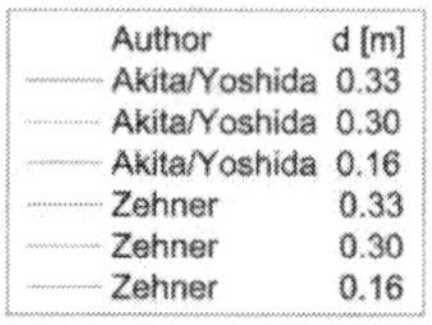

$$\varepsilon = \frac{u_g/u_s}{\left(1 + 4\frac{u_g}{u_s}^{2/3}\frac{u_{l0}}{u_s}\right)^{1/2}}$$

$$u_{l0} = \left(\frac{1}{2.5}\frac{\rho_l - \rho_g}{\rho_l}u_g g D\right)^{1/3}$$

[5] K. Akita, F. Yoshida, *Industrial & Engineering Chemistry Process Design and Development 1973, 12 (1), 76-80. DOI: 10.1021/i260045a015.*

[6] P. Zehner, *DECHEMA-Monogr. 1989, 114 (Copyright (C) 2014 American Chemical Society (ACS). All Rights Reserved.), 215-32.*

2. Influence of scale

EVONIK INDUSTRIES

- Verified decreasing gas holdup with column diameter (about 1.5 to 2 %)
- Zehner and Akita models unable to predict gas holdup in both cases

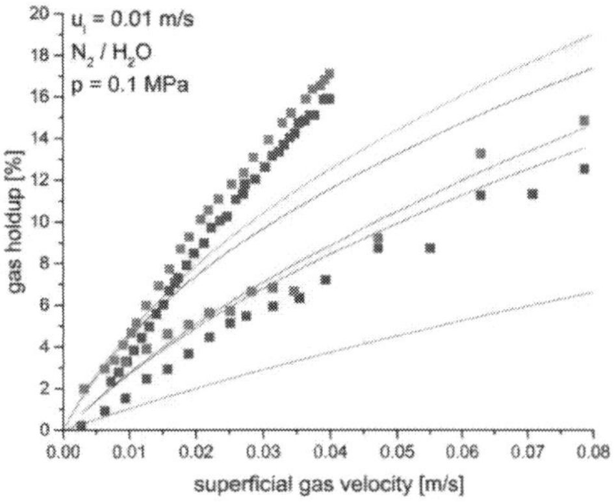

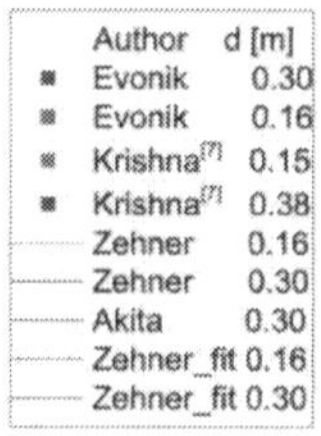

$$\varepsilon = \frac{u_g/u_s}{\left(1 + 4\frac{u_g}{u_s}^{2/3}\frac{u_{l0}}{u_s}\right)^{1/2}}$$

- Fitting parameter:
  - $u_s = f(\rho_l, \rho_g, \sigma)$

- gas holdup reduction with increasing diameter caused by intensified liquid circulation

[7] R. Krishna, J. M. v. Baten, M. I. Urseanu, *Chemical Engineering & Technology 2001, 24 (5), 451-458. DOI: 10.1002/1521-4125(200105)24:5<451::aid-ceat451>3.0.co;2-8.*

3. Effect of pressure

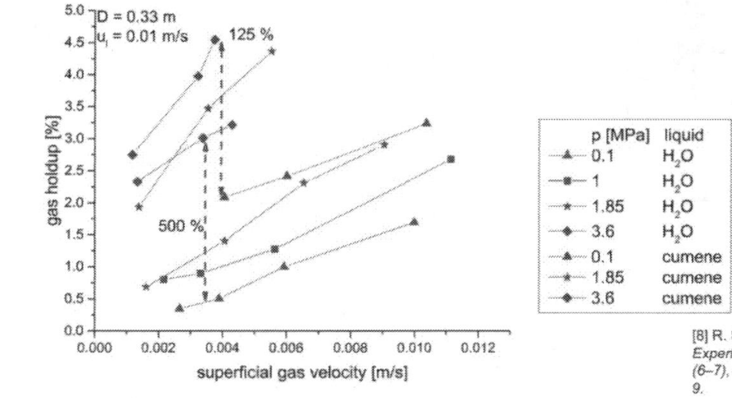

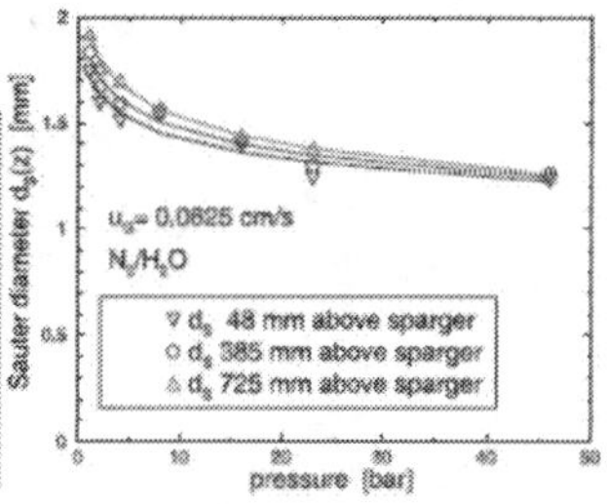

[8] R. Schäfer, C. Merten, G. Eigenberger, *Experimental Thermal and Fluid Science 2002, 26 (6–7), 595-604. DOI: 10.1016/s0894-1777(02)00189-9.*

- Higher gas density causes increase in gas holdup
- Relative increase in cumene lower because of smaller initial bubble sizes

3. Effect of pressure

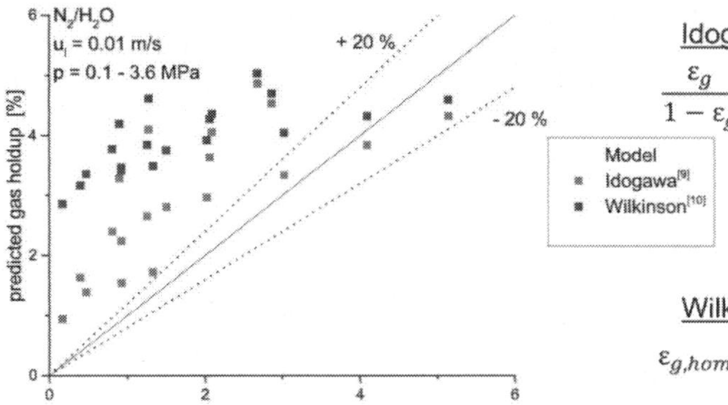

Idogawa[9]:

$$\frac{\varepsilon_g}{1-\varepsilon_g} = 0.059 u_g^{0.8} \rho_g^{0.17} \left(\frac{\sigma_l}{72}\right)^{-0.22\exp(-p)}$$

Wilkinson[10]:

$$\varepsilon_{g,hom} = \frac{u_g}{u_{sb}}\,;\ \varepsilon_{g,het} = \frac{u_{tr}}{u_{sb}} + \frac{u_g - u_{tr}}{u_{lb}}$$

- Idogawa tends to overestimate holdups
  - ➢ Equation derived from very small column diameter (0.0508 m)
- Wilkinson seems to underestimate transition velocity

[9] K. Idogawa, K. Ikeda, T. Fukuda, S. Morooka, *International chemical engineering 1987, 27 (1), 93-99.*
[10] P. Wilkinson, A. Spek, L. van Dierendonck, *AIChE Journal 1992, 38 (4), 544-554.*

4. Radial distribution – measurement technique

RUB

9

- Underestimation of holdup
  - Small bubbles are not detected
- Overestimation of holdup
  - Circulating bubbles are detected more than once
- Valid measuring technique for multiphase flows

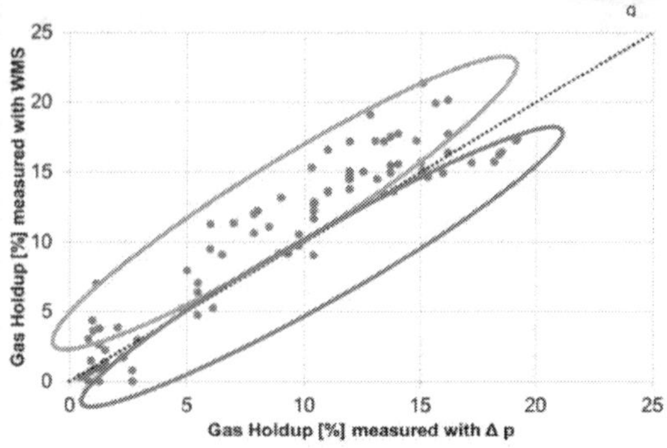

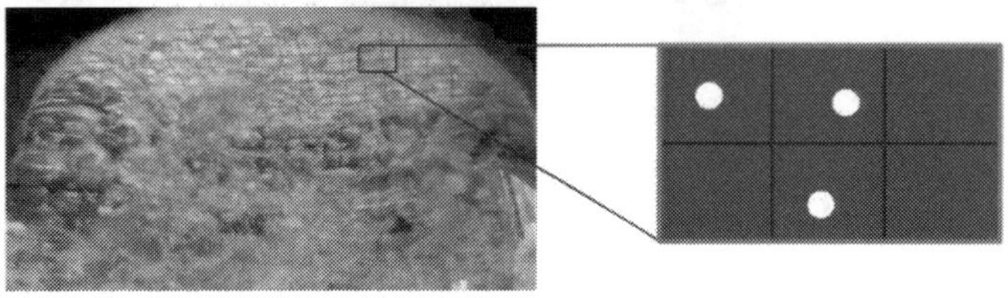

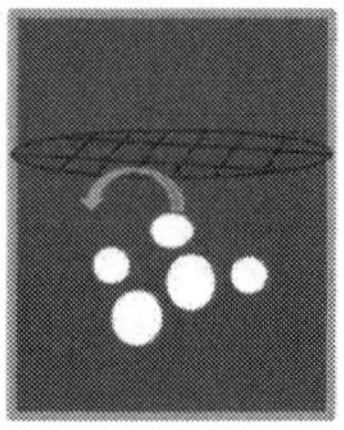

* Details about the Wire Mesh Sensor
2 pm ‚Liquid Phase Backmixing in Bubble Columns'

4. Radial distribution – comparison with literature

RUB

10

- **Comparison with literature**
- Differences
  - Measurement system
  - Gas/ liquid properties
  - Sparger design
  - Position of measuring technique

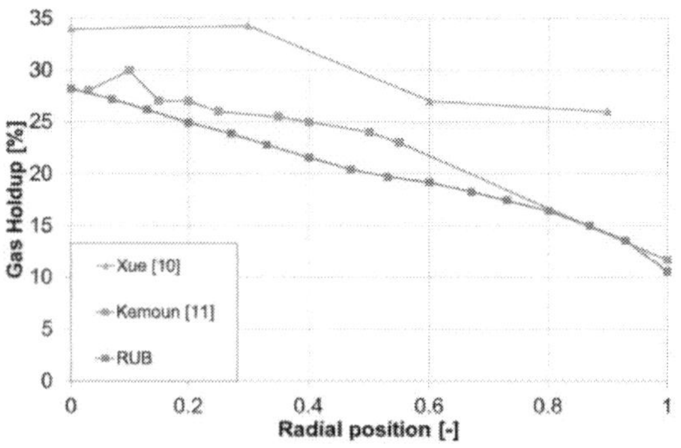

	Geometry	Measuring Technique	Superficial gas velocity [m/s]	Gas Liquid System
Xue et al. [10]	d_i= 162mm	Four point optical fibre probe	0.08	Water/ air
Kemoun et al.[11]	d_i= 162mm	CT	0.14	Water/ air
RUB	d_i= 300mm H/d = 8.3	WMS	0.091	Deionized water/ air

[10] J. Xue , M. Al- Dahhan, M. P. Dudukovic, *AIChE Journal 2008, 54 (2), 350-363.*
[11] *A. Kemoun, B. C. Ong, P. Gupta, M. H. Al-Dahhan, M. P. Dudukovic, International Journal of Multiphase Flow 2001, 27, 929-946*

4. Radial distribution – influence of superficial gas velocity

RUB
11

- **Influence of superficial gas velocity**
  - Increase of holdup in the center of the column
  - Due to flow regimes

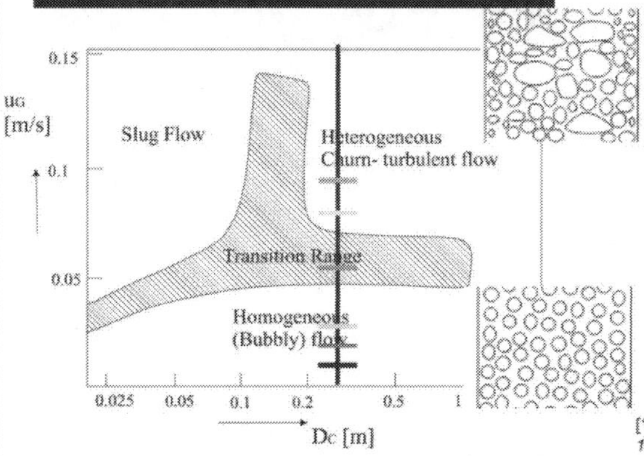

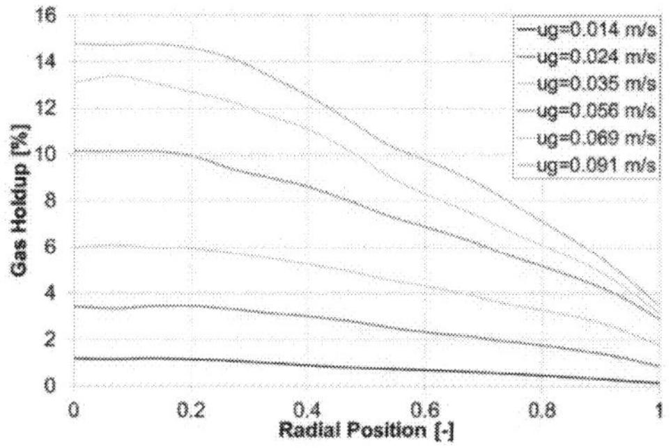

[1] *Y. T. Shah, B. G. Kelkar, S. P. Godbole, W. D. Deckwer, AIChE Journal 1982, 28 (3), 353-379.*

4. Radial axial distribution

RUB
12

- **Gas holdup at different axial positions**
  - Increase of holdup in the upper region of the column

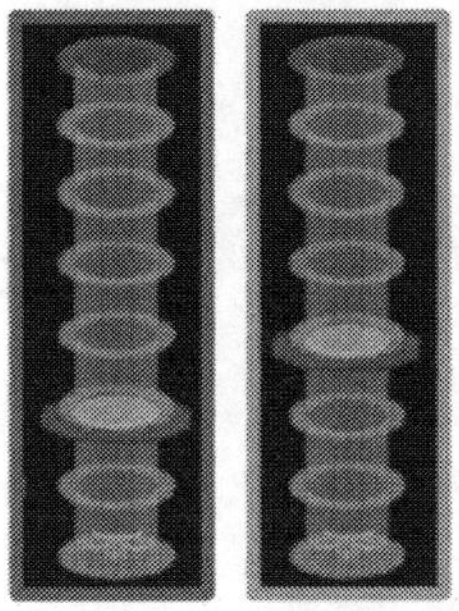

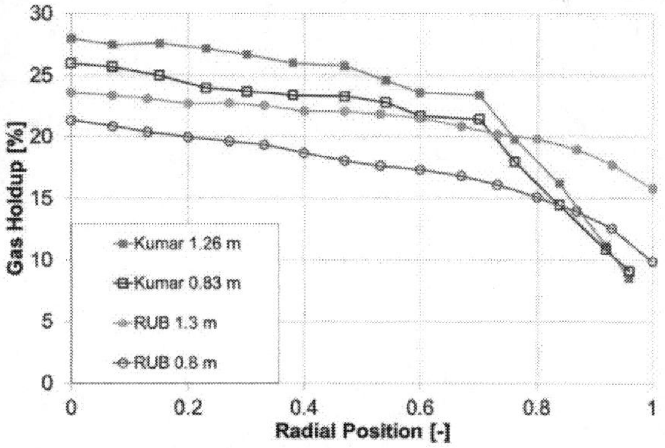

	Geometry	Height measuring system [m]	Superficial gas velocity [m/s]
Kumar et al. [12]	d_i= 260 mm	0.83 1.26	0.05
RUB	d_i= 300 mm H/d = 8.3	0.8 1.80 above sparger	0.056

[12] S. B. Kumar, D. Moslemian, M. P. Dudukovic, AIChE Journal 1997, 43 (6), 1414- 1425

4. Radial distribution – effect of sparger design

- The higher the proportion of the gas distributor with regard to the cross section the more homogeneous the radial distribution of the gas phase

- **Spider Sparger**
  - d_s=271 mm
  - d_h= 1 mm
  - n_h=80

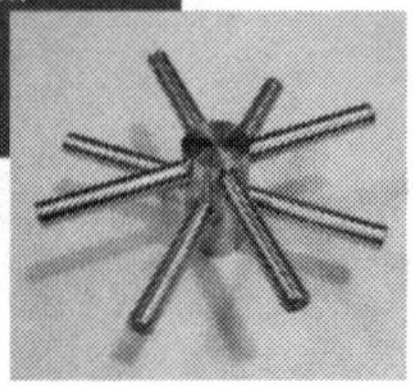

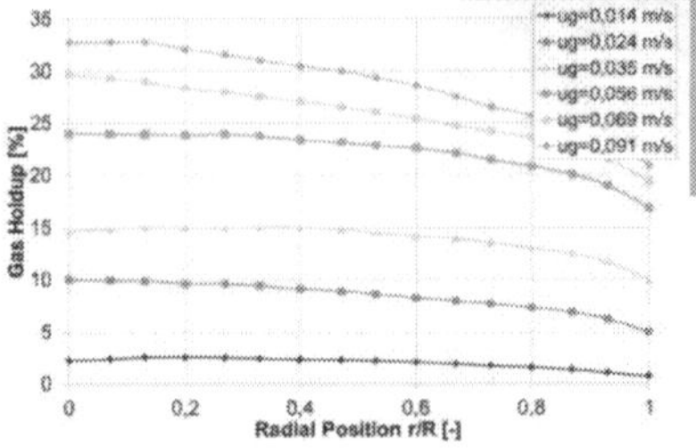

- **Perforated Plate**
  - d_s=204 mm
  - d_h= 1 mm
  - n_h=352

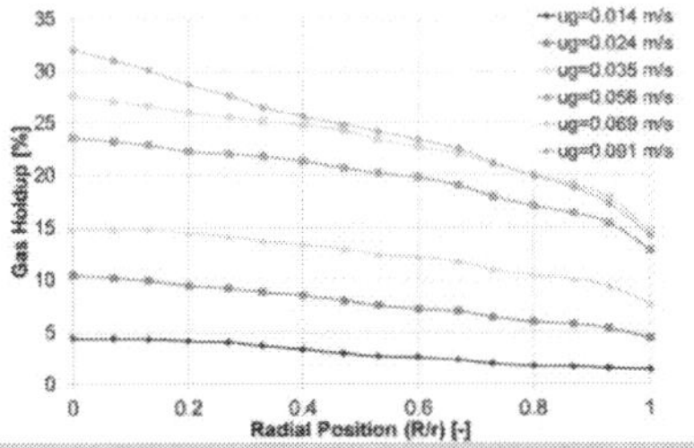

Radial distribution – sparger design effect

- **Gas holdup increases**
  - With column height
  - With superficial gas velocity

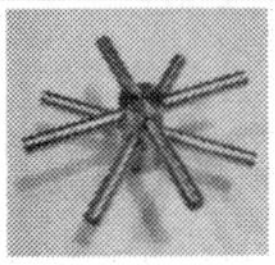

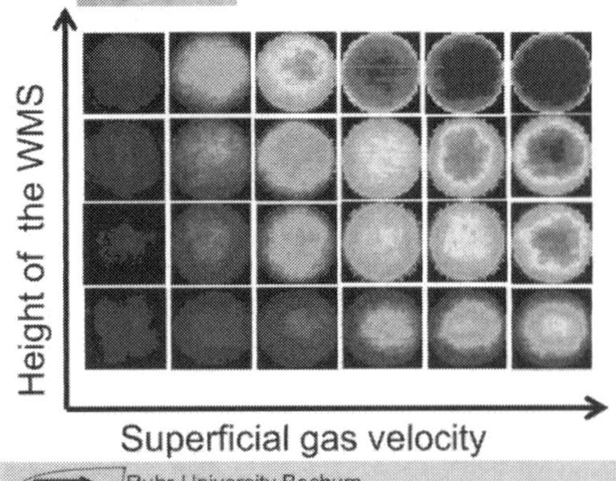

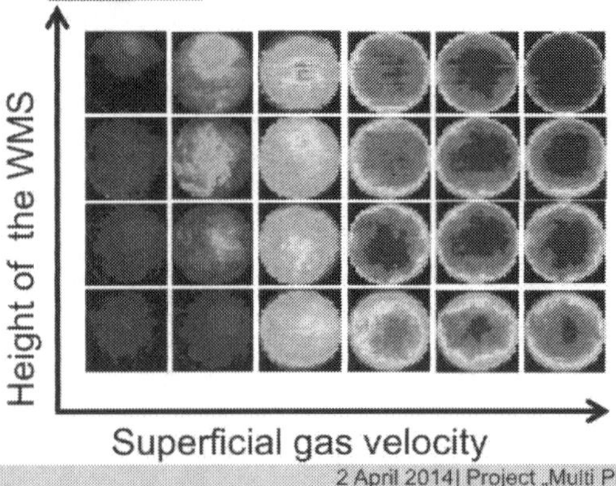

4. Radial distribution – pressure influence

RUB 15

- **Pressure influence**
  - ➢ Holdup increases with increase of pressure

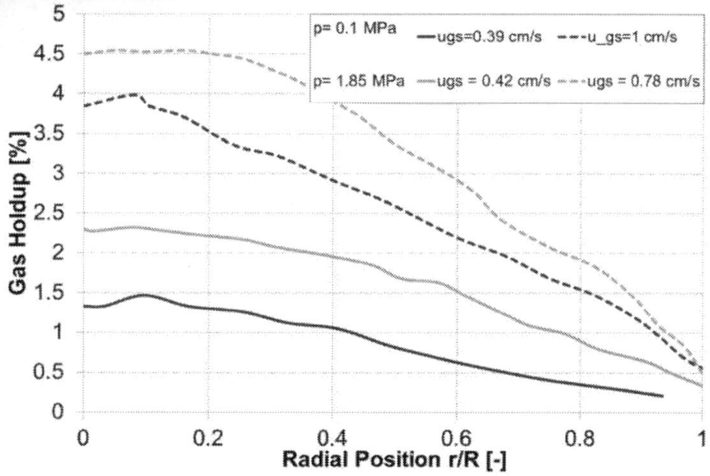

Summary

RUB 16

Overall gas holdup

- ➢ Further model validation necessary for precise prediction of gas holdups

Local gas holdup

- ➢ WMS; valid measuring technique in industrial scale and conditions
- ➢ additional experiments
  - ➢ with organic solvents
  - ➢ In different scales

Systematic approach with standardized equipment

- Sparger design
- Column geometry
- Liquid and gas velocities
- Identical liquid properties

SPONSORED BY THE

Federal Ministry
of Education
and Research

The Multi-Phase project (full title: Increasing Energy Efficiency and Reducing Greenhouse Gas Emissions by Multiscale Modeling of Multiphase Reactors) is supported by the BMBF under support no. 033RC1102A.

EVONIK
INDUSTRIES

RUB Ruhr-University Bochum,
Laboratory of Fluid Separations

Appendix A – references correlations

EVONIK INDUSTRIES

[1] K. Akita, F. Yoshida, *Industrial & Engineering Chemistry Process Design and Development 1973, 12 (1), 76-80. DOI: 10.1021/i260045a015.*

[2] H. Hikita et al., *The Chemical Engineering Journal 1980, 20 (1), 59-67.*

[3] G. A. Hughmark, *Ind. Eng. Chem. Process Des. Dev. 1967, 6 (Copyright (C) 2014 American Chemical Society (ACS). All Rights Reserved.), 218-20. DOI: 10.1021/i260022a011.*

[4] J. B. P. Joshi, U. V. ; Prasad, C. V. S. ; Phanikumar, D. V. ; Deshpande, N. S. ; Thorat, B. N., *Proceedings of the Indian National Science Academy 1998, 64A (4), 441-567.*

[5] A. Mersmann, *Chemie Ingenieur Technik 1977, 49 (9), 679-691. DOI: 10.1002/cite.330490902.*

[6] I. G. Reilly et al., *The Canadian Journal of Chemical Engineering 1986, 64 (5), 705-717. DOI: 10.1002/cjce.5450640501.*

[7] J. B. Joshi, M. M. Sharma, *Trans. Inst. Chem. Eng. 1979, 57 (Copyright (C) 2014 American Chemical Society (ACS). All Rights Reserved.), 244-51.*

[8] P. Wilkinson, A. Spek, L. van Dierendonck, *AIChE Journal 1992, 38 (4), 544-554.*

[9] K. Idogawa, K. Ikeda, T. Fukuda, S. Morooka, *International chemical engineering 1987, 27 (1), 93-99.*

Appendix B – pressurized column

EVONIK INDUSTRIES

Results obtained via:

(1,2) radial gas holdup profiles (Gamma-CT, WMS)

(3) Axial gas holdup profiles (delta-P)

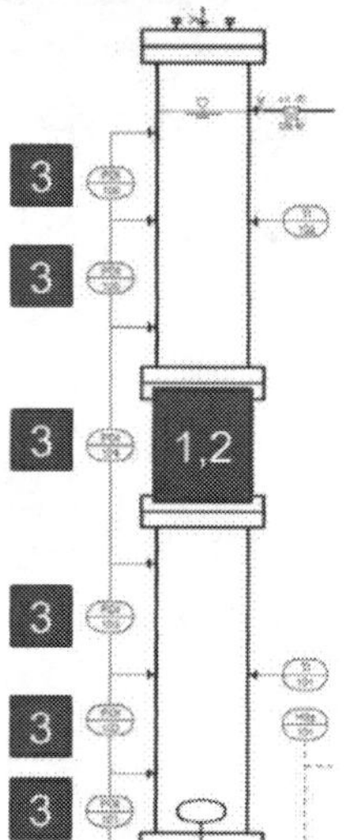

Washington University in St. Louis — Engineering

Time-Series Analysis of Optical Measurements in Gas-Liquid Stirred tanks

Boung Wook Lee, Yujian Sun, M.P. Dudukovic

Chemical Reaction Engineering Laboratory (CREL)
Energy, Environment, and Chemical Engineering Department
Washington University in St. Louis

2014 AIChE Spring Meeting
New Orleans, LA
4/1/2014

Outline

- Introduction (Optical probe & Time-series analysis)
- Experimental (Stirred tank)
- Data Processing (Theory & Methods)
- Results and Discussions
- Conclusions

Introduction | Experimental | Data Processing | Result & Discussion | Conclusions

Optical probes

- Various types exist
  - Flat end [1]
  - Conical end [2]
- Some advantages [2]
  - Small size
  - Tolerance towards harsh environment
  - High signal to noise ratio
- Some disadvantages
  - Intrusiveness
  - Directional sensitivity

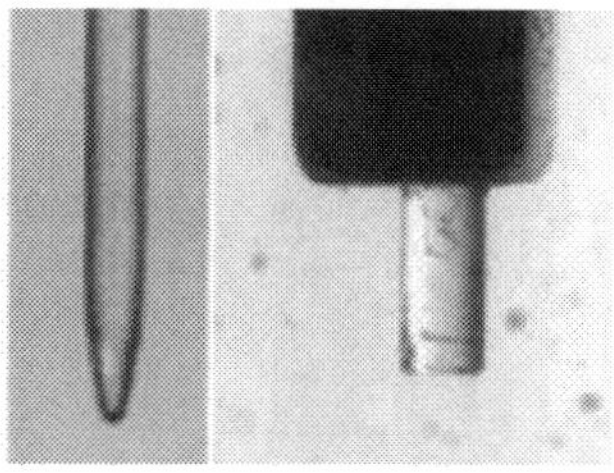

[1] Avdeev et al. (2004)
[2] Xue, J. (2004)

2

Introduction | Experimental | Data Processing | Result & Discussion | Conclusions

Time-series analysis

- Describes the systems of dynamic and chaotic behavior in three domains
  - Time domain (mean, standard deviation, etc.)
  - Frequency domain (Fourier & wavelet transform)
  - State space domain
- Applied on measurement data from various reactor types
  - Pressure fluctuation: fluidized bed and bubble column reactors [3, 4]
  - Pressure, torque, and conductance: gas-liquid stirred tank reactor [5, 6]
  - Flat-end optical probe: fluidized bed reactor [7]
  - Computer Aided Radioactive Particle Trajectory (CARPT): Bubble column reactors [8]

[3] Drahos et al. (1992)
[4] Johnsson et al. (1990)
[5] Khopkar et al. (2005)
[6] Paglianti et al. (2000)
[7] Werther et al. (1996)
[8] Fraguio et al. (2008)

3

Introduction | Experimental | Data Processing | Result & Discussion | Conclusions

Gas-liquid stirred tank

- Gas: air, Liquid: water
- Central Rushton impeller at 1/3 height of the tank
- Gas sparger below the impeller
- Four baffles on the reactor wall
- D_T (tank diameter) = H_T (liquid height)
- Impeller diameter = $D_T/3$
- Baffle width = $D_T/10$

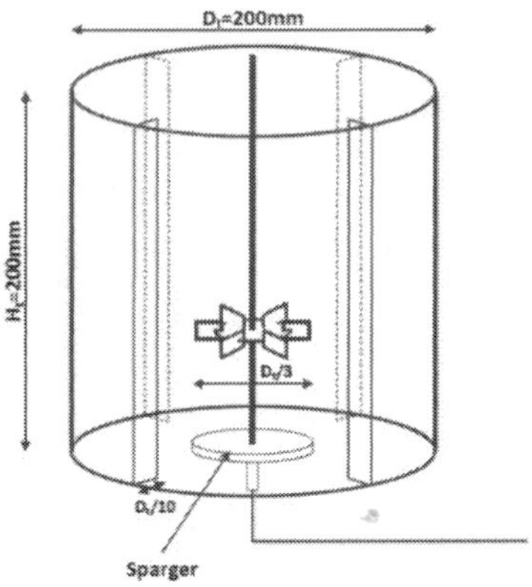

4

Introduction | Experimental | Data Processing | Result & Discussion | Conclusions

Flow regime

- Commonly characterized with two dimensionless numbers
  - Froude number (Fr) : Impeller driven acceleration to gravity ratio
  - Flow number (Fl) : Gas flow rate to impeller driven flow ratio
- Linear relationship between the Fr and dispersion parameters observed at r = 0.6R on impeller discharge plane [9]

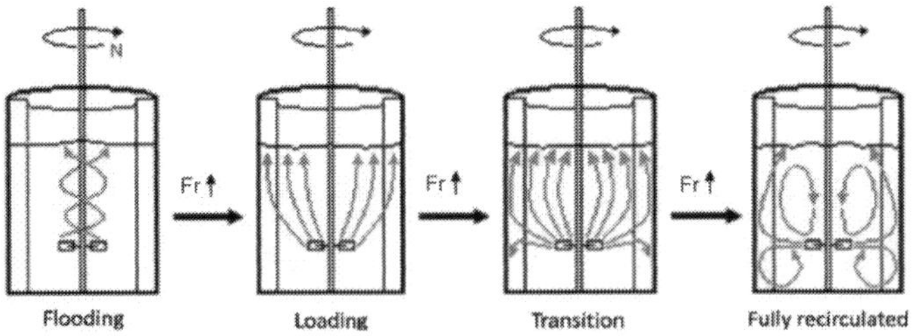

Major bubble trajectories. Figure modified from Mueller and Dudukovic [10]

[9] Lee & Dudukovic. (2014)
[10] Mueller & Dudukovic. (2010)

5

Measurements

- Due to impeller motion, bubbles change their behavior from deterministic to chaotic on impeller discharge plane
- Measurements taken at 0.4R, 0.5R, 0.6R, 0.7R, and 0.8R
- Probes placed on windward side with their tips pointing towards the center
- Voltage signals recorded at 40 kHz for 138.24 seconds

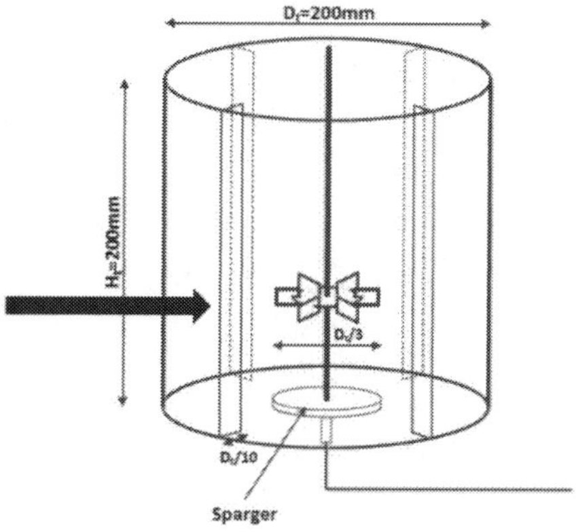

Data processing

- Gas holdup: the Ergodic theorem

$$\varepsilon_{Gas,local} = \frac{V_{Gas,local}}{V_{Liquid,local} + V_{Gas,local}} = \frac{t_{Gas}}{t_{Gas} + t_{Liquid}}, \qquad (4)$$

- Bubble count: Number of jumps from the liquid phase to the gas phase per unit time
- Signal self-resemblance: autocorrelation sequence

$$r_{xx}(m) = \sum_{n=0}^{N-m-1} x(n)x(n+m) \quad (5) \quad R_{xx}(m) = \frac{r_{xx}(m)}{r_{xx}(0)}. \qquad (6)$$

- Power spectral density (PSD) estimate: Welch's method

$$P_{xx}(f) = \sum_{n=1}^{N_s} P_{xx}^i(f). \quad P_{xx}^i(f) \; (8) = \frac{1}{N_s U}\left|\sum_{n=1}^{N_s} x_i(n)w(n)\exp(-j2\pi fn)\right|^2, \quad U = (7) \; \frac{1}{N_s}\sum_{n=1}^{N_s} w^2(n)$$

Gas holdup and bubble count profiles

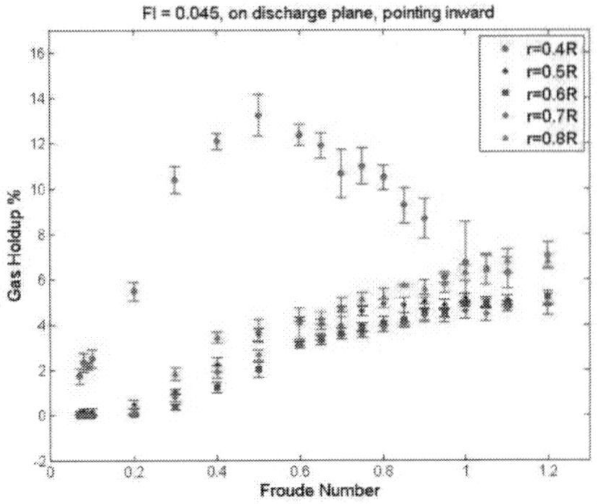

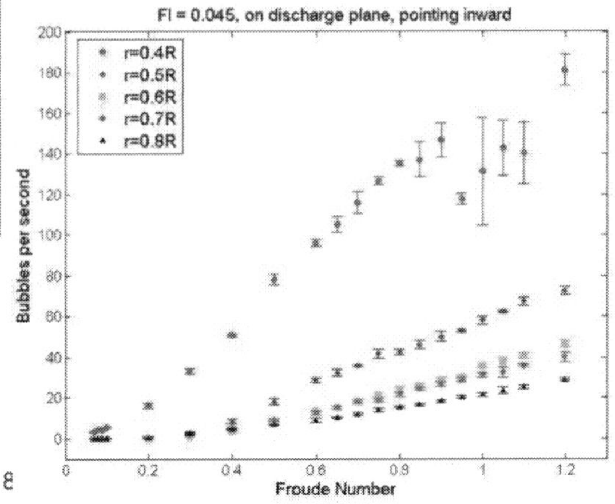

Gas holdup and bubble count profiles

- Linear relationships between the Fr and the two parameters (gas holdup and bubble count) observed at r ≥ 0.5R
- Rise-dip-rise pattern for the two parameters observed at r = 0.4R
- Comparatively higher gas holdups near the impeller observed
  - Also observed by other means of measurements [11, 12]
  - Due to extension of cavities, dispersions by the velocity field near the impeller, and bubbles leaving the radial jet stream before reaching the outer reactor regions
- The gas holdup values observed to be relatively constant in the outer region regardless of the operating condition
  - In agreement to other measurements [11, 12]
- As the flow regime transitioned into the fully recirculated regime, the distinction between inner and outer reactor disappeared

[11] Ford et al. (2008)
[12] Kong et al. (2012)

Autocorrelation sequence

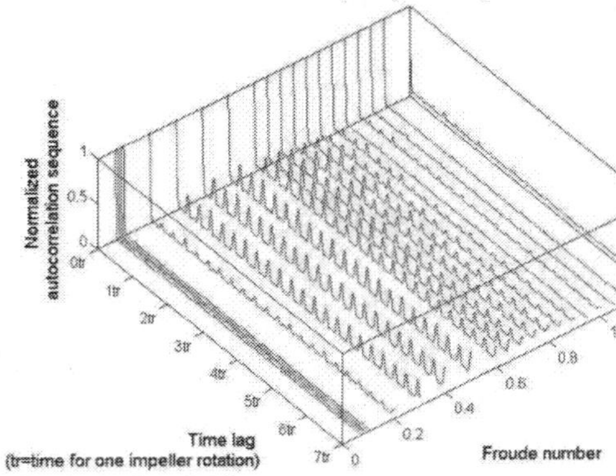

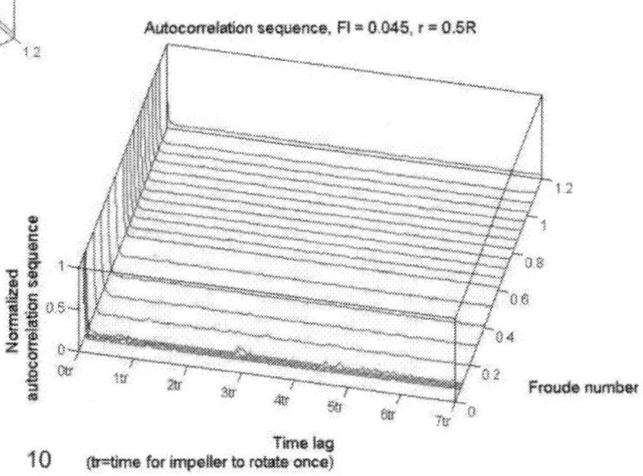

10

Autocorrelation sequence

- Time lags are in terms of impeller movement
  - Each line represents different time scales
- Periodic sequences were observed at multiples of 1/3 t_r (t_r = time for one impeller rotation) only at r = 0.4R at Fr ≤ 0.9
  - At our operating conditions, transition to the fully recirculated regime occurs at Fr = 0.91
  - The magnitude of the peaks increased up to Fr = 0.5, decreased up to Fr = 0.9
- For all operating conditions, a fast decay near zero time lag was followed by near constant autocorrelation sequences
- PSD estimates revealed similar information

11

PSD estimate

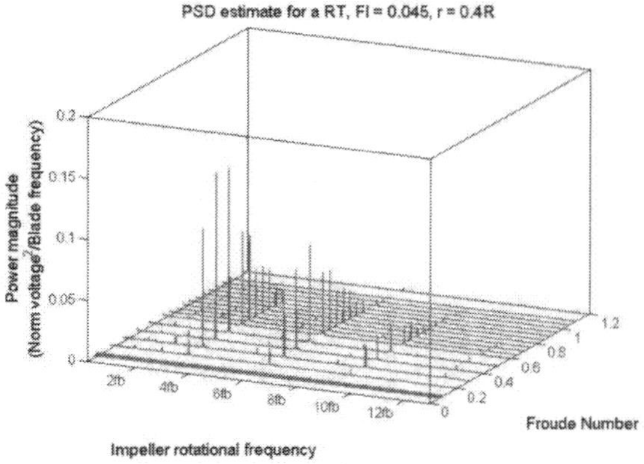

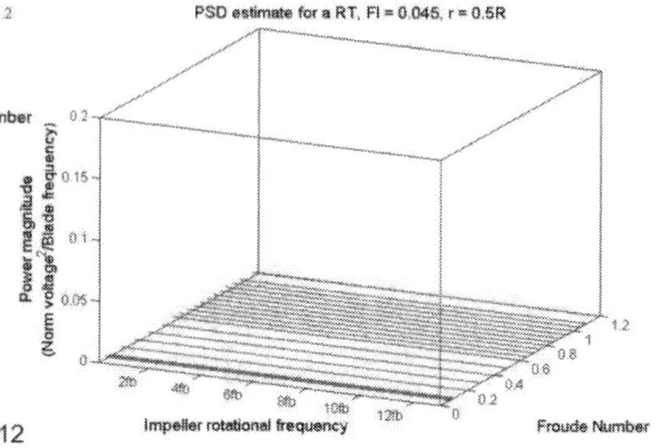

12

PSD estimate

- Dominant frequencies observed at r = 0.4R in multiples of $3f_b$ (f_b = impeller rotational frequency)
  - Consistent with sub-harmonics observed by Bombac et al (1997) [13]
- Flat PSD estimates observed for r ≥ 0.5R
  - Bubbles are detected at random frequencies in the outer region
- The peak height was proportional to the gas holdup values
- As the flow regime transitioned to the fully recirculated regime (Fr ≥ 0.91), no dominant frequencies were observed anywhere within our measurements

13

[13] Bombac et al. (1997)

Conclusions

- Treatment of optical probe measurements in gas-liquid STR equipped with a Rushton turbine revealed inner and outer region within the impeller discharge plane
- The two regions were identifiable in both the time and frequency domain analyses
- As the flow regime transitioned to the fully recirculated regime, the separation between the two regions disappeared
- Further investigation of intrusive probes needed for better description of gas dispersions

14

Acknowledgement

- Past and current CREL members
- EECE Department, Washington University
- Funding: NSF Grant 0933780

15

Introduction > Experimental > Data Processing > Result & Discussion > Conclusions

References

[1] Avdeev MV, Konovalov AN, Bagratashvili VN, et al. The fibre optic reflectometer: A new and simple probe for refractive index and phase separation measurements in gases, liquids and supercritical fluids. *Physical Chemistry Chemical Physics*. 2004;6(6):1258-1263.

[2] Xue J. *Bubble Velocity, Size and Interfacial Area Measurements in Bubble Columns*. Saint Louis: Department of Chemical Engineering, Washington University; 2004.

[3] Drahoš J, Bradka F, Punčochář M. Fractal behaviour of pressure fluctuations in a bubble column. *Chemical Engineering Science*. 1992;47(15–16):4069-4075.

[4] Johnsson F, Zijerveld RC, Schouten JC, van den Bleek CM, Leckner B. Characterization of fluidization regimes by time-series analysis of pressure fluctuations. *International Journal of Multiphase Flow*. 2000;26(4):663-715.

[5] Khopkar AR, Panaskar SS, Pandit AB, Ranade VV. Characterization of Gas−Liquid Flows in Stirred Vessels Using Pressure and Torque Fluctuations. *Industrial & Engineering Chemistry Research*. 2005/04/01 2005;44(9):3298-3311.

[6] Paglianti A, Pintus S, Giona M. Time-series analysis approach for the identification of flooding/loading transition in gas–liquid stirred tank reactors. *Chemical Engineering Science*. 2000;55(23):5793-5802.

[7] Werther J, Hage B, Rudnick C. A comparison of laser Doppler and single-fibre reflection probes for the measurement of the velocity of solids in a gas-solid circulating fluidized bed. *Chemical Engineering and Processing: Process Intensification*. 1996;35(5):381-391.

[8] Fraguio MaS, Cassanello MC, Degaleesan S, Dudukovic M. Flow Regime Diagnosis in Bubble Columns via Pressure Fluctuations and Computer-Assisted Radioactive Particle Tracking Measurements. *Industrial & Engineering Chemistry Research*. 2009/02/04 2008;48(3):1072-1080.

[9] Lee BW, Dudukovic MP. Determination of flow regime and gas holdup in gas–liquid stirred tanks. *Chemical Engineering Science*. 2014;109(0):264-275.

[10] Mueller SG, Dudukovic MP. Gas Holdup in Gas−Liquid Stirred Tanks. *Industrial & Engineering Chemistry Research*. 2010/11/03 2010;49(21):10744-10750.

[11] Ford JJ, Heindel TJ, Jensen TC, Drake JB. X-ray computed tomography of a gas-sparged stirred-tank reactor. *Chemical Engineering Science*. 2008;63(8):2075-2085.

[12] Kong LN, Li W, Han LC, et al. On the measurement of gas holdup distribution near the region of impeller in a gas-liquid stirred Rushton tank by means of gamma-CT. *Chemical Engineering Journal*. Apr 2012;188:191-198.

[13] Bombač A, Žun I, Filipič B, Žumer M. Gas-filled cavity structures and local void fraction distribution in aerated stirred vessel. *AIChE Journal*. 1997;43(11):2921-2931.

16

Introduction > Experimental > Data Processing > Result & Discussion > Conclusions

Questions?

17

352154 Liquid Phase Back Mixing in Bubble Columns

Tuesday, April 1, 2014: 2:00 PM
Prince of Wales (Hilton New Orleans Riverside)

Linda Schlusemann and Marcus Grünewald, Thermo- and Fluiddynamics, Ruhr University Bochum, Bochum, Germany

Liquid phase back mixing in bubble columns

Linda Schlusemann*[1], Marcus Grünewald[1]
* schlusemann@fluidvt.rub.de
[1] Ruhr-University Bochum, Chair of fluid processing, 44801 Bochum, Germany

1. Motivation

Due to their simple design bubble columns are widely used in chemical, petrochemical industry and in biotechnology. Despite the simple geometry, the measurement of phase distribution and flow structure in multiphase columns is a major challenge. Especially back mixing effects have mainly been observed in computational fluid dynamic (CFD) simulations rather than in experimental studies. Therefore it is reasonable to invent an experimental setup to measure the back mixing and flow structure of the liquid phase in bubbles columns in industrial scale.

2. Experimental studies

For the experimental development of the hydrodynamic a bubble column made of Plexiglas ® with an inner diameter of 290 mm and a height to diameter ratio of 14 (see fig. 1) is used. The column can operate in semi- batch and co- current flow. Water is used as continuous/liquid phase and nitrogen as gas phase.

The measurement technique is a wire mesh sensor consisting of 32x32 measurement points in the cross section of the column. The flow structure of the liquid phase is investigated by adding acetic acid as a tracer to the liquid flow on the bottom of the column. The wire mesh sensor can measure the resulting increase in conductivity at each mesh point. This is how the flow structure in the cross section of the column can be determined.

3. Goals

These experiments should help to build a model for the flow structure of the liquid phase in bubble columns.

This presentation includes a detailed description of the measurement technique and the experimental setup. Furthermore the first results of own experimental studies are presented and discussed.

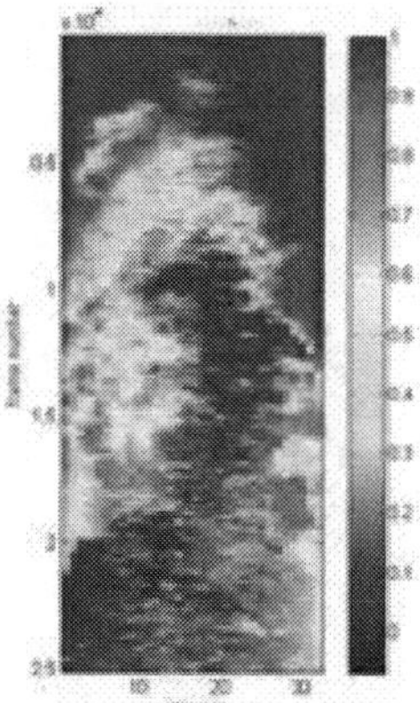

Dimensional Analysis of Gas Flow Through Packed Beds

Meng Zhang[1], Jonathan H. Worstell[2],
and Jacob Arredondo[3]

[1] University of Houston
[2]Worstell and Worstell, Consultants and University of Houston
[3]Rice University

Overview

Why the interest in flow through packed beds.

Issues concerning packed beds.

Derive the Ergun equation for packed beds.

Establish dimensionless parameter function for packed beds.

Discuss data from packed beds.

Conclusions

Summary

Why Interest in Packed Beds?

Absorption / Extraction / Stripping

Adsorption

Distillation

Drying

Reaction

Packing, in one form or another,
is present in most chemical plants.

Why So Prevalent?

Increases the surface area of the active agent.

Absorption / Extraction / Stripping
Gas-liquid processes
Increases liquid surface area.

Adsorption
Gas or liquid processes
Increases solid surface area = increases number of active sites
Porous solids

Distillation
Vapor-liquid processes
Increases liquid surface area = enhances attainment of equilibrium
Shaped solids

Why So Prevalent?

Packed Bed Reactors
Gas, liquid, or gas-liquid processes
Increases solid surface area = increases number of active sites
Porous solids

Issues

Primary

ΔP

Increased surface area = increased friction along surface

Secondary

Bulk Strength
Will solid support its own weight?
Attrition Resistance
Will solid resist fragmentation?

Design Concerns

Main design concern = predicting ΔP

Experimental Prediction
Fill circular column with packing;
Pump fluid through it;
Measure ΔP between inlet and outlet.

Issues
D_{Solids} / D_{column}
Wall effects

Requires large diameter column to be accurate
Requires large fluid quantities

Design Concerns

Calculation Prediction
Ergun equation

$$\frac{P_{In} - P_{Out}}{Z} = \frac{\Delta P}{Z} = 150\mu\left(\frac{v_S}{D_P^2}\right)\frac{(1-\varepsilon)^2}{\varepsilon^3} + 1.75\rho\left(\frac{v_S^2}{D_P}\right)\frac{(1-\varepsilon)}{\varepsilon^3}$$

where v_S is superficial fluid velocity and D_P is diameter of solid.

Dimensionless form of Ergun equation

$$\frac{\varepsilon^3 D_P}{(1-\varepsilon)\rho v_S^2}\frac{\Delta P}{Z} = 150\left(\frac{\mu}{\rho D_P v_S}\right)(1-\varepsilon) + 1.75$$

Some say it works, others say not.

Derivation of Ergun Equation

Assume interstices of packed bed form small diameter pipes/tubes

Consider Hagen-Poiseuille equation for interstitial flow

$$\Delta P = \frac{2\mu Z v_I}{R_H^2} = \frac{2\mu Z \left(v_S / \varepsilon \right)}{R_H}$$

where R_H is the "hydraulic radius".

$$R_H = \frac{D_P}{6}\left(\frac{\varepsilon}{1-\varepsilon}\right)$$

R_H is a measure of channel flow efficiency.

Derivation of Ergun Equation

Making the appropriate substitution

$$\Delta P = \frac{2\mu Z \left(v_S / \varepsilon \right)}{R_H^2} = \frac{2\mu Z \left(v_S / \varepsilon \right)}{\frac{D_P}{6}\left(\frac{\varepsilon}{1-\varepsilon}\right)} = 72\left(\frac{\mu Z v_S}{D_P^2}\right)\frac{(1-\varepsilon)^2}{\varepsilon^3}$$

But flow through packed bed is longer than physical length L of bed.

Empirical correction factor

$$Z = \left(25/12\right)L$$

Derivation of Ergun Equation

Making the appropriate substitution

$$\Delta P = 150\left(\frac{\mu L v_S}{D_P^2}\right)\frac{(1-\varepsilon)^2}{\varepsilon^3}$$

Blake-Kozeny Equation
Laminar Flow

Derivation of Ergun Equation

For turbulent flow

$$\Delta P = f\left(\frac{L}{D}\right)\frac{\rho v_l^2}{2}$$

Let

$$D = 4R_H = \frac{4}{6}D_P\frac{1-\varepsilon}{\varepsilon}$$

Then

$$\Delta P = \frac{6}{4}f\left(\frac{L}{D_P}\right)\frac{\rho\left(v_S/\varepsilon\right)^2}{2}\frac{1-\varepsilon}{\varepsilon} = \frac{3f}{2}\frac{\rho L v_S^2}{2D_P}\frac{1-\varepsilon}{\varepsilon^3}$$

Derivation of Ergun Equation

But, generally assume (3/2) f = 3.5; thus

$$\Delta P = 3.5\frac{\rho L v_S^2}{2}\frac{1-\varepsilon}{\varepsilon^3} = 1.75\frac{\rho L v_S^2}{D_P}\frac{1-\varepsilon}{\varepsilon^3}$$

Burke-Plummer Equation

But, what about the transition zone between laminar and turbulent flow?

Ergun filled that void.

Derivation of Ergun Equation

Flow in Transition Region

Ergun simply added the Blake-Kozeny and Burk-Plummer equations.

$$\Delta P = 150\left(\frac{\mu L v_S}{D_P^2}\right)\frac{(1-\varepsilon)^2}{\varepsilon^3} + 1.75\frac{\rho L v_S^2}{D_P}\frac{1-\varepsilon}{\varepsilon^3}$$

Rearranging gives the dimensionless form of the equation.

$$\frac{\varepsilon^3}{1-\varepsilon}\left(\frac{D_P}{\rho v_S^2}\right)\left(\Delta P/L\right) = 150\left(\frac{\mu}{\rho D_P v_S}\right) + 1.75$$

Plot of Gas Flow Through a Packed Bed

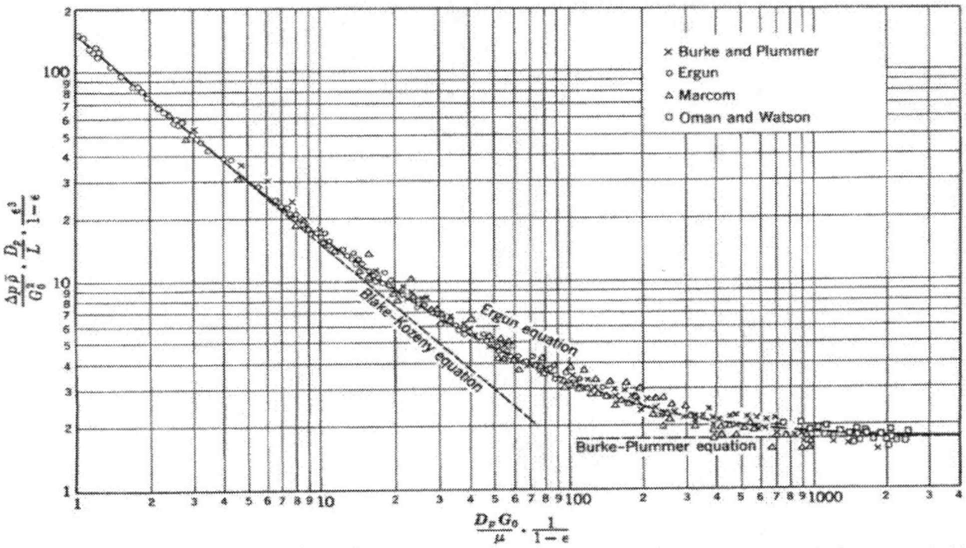

Plot of Gas Flow Through a Packed Bed

Plot is for solid spheres.

Shaped solids not necessarily spherical.

Convert to "spheres" with a sphericity factor.

Packing Material Tested

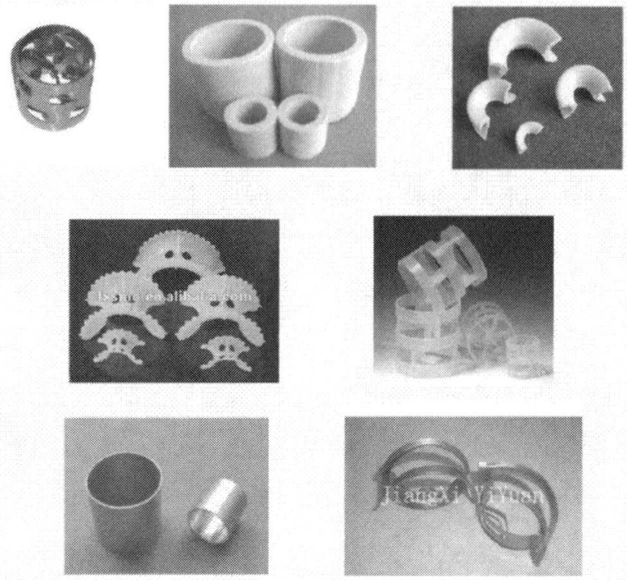

17

Why Divided Opinion on Ergun Equation?

Mathematical addition of two limiting cases.

Tortuosity estimated as 25/12.

Friction factor estimated as (3/2) f = 3.5.

Uses $\Delta P/L$ instead of dP/dL for large pressure drops.

Plotted as $\frac{\varepsilon^3}{1-\varepsilon}\left(\frac{D_P}{\rho v_S^2}\right)\left(\Delta P/L\right)$

versus Re / (1 - ε) instead of (1 – ε) / Re.

Length is L, the flow length, not Z, the physical length.

Another Concern about Ergun Equation

Does Ergun equation apply to non-porous solids.

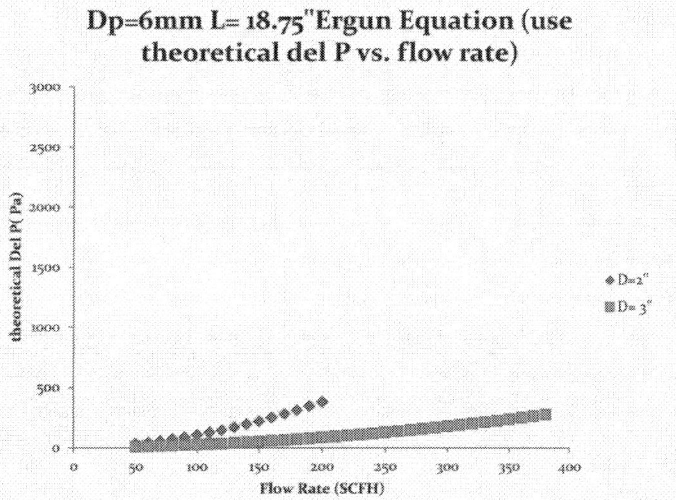

Another Concern about Ergun Equation

Does Ergun equation apply to non-porous solids.

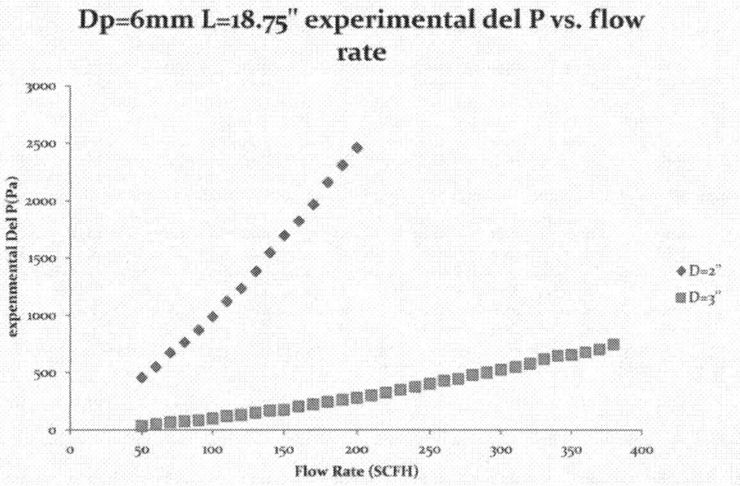

Another Concern about Ergun Equation

Does Ergun equation apply to porous solids.

D = 2" L= 18.75" Ergun Equation (theoretical del P vs. flow rate)

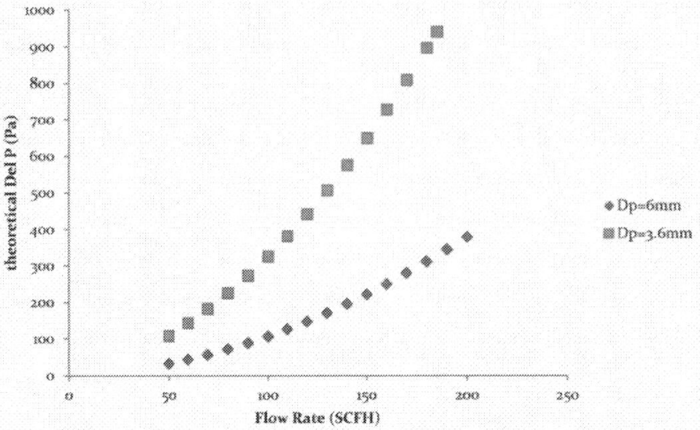

Another Concern about Ergun Equation

Does Ergun equation apply to porous solids.

D=2" L=18.75" Experimental del P vs. flow rate

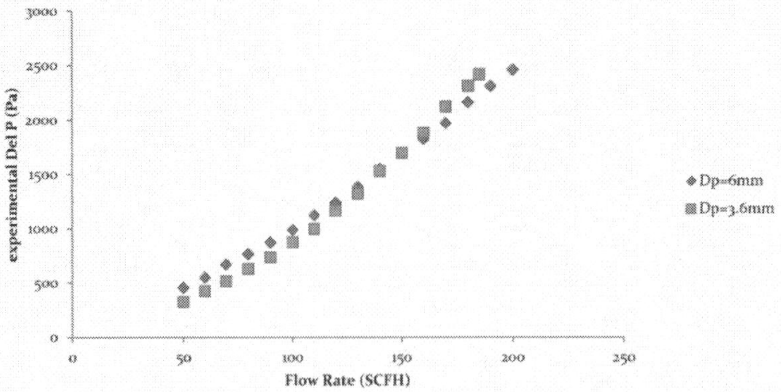

Fluid Flow through a Packed, Circular Pipe

Described in 1952 by S. Ergun.

$$f(v, D, L, dP/Z, \rho, \mu) = 0$$

Units	dP/Z	D	L	v	ρ	μ
L	-2	1	1	1	-3	-1
M	1	0	0	0	1	1
T	-2	0	0	-1	0	-1
Variable Index	k_1	k_2	k_3	k_4	k_5	k_6

$$\begin{bmatrix} -2 & 1 & 1 & 1 & -3 & -1 \\ 1 & 0 & 0 & 0 & 1 & 1 \\ -2 & 0 & 0 & -1 & 0 & -1 \end{bmatrix}$$

23

Definition of Characteristic Length

When we load a bed, what do we know?
The weight of packing put into the circular column.
Converting its $W_{packing}$ to volume, we obtain

$$W_{packing}/\rho_{LBD} = V_{Packing} = A_{CS} Z_{Packing}$$

The area available for fluid flow is εA

$$L_{Char} = \frac{\left(W_{packing}/\rho_{LBD}\right)}{\varepsilon A_{CS}} = \frac{A_{CS} Z_{Packing}}{\varepsilon A_{CS}} = \frac{Z_{Packing}}{\varepsilon} = L$$

If Z = 3 m and ε = 0.4, then L = 4.5 m; thus, L is an indication of tube length; i.e., flow path length through the packing.

24

Fluid Flow through a Packed, Circular Pipe

Rank Calculation

$$\begin{vmatrix} 1 & -3 & -1 \\ 1 & 1 & 1 \\ -1 & 0 & 1 \end{vmatrix} = 1$$

$$N_P = N_V - R$$

$$N_P = 6 - 3 = 3$$

$$\begin{bmatrix} k_1 \\ k_2 \\ k_3 \\ k_4 \\ k_5 \\ k_6 \end{bmatrix} = \begin{bmatrix} I & 0 \\ -R^{-1}B & R^{-1} \end{bmatrix} \begin{bmatrix} k_1 \\ k_2 \\ k_3 \\ a \\ b \\ c \end{bmatrix}$$

25

Fluid Flow through a Packed, Circular Pipe

$$\begin{bmatrix} k_1 \\ k_2 \\ k_3 \\ k_4 \\ k_5 \\ k_6 \end{bmatrix} = \begin{bmatrix} 1 & 0 & 0 & 0 & 0 & 0 \\ 0 & 1 & 0 & 0 & 0 & 0 \\ 0 & 0 & 1 & 0 & 0 & 0 \\ -3 & 1 & 1 & -1 & -3 & -2 \\ -2 & 1 & 1 & -1 & -2 & -1 \\ 1 & -1 & -1 & 1 & 3 & 1 \end{bmatrix} \begin{bmatrix} k_1 \\ k_2 \\ k_3 \\ 0 \\ 0 \\ 0 \end{bmatrix} \begin{matrix} dP/dZ \\ D \\ L \\ v \\ \rho \\ \mu \end{matrix}$$

$$\Pi_1 = \frac{k_1 k_6}{k_4^3 k_5^2} = \frac{\left(dP/Z\right)\mu}{v^3 \rho^2}$$

$$\Pi_2 = \frac{k_2 k_4 k_5}{k_6} = \frac{D\, v\, \rho}{\mu} = Re_D$$

$$\Pi_3 = \frac{k_3 k_4 k_5}{k_6} = \frac{L\, v\, \rho}{\mu}$$

26

Fluid Flow through a Packed, Circular Pipe

$$\frac{\Pi_3}{\Pi_2} = \frac{L v \rho / \mu}{D v \rho / \mu} = \frac{L}{D}$$

$$f(\Pi_1, \Pi_2\, \Pi_{3/2}) = 0$$

$$\Pi_1 = \kappa f(\Pi_2\, \Pi_{3/2})$$

27

Fluid Flow through a Packed, Circular Pipe

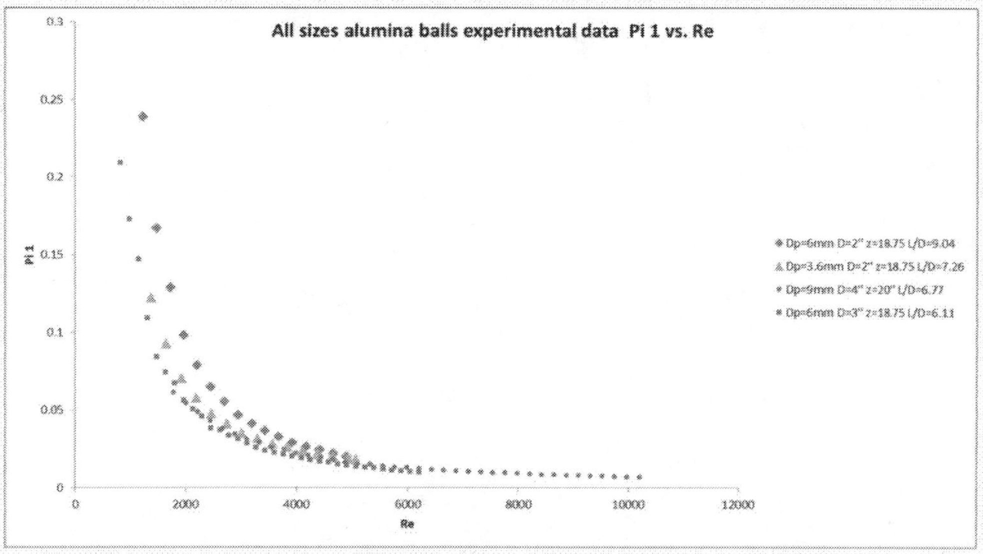

28

Fluid Flow through a Packed, Circular Pipe

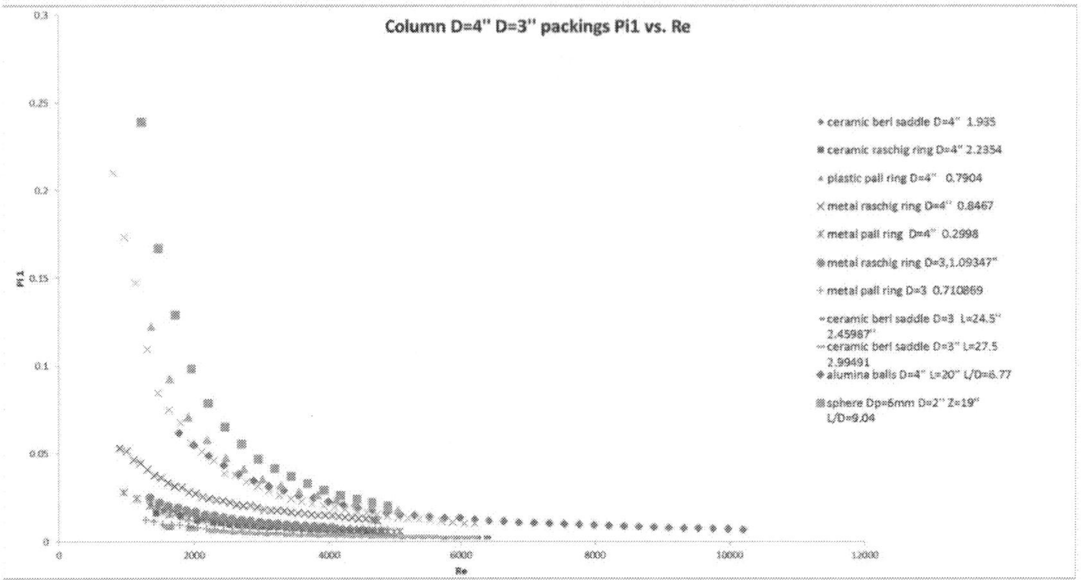

29

Presenting the Ergun Equation

Ergun Equation

$$\Delta P = 150\left(\frac{\mu L v_S}{D_P^2}\right)\frac{(1-\varepsilon)^2}{\varepsilon^3} + 1.75\,\frac{\rho L v_S^2}{D_P}\,\frac{1-\varepsilon}{\varepsilon^3}$$

Rearranging gives the dimensionless form of the equation.

$$\frac{\varepsilon^3}{1-\varepsilon}\left(\frac{D_P}{\rho v_S^2}\right)\left(\Delta P/L\right) = 150\left(\frac{\mu}{\rho D_P v_S}\right) + 1.75$$

Jim Sturnfield, Dow Chemical Co, suggested this linear form

$$\frac{\left(\Delta P/L\right)\mu}{v_S^2\,\rho^2}\left(\frac{\rho D_P v_S}{\mu}\right)^2 = \frac{1.75(1-\varepsilon)}{\varepsilon^3}\left(\frac{\rho D_P v_S}{\mu}\right) + \frac{150(1-\varepsilon)^2}{\varepsilon^3}$$

Presenting the Ergun Equation

The Sturnfield linearized Ergun Equation

$$\frac{\left(\Delta P/L\right)\mu}{v_S^2\,\rho^2}\left(\frac{\rho\,D_P\,v_S}{\mu}\right)^2 = \frac{1.75(1-\varepsilon)}{\varepsilon^3}\left(\frac{\rho\,D_P\,v_S}{\mu}\right) + \frac{150(1-\varepsilon)^2}{\varepsilon^3}$$

is equivalent to

$$\Pi_1 * \Pi_2^2 = \left(\frac{1.75(1-\varepsilon)}{\varepsilon^3}\right)\Pi_2 + \frac{150(1-\varepsilon)^2}{\varepsilon^3}$$

if Ergun L = Z and D_P = D.

Linearized for of Dimensional Analysis Results

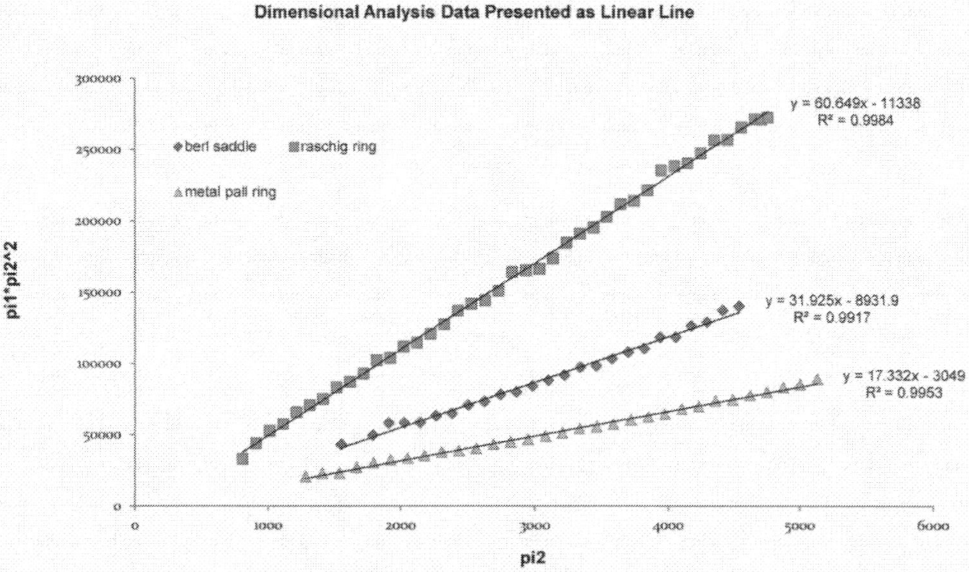

32

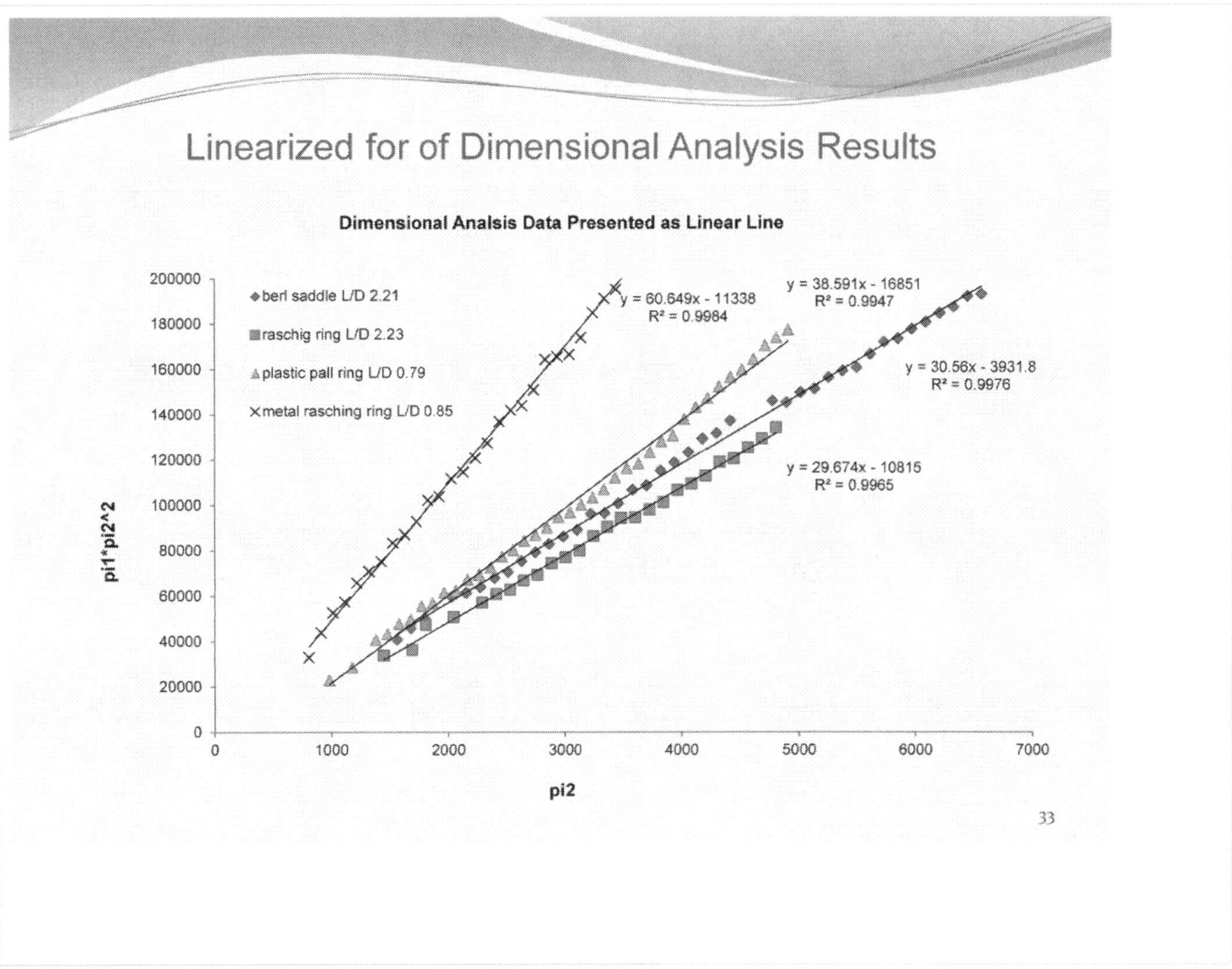
Linearized for of Dimensional Analysis Results
Dimensional Analsis Data Presented as Linear Line
berl saddle L/D 2.21
raschig ring L/D 2.23
plastic pall ring L/D 0.79
metal rasching ring L/D 0.85
y = 60.649x - 11338
R² = 0.9984
y = 38.591x - 16851
R² = 0.9947
y = 30.56x - 3931.8
R² = 0.9976
y = 29.674x - 10815
R² = 0.9965
pi1*pi2^2
pi2
33

Reactive Distillation:

Using The Reactor as a Reboiler

Thomas A. Kenat, Ph.D., P.E.
KenaTech Process Engineering
Medina, Ohio www.kenatech.com

KenaTech AIChE National Meeting, New Orleans, LA - Spring, 2014 3

KenaTech AIChE National Meeting, New Orleans, LA - Spring, 2014 4

Reactive Distillation:

Using The Reactor as a Reboiler

Thomas A. Kenat, Ph.D., P.E.
KenaTech Process Engineering
Medina, Ohio   www.kenatech.com

What is Process Intensification?

Combining or modifying processes in unique or creative ways to:

- Increase yields or product quality.
- Reduce equipment or operating costs.
- Save or recover energy.
- Reduce undesirable byproducts.
- Reduce emissions or destroy pollutants.
- Make the process more sustainable.

What is Reactive Distillation?

Combining distillation and chemical reaction in unique or creative ways to:

- Use the reaction to improve the separation.
- Reduce column size to reduce capital cost.
- Reduce energy cost by improved separation.
- Reduce energy cost by using the heat of reaction to replace the column reboiler.

KenaTech AIChE National Meeting, New Orleans, LA - Spring, 2014 7

The Vinyl Chloride (VCM) Process
(Nearly a Century of Process Intensification)

- Vinyl Chloride Monomer (VCM) is the monomer used to produce Polyvinyl Chloride (PVC), one of the most widely used plastics in the world.
- Since the first VCM process was invented in 1912 and commercialized in the 1920's, the process for the production of VCM has undergone many improvements, most of which would be regarded as process intensification.

KenaTech AIChE National Meeting, New Orleans, LA - Spring, 2014 8

The Acetylene VCM Process

$CaO + 3\,C \rightarrow CaC_2 + CO$
(Electric Arc Furnace – Graphite Electrodes)

$CaC_2 + H_2O \rightarrow C_2H_2 + Ca(OH)_2$

$2NaCl + 2H_2O \rightarrow 2NaOH + H_2 + Cl_2$
(H_2 and Cl_2 are produced by electrolysis)

$H_2 + Cl_2 \rightarrow HCl$

$C_2H_2 + HCl \rightarrow C_2H_3Cl$ (Vinyl Chloride)
(Hydrochlorination of Acetlyene)

The Acetylene VCM Process

- Most widely used in the 1930's and 1940's.
- Very high electrical energy consumption.
- High levels of toxic emissions.
- Still the most commonly used process in China.

The First “Balanced” Process

Cl_2 + $C_2H4 \rightarrow C_2H_4Cl_2$ Ethylene Dichloride (EDC)
(Direct Chlorination of Ethylene – Liquid Phase)

$C_2H_4Cl_2 \rightarrow C_2H_3Cl$ + HCl
(Thermal Cracking)

C_2H_2 + HCl $\rightarrow C_2H_3Cl$
(Hydrochlorination of Acetlyene)

The First “Balanced” Process

- Takes advantage of lower-cost ethylene, and cheap natural gas of the 1950’s for cracking furnaces.
- Eliminates energy-wasting HCl production, using HCl from cracking to react with ethylene.
- Direct chlorination is a low-temperature, liquid phase process with high yields and high purity.

The Second “Balanced” Process

Cl_2 + C_2H4 → $C_2H_4Cl_2$ Ethylene Dichloride (EDC)
(Direct Chlorination of Ethylene – Liquid Phase)

C_2H_4 + 2HCl + ½O_2 (Air) → $C_2H_4Cl_2$ (EDC) + H_2O
(Oxyhydrochlorination of Ethylene)

$2C_2H_4Cl_2$ → $2C_2H_3Cl$ + 2HCl
(Thermal Cracking)

The Second “Balanced” Process

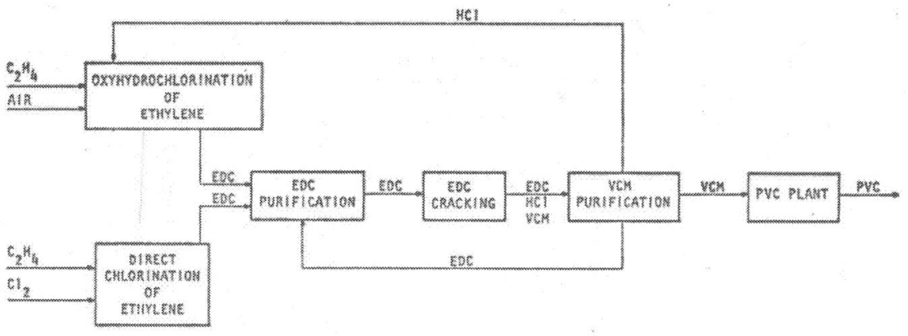

The Second "Balanced" Process

- Takes full advantage of lower-cost ethylene, and completely eliminates acetylene from the process.
- Uses HCl from cracking to react with ethylene in the oxychlorination reactor to produce EDC.
- Recovers heat of reaction as high-pressure steam.
- Oxychlorination produces more byproducts than previous process, but overall costs are lower.

The Second "Balanced" Process

- Opportunities for further improvement:
- The direct chlorination of ethylene is a highly exothermic liquid-phase reaction consuming a large quantity of cooling water and wasting heat.
- Distillation of EDC consumes a lot of steam.
- Can we use this heat of reaction to distill EDC?

The Direct Chlorination Reactors

The Boiling Reactor Concept

- The Direct Chlorination process produces a pure product with high yield, but is highly exothermic, and requires cooling to remove the heat of reaction.
- The challenge was to find a way to use this heat to make the process more efficient and economical.
- The Boiling Reactor would become the reboiler of the EDC distillation column, using this heat of reaction to distill the EDC, rather than wasting it by cooling.
- But is there enough heat available to do the job?

Let's Look at the Thermodynamics

- For the direct chlorination of ethylene:
- Cl_2 + C_2H4 $\rightarrow$ $C_2H_4Cl_2$ (EDC)
- The heat of reaction is 794 Btu / lb
- The heat of vaporization of EDC is 139 Btu/lb
- A factor of 5.7 is more than adequate to distill the EDC from both the boiling reactor and oxychlorination.

KenaTech AIChE National Meeting, New Orleans, LA - Spring, 2014 19

Boiling Reactor R&D Program

- Build a small microplant in the laboratory:
- This will test the concept, to see how much height is needed for the reaction to be complete before boiling begins, so that efficiency is optimized.
- Build a pilot plant at the production facility:
- Its product is fed into the distillation column, and data can be gathered at semi-production scale.

KenaTech AIChE National Meeting, New Orleans, LA - Spring, 2014 20

The Boiling Reactor Microplant

KenaTech AIChE National Meeting, New Orleans, LA - Spring, 2014 21

Detail of Microplant Feed Zone

KenaTech AIChE National Meeting, New Orleans, LA - Spring, 2014 22

Microplant Conclusions

- The concept was successful, and the efficiencies were close to those in the low-temperature reactors.
- The microplant data provided:
  - Proof that the concept was workable.
  - Design and scale-up data for the pilot plant.

The Boiling Reactor Pilot Plant (1) (Calvert City, KY)

KenaTech | AIChE National Meeting, New Orleans, LA - Spring, 2014 | 25

The Boiling Reactor Pilot Plant (2)

KenaTech | AIChE National Meeting, New Orleans, LA - Spring, 2014 | 26

Pilot Plant Feed Zone

KenaTech | AIChE National Meeting, New Orleans, LA - Spring, 2014 | 27

Pilot Plant Flow Control Valve

KenaTech | AIChE National Meeting, New Orleans, LA - Spring, 2014 | 28

Pilot Plant Conclusions

- Efficiencies are very close to those of the existing low-temperature direct chlorination reactors.
- The pilot plant data provided:
  - Design and scale-up data for a commercial unit.
  - Materials of construction data.
  - Sufficient information to modify an existing boiling reactor at another BFG plant in Texas.

Existing Boiling Reactor, LaPorte, TX

(Plant purchased from Diamond Shamrock – Stauffer Design)

Existing Boiling Reactor, LaPorte, TX

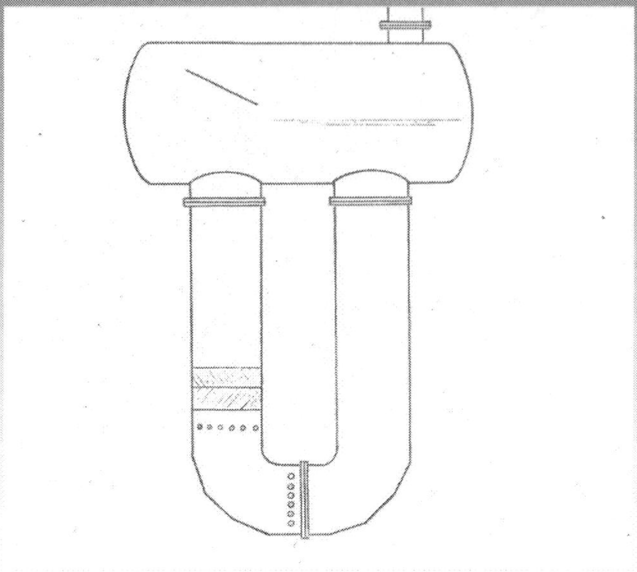

KenaTech | AIChE National Meeting, New Orleans, LA - Spring, 2014 | 31

Modified Boiling Reactor, LaPorte, TX
Increased Capacity from 300 to 900 Tons/day

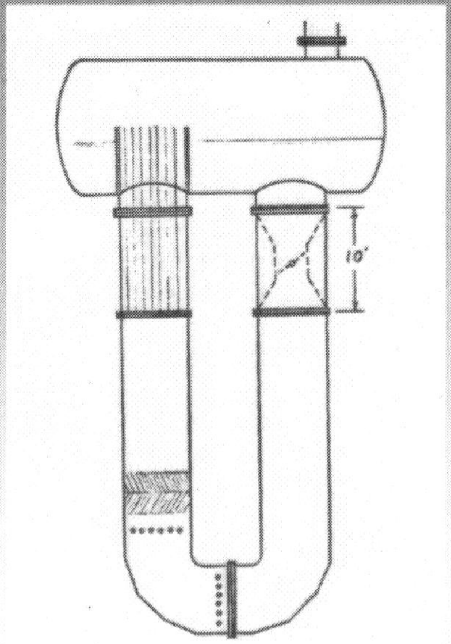

KenaTech | AIChE National Meeting, New Orleans, LA - Spring, 2014 | 32

Boiling Reactor Drum Removed

KenaTech AIChE National Meeting, New Orleans, LA - Spring, 2014 33

Extensions and New Deck Installed

KenaTech AIChE National Meeting, New Orleans, LA - Spring, 2014 34

Replacing Drum

KenaTech AIChE National Meeting, New Orleans, LA - Spring, 2014 35

Boiling Reactor Modifications Complete

KenaTech AIChE National Meeting, New Orleans, LA - Spring, 2014 36

Results of Modifications:

- Capacity increased from 300 to 900 tons/day.
- Surging was eliminated by adding damper valve.
- Conversion was significantly increased.
- Ethylene and chlorine efficiency were increased.

KenaTech AIChE National Meeting, New Orleans, LA - Spring, 2014 37

Conclusions:

- Using the direct chlorination reactor as the reboiler to the EDC distillation column resulted in:
  - Capital savings through tripling reactor capacity.
  - Significant energy savings.
- The Boiling Reactor, in some form, is now used in most state-of-the-art vinyl chloride processes.

KenaTech AIChE National Meeting, New Orleans, LA - Spring, 2014 38

Conclusions (cont'd):

- This concept can be used in any process where:
  - The reaction product must be distilled.
  - The reaction is exothermic, and the heat of reaction far exceeds the heat of vaporization.
  - The reaction can be carried out at a temperature near the product's boiling point.

351420 Heat Integrated Reactive Distillation Using External Side Reactors for Synthesis of Tri-Ethyl Citrate

Abstract:

Organic acids, typically produced by fermentation of carbohydrate feedstocks, constitute an important class of biorenewable platform chemicals that can be further converted to useful products. Tri-ethyl citrate (TEC), produced from citric acid and ethanol, is currently gaining of attention as a non-toxic, biocompatible plasticizer that can be used in place of petroleum-based phthalate compounds. Phthalates have potential carcinogeneity and overall health concerns to the extent that they have been banned in Europe from use in children's toys and human contact applications. At present, the potential of TEC is limited by the lack of large-scale, energy efficient and economic production.

Esters are advantageously produced using reactive distillation, where simultaneous reaction and separation take place in the same process unit. However, the hardware design of standard reactive distillation columns exhibits challenges for slow chemical reactions such as tri-ethyl citrate formation, because high liquid holdup and large quantities of catalyst are required inside the reactive distillation column to obtain complete conversion. To overcome this problem and make the synthesis of citrate esters economically competitive with their petroleum-based counterparts, we have developed and demonstrated advanced reactive distillation concepts involving heat integration and external side reactors. AspenPlus process analysis has been performed using available kinetic data for esterification of citric acid with ethanol to integrate a pre-reactor as well as side reactors, which can have larger catalyst volumes and operate at any desired temperature, with a fractionation column. Heat integration is examined for energy efficiency and maintaining temperatures in side reactors to optimize catalyst utilization. The concept of integrating side reactors with distillation has been validated experimentally using the reactive distillation facility at Michigan State University with a pre-reactor and one side reactor coupled to a distillation column. An 85% conversion of acid groups was obtained in this configuration when the pre-reactor and side reactor were operated at 110°C using Amberlyst-15 as a catalyst. Using AspenPlus simulation for the case of multiple side reactors, greater than 98% conversion of citric acid groups can be obtained.

351657 Reactive Distillation Process of Methyl Acetate Hydrolysis Directly Intensified by Auxiliary Reaction of Methanol Dehydration

Abstract:

The green novel technology reactive distillation (RD) has provided with great potentials. However, it is limited due to physical properties of the reaction system, such as multiple liquid phases restricting solubility of catalyst and miscibility of reactants, and formulation of azeotropic mixtures entraining reactant or product and making downstream separation very difficult. Typical industrial examples include complete recovery of methyl acetate (MeOAc) from waste of polyvinyl alcohol processes. For MeOAc hydrolysis, the chemical equilibrium constant is very low. In order to reach high conversion, excessive water is needed which causes high energy consumption in downstream separation.

In order to make use of the advantages of reactive distillation, process intensification by auxiliary reactions is proposed in this contribution. Methanol is introduced as subsidiary reactant and provides water by its dehydration reaction. Water is supplied for hydrolysis reaction of MeOAc, which produces methanol and product acetic acid (HOAc). Since methanol is lighter and water is heavier, they are moved by distillation effect to their reactive zones respectively once formed. As a result, extra water is not needed and methanol is in-situ transferred to product dimethyl ether (DME). The reaction of methyl acetate hydrolysis is thus directly intensified by methanol dehydration.

A residue curve maps (RCMs) is developed to analyze the mechanism of intensification for the case with two reactions and multiple components. The analysis of thermodynamics and RCMs reveals that the reaction of MeOH dehydration is the controlling step which limits the conversion of MeOAc. Based on the chemical equilibrium and kinetically controlled design, a novel process is developed to hydrolyze MeOAc and MeOH in the process of polyvinyl alcohol. Compared to traditional processes, no additional water is required to feed into the RD column and the process is significantly simplified. With equal mole of MeOAc and MeOH as feed and a set of pre-reactor, closely 100% conversion of MeOAc and MeOH can be achieved and high purity dimethyl ether and acetic acid are the products in the RD column.

Simulation and Optimization of Different Extractive Distillation Schemes for Aromatics Recovery from Pyrolysis Gasoline

***Mansour Emtir*[1] *, Elmahboub Edreder*[2]**

[1]*Libyan Petroleum Institute, Tripoli, Libya*
[2]*National Oil Corporation, Tripoli, Libya*

2014 AIChE Spring Meeting and Global Congress on Process Safety
March 30 - April 3, 2014
Hilton New Orleans-Riverside, New Orleans, LA

Outlines

- Overview
- Why extractive distillation (ED)?
- Objectives
- A case study
- Assumptions and techniques
- Investigated ED schemes
- Composition profiles
- Optimization results
- Conclusions

2

Overview

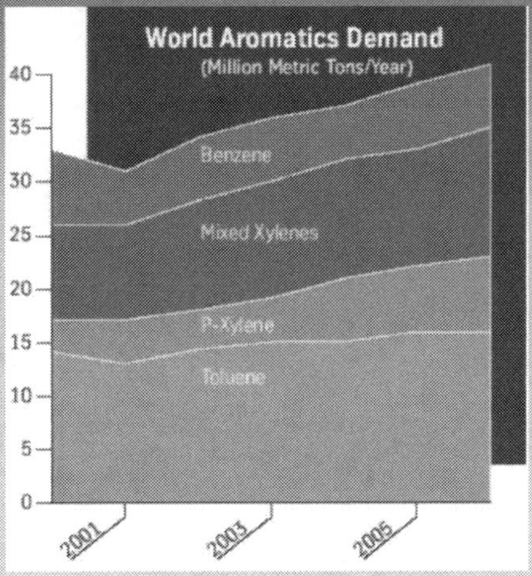

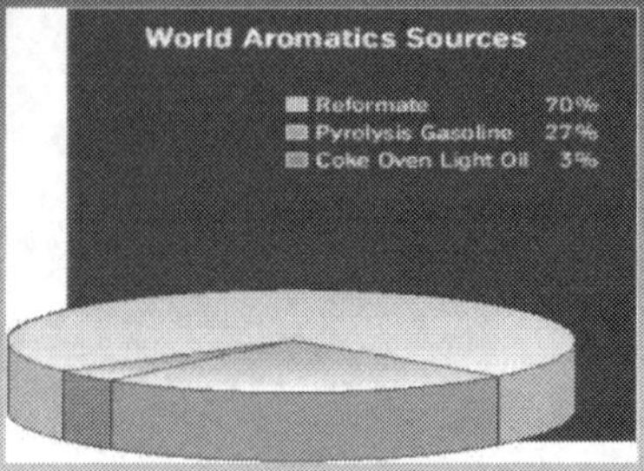

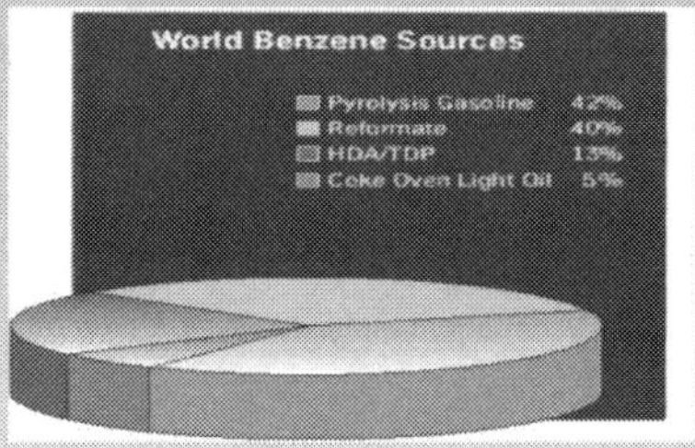

Why Extractive Distillation (ED)?

- The aromatic components usually found with other close-boiling hydrocarbons, some of which form azeotropes
- The working principle of extractive distillation (ED) is the alteration of the relative volatility of components in the presence of highly selective solvent

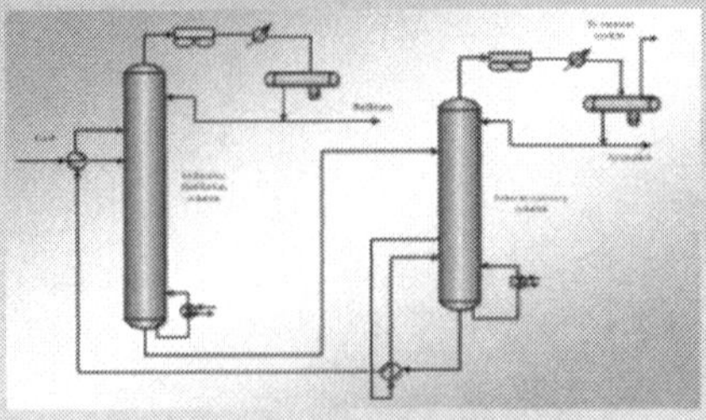

Objectives

- ✓ Design and simulation of conventional extractive distillation, complex ED schemes and heat-integrated ED scheme for recovery of aromatics from pyrolysis gasoline
- ✓ Optimizing of the studied ED schemes based on total annual cost (TAC) as the objective function
- ✓ Investigating a various solvent types on the economic optimization of studied extractive distillation schemes
- ✓ Study the effect of solvent stream temperature and solvent/feed ratio parameters on TAC

5

A case study

- Feed stream = 45100 kg/hr
- Selected solvents: N-methyl pyrolidone [NMP], Di methylene formamide [DMF] and Sulfolane and Tetra ethylene glycol [TEG]

Components	Streams (kg/hr)				
	Feed	Solvent	Non Arom.	Aromatic	Solvent R.
Cyclo pentane	2937	0	2936.6	0.39	0
Benzene	10583	0	0	10583	0
1-Hexene	1359	0	0	1359	0
Toluene	6758	0	0	6758	0
m-xylene	1324.1	0	0	1324.1	0
1M-3EBenzene	1005	0	0	1002.8	2.17
n-m-pyrolidone	0	162181.59	0	39.31	162143.3

6

Assumptions and Techniques

- HYSYS process simulator package is used.
- The amount of benzene in non-aromatic stream should never exceed 1.0 vol. %.
- Feed and products are saturated liquids at atmospheric pressure.
- Top vapor temperature ≥ 47 °C.
- Valve tray (Glitsch type) is selected as column internal, 70-75 % maximum flooding velocity.
- Plant life = 10 years, operating hours are 8000hr/yr.
- The objective function is minimum total annual cost (TAC).
- Optimization parameters are: S/F ratio, number of trays, feed location, solvent type and temperature.

7

Investigated ED schemes

- Conventional ED scheme (base case)
- Petlyuk column
- Divided wall column
- Conventional heat-integrated ED scheme

8

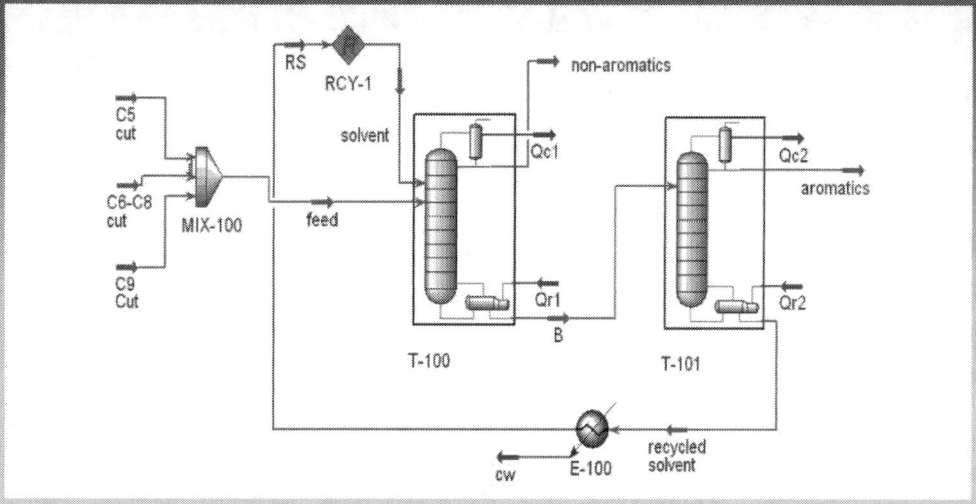

Conventional extractive distillation (base case)

9

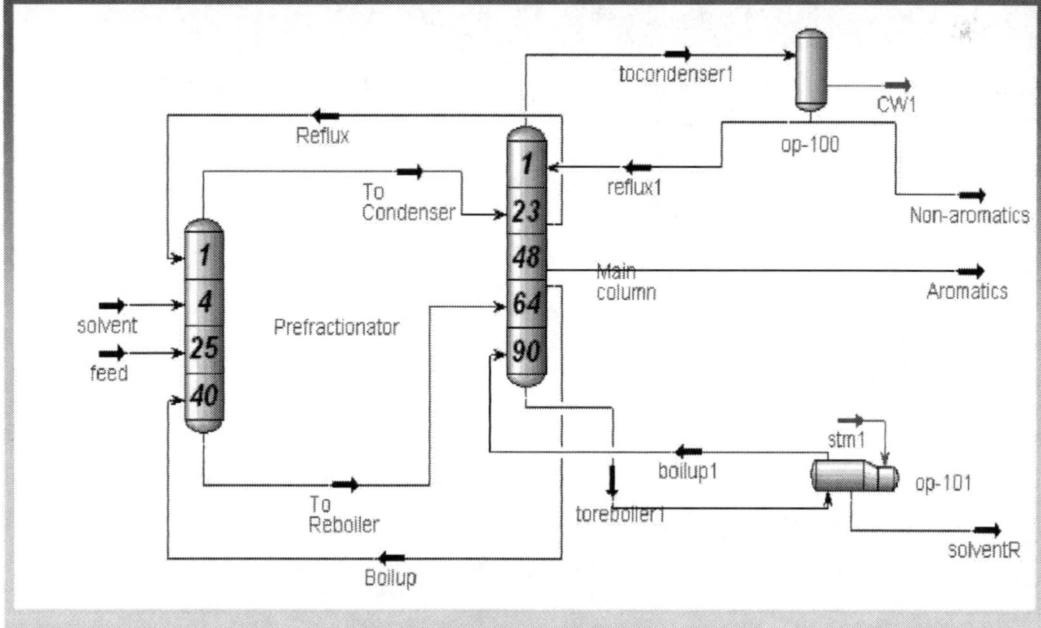

Petlyuk column

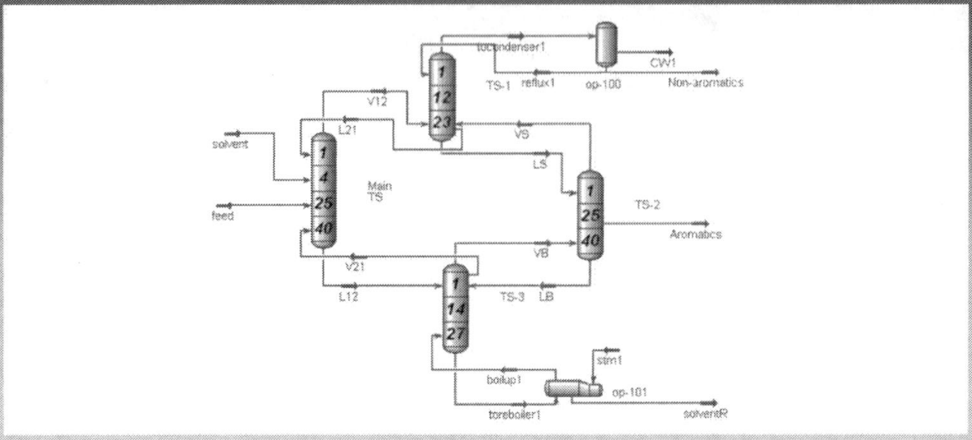

Divided-wall column

11

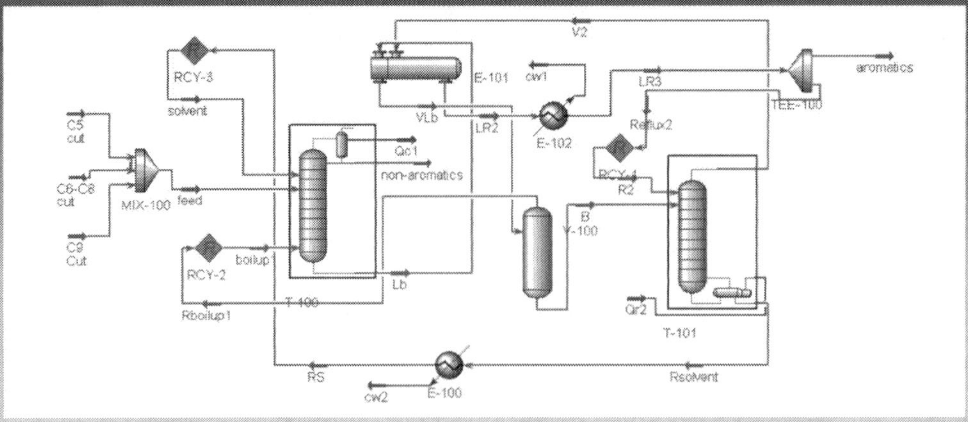

Conventional ED with backward heat integration

12

Benzene compositions profiles

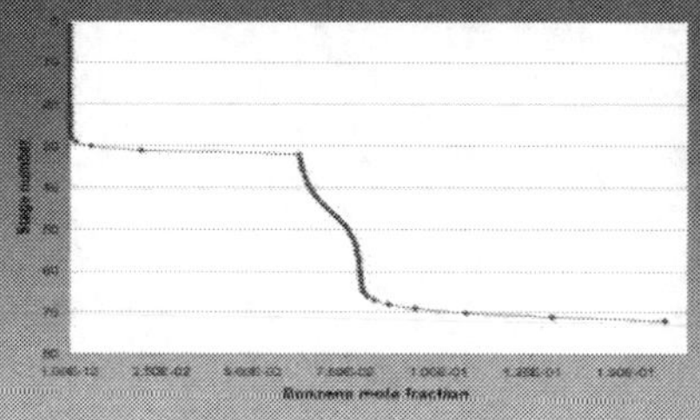

Composition profile along the extraction column

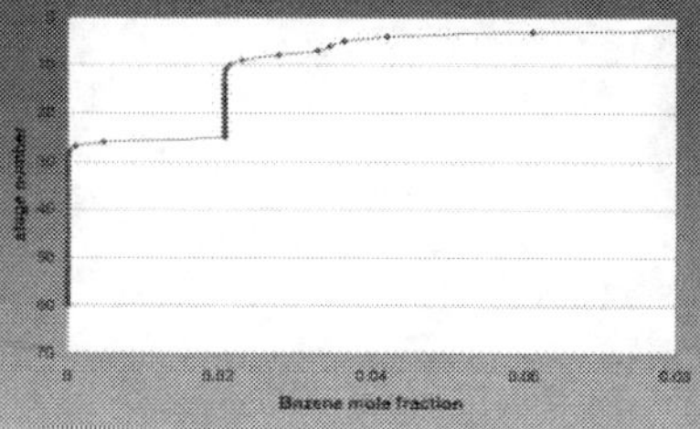

Composition profile along the stripping column

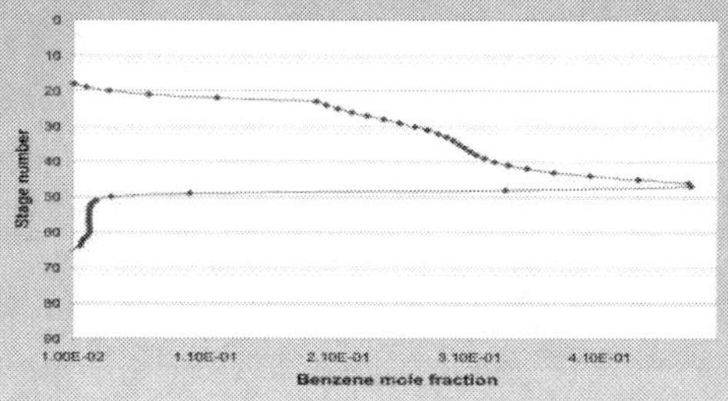

composition profile along the DWC

13

Conventional and complex ED optimization results

	Conventional ED			Petlyuk column		
S/F	3.0	2.5	2.0	3.0	2.5	2.0
Annual capital cost ($/year)*$10^5$	5.08	4.75	4.54	3.94	3.41	3.34
Annual operating cost ($/year)*$10^6$	3.48	3.25	3.04	2.36	2.39	1.85
TAC ($/year)*$10^6$	3.99	3.73	3.49	2.75	2.73	2.19

14

Conventional and heat-integrated ED optimization results

Solvent type	NMP	TEG	Sulfolane	DMF
Recycled solvent temperature C°	202.3	278.3	287	153.6
Solvent recovery %	0.999	0.999	0.988	0.98
Annual capital cost ($/year)*$10^5$	4.75	6.79	7.04	4.24
Annual operating cost ($/year)*$10^6$	3.25	5.80	5.87	3.53
TAC ($/year)*$10^6$	3.73	6.48	6.57	3.95

Solvent type	NMP	TEG	Sulfolane	DMF
TAC ($/year)*$10^6$	2.61	5.82	5.00	2.36
TAC saving %	30	10	24	40

15

Conclusions

- Energy-integrated extractive distillation configurations (Petlyuk column, DWC and heat-integrated ED) proved to be an attractive alternative configurations to conventional extractive distillation by showing a higher savings of TAC ranging from 10 to 40 % in TAC
- The ranking of solvents in terms of TAC savings achievement are DMF, NMP, Sulfolane and TEG respectively
- DMF entrainer prove to be the best choice regarding TAC savings of 40 % which is attributed to low recycled entrainer temperature that allows for the usages of LP-steam and less cooling rates

16

THANK YOU FOR YOUR ATTENTION !

A Reactive Distillation Column with Two Reaction Zones for the Cyclohexanol Production from Cyclohexene

Xinhui Yao, Kejin Huang, Kun Wang, Ningning Wei,
Mingyao Li and Shaofeng Wang

College of Information Science & Technology
Beijing University of Chemical Technology
People's Republic of China

Highlights and objectives

- **Overview of cyclohexanol production**
- **Characters of conventional coupled reactive distillation columns (CRDC)**
- **A novel reactive distillation column with two reactive sections (RDC-TRS)**
- **Results and analysis**
- **Conclusions**

Overview of cyclohexanol production

Oxidation of cyclohexane

- Safety risks
- Low selectivity

and

Hydrogenation of phenol

- High prices of material

Asahi Chemical Co.

Direct hydration of cyclohexene

- Low reaction rate
- Large amounts of catalyst

Steyer et al.

Indirect hydration of cyclohexene

- High reaction rate with formic acid as an entrainer

2014 AIChE Spring Meeting &
10th Global Congress on Process Safety

Production of cyclohexanol by indirect hydration of cyclohexene — coupled reactive distillation columns (CRDC)

➢ Reaction

- R1: ENE + FA ↔ FCE
- R2: FCE + H2O ↔ NOL + FA
- R3: ENE + H2O ↔ NOL

➢ Separation

- ANE fed as an impurity
- Boiling points at 10 kpa
  ANE < ENE < FA < H2O < FCE < NOL

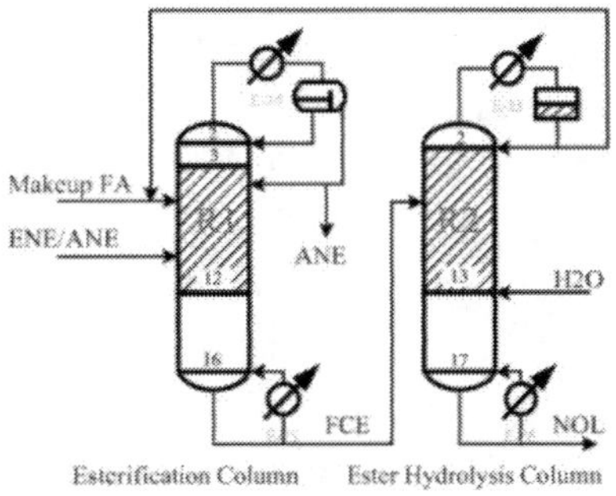

[Katariya et al., Ind Eng Chem Res, 48, 9534–9545, 2009]

Beijing University of Chemical Technology (BUCT)

Characters of CRDC

- Complicated configuration
  - Two reactive distillation columns involved
  - Recycle from the second column to the first column
  - Bottom product of the first column fed to the second column

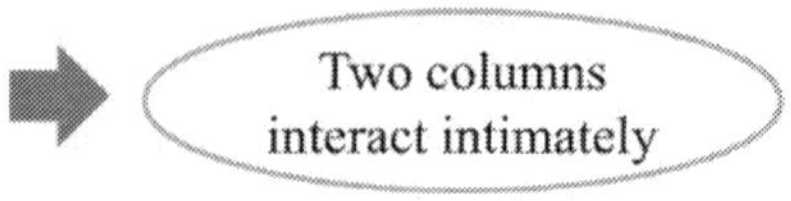

- Opportunity of further process intensification

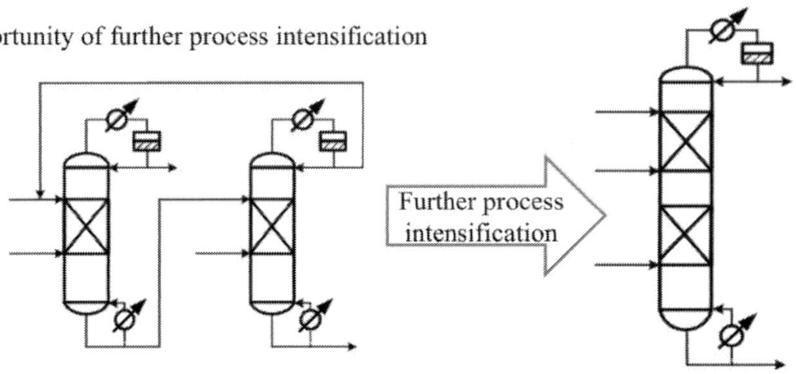

2014 AIChE Spring Meeting &
10th Global Congress on Process Safety

A novel reactive distillation column with two reactive sections (RDC-TRS)

- Potential advantages
  - Intensifying internal mass and energy integration
  - Increasing process efficiency
  - Reducing capital investment and operating cost
- Challenges to be solved
  - More components involved
  - More variables to be coordinated
  - Strong interaction between reaction and separation operations

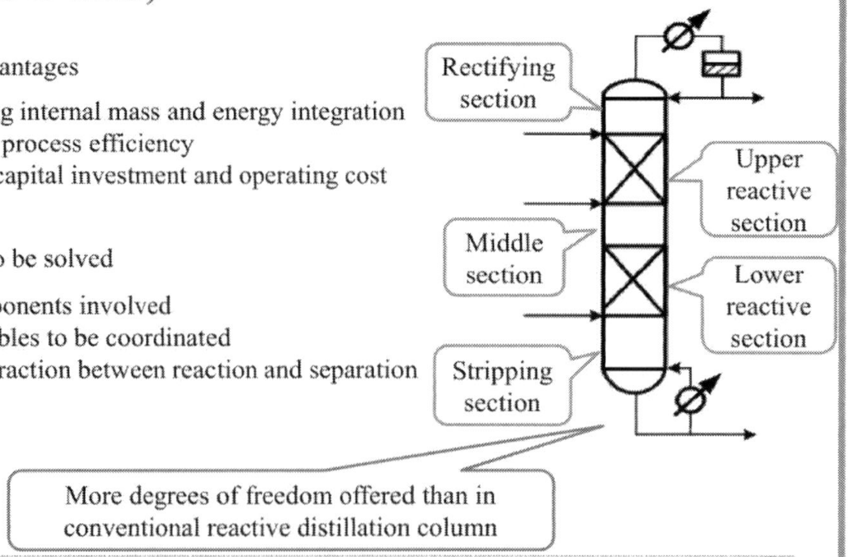

Beijing University of Chemical Technology (BUCT)

Procedure for synthesis and design of RDC-TRS

Step 1

Guess initial process design according to coupled reactive distillation columns

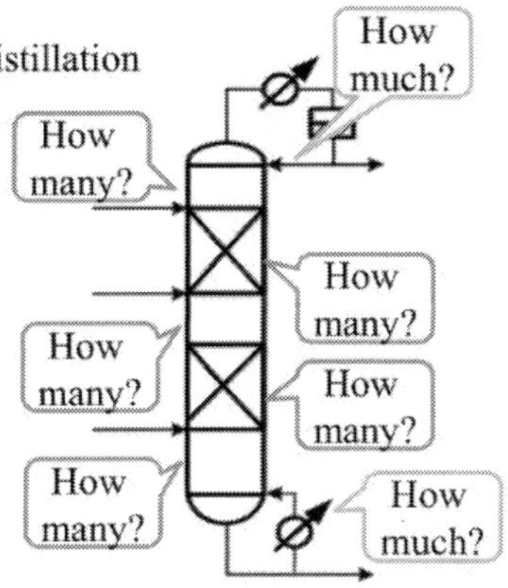

Step 2

Searching for the optimum value of structural variables in a nest-loop manner

- Structural variables: Nrec, Ns, Nrs1, Nrs2, Nm
- Objective function: total annual cost (TAC)

Step 3

Searching for the optimum value of operating variables in a sequential manner

•Operating variables: reflux flow rate and reboiler heat duty

•Objective function: TAC

In case of products specification unsatisfied, external recycle is frequently employed

Xinhui Yao et al., Ind. Eng. Chem. Res., 2013, 52, 16870 - 16879;
Haisheng Chen et al., AIChE J., 2013, 59, 2015 - 2032;
Haisheng Chen et al., Ind. Eng. Chem. Res., 2012, 51, 14473 - 14488.

Relationships between TAC and relevant decision variables

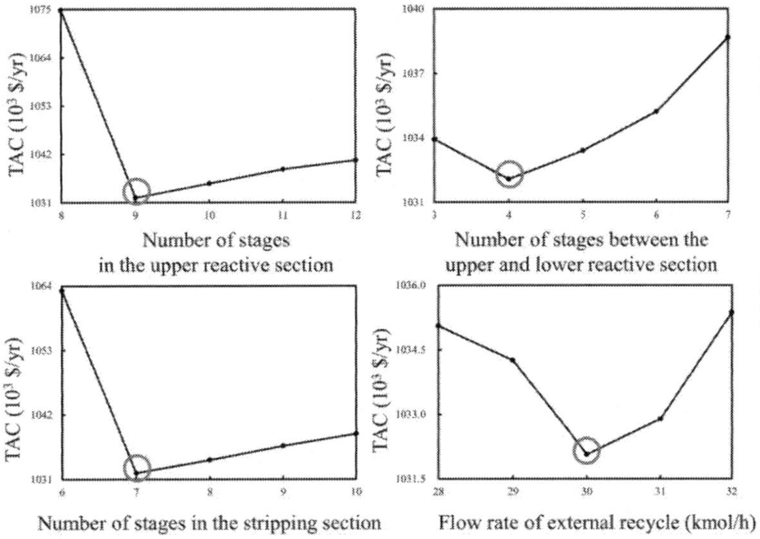

- All parabolic curves
- Lowest points

- Valid procedure
- RDC-TRS potentially energy efficient

2014 AIChE Spring Meeting &
10th Global Congress on Process Safety

Results and analysis

- Two reactive sections divided by four stages
- External recycle

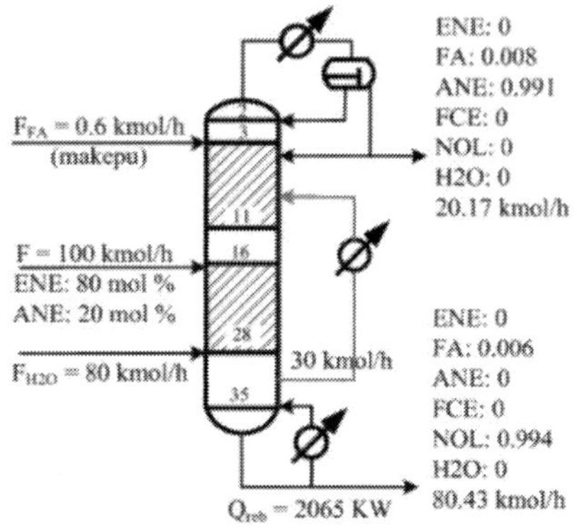

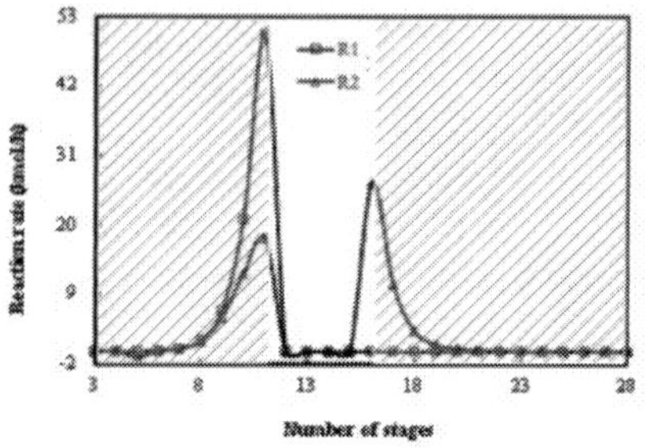

Beijing University of Chemical Technology (BUCT)

2014 AIChE Spring Meeting &
10th Global Congress on Process Safety

Comparison of liquid composition profiles

- different liquid composition profiles from esterification column
- Similar liquid composition profiles with ester hydrolysis column

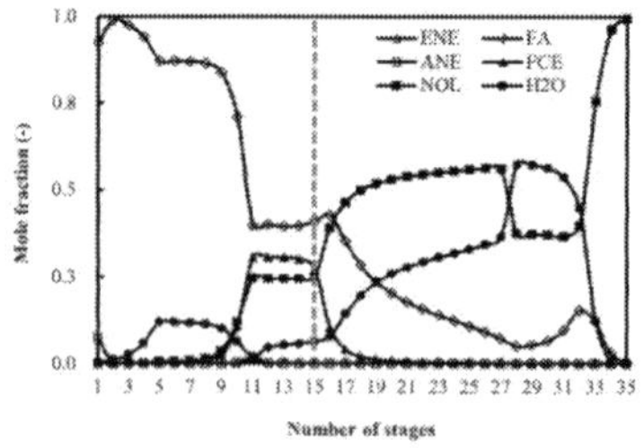

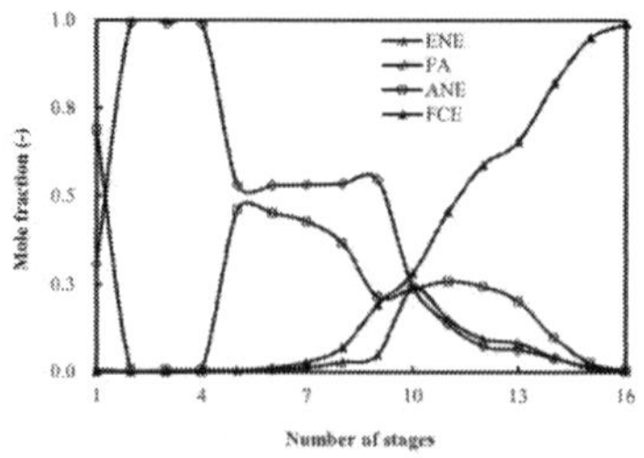

Esterification column

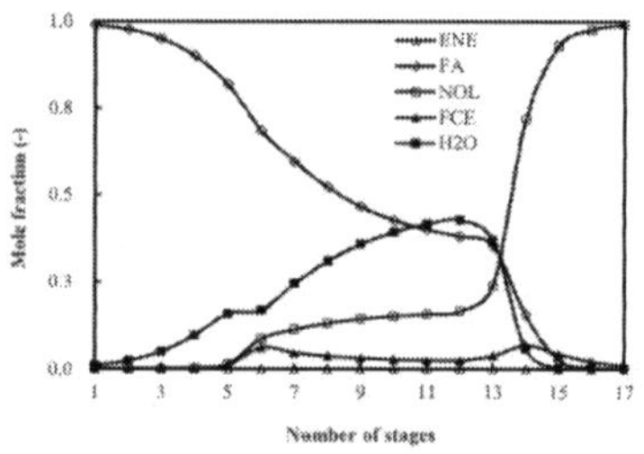

Ester hydrolysis column

Beijing University of Chemical Technology (BUCT)

2014 AIChE Spring Meeting & 10th Global Congress on Process Safety

Comparison between RDC-TRS and CRDC

Structure		CI (10^3 \$/yr)	OC (10^3 \$/yr)	TAC (10^3 \$/yr)	Comparison of CI (%)	Comparison of OC (%)	Comparison of TAC (%)
CRDC	Esterification Column	146.11	346.99	493.10	100	100	100
	Ester Hydrolysis Column	326.43	1004.59	1331.02			
RDC-TRS		362.09	669.97	1032.06	76.6	49.6	56.6

RDC-TRS leads to lower capital investment (CI), operating cost (OC) and TAC

Beijing University of Chemical Technology (BUCT)

2014 AIChE Spring Meeting & 10th Global Congress on Process Safety

Conclusions

- **We proposed a reactive distillation column with two reactive sections and gave the design method**
- **The arrangement of two separate reactive sections could enhance internal mass integration and/or internal energy integration**
- **External recycle played an important role in satisfying product specifications**
- **RDC-TRS could cut considerably CI, OC and TAC in comparison with CRDC for cyclohexanol production**

Beijing University of Chemical Technology (BUCT)

2014 AIChE Spring Meeting &
10th Global Congress on Process Safety

Thank you for your listening!

Beijing University of Chemical Technology (BUCT)

353790 A Recipe to Excel: Beyond Technical Competency

Abstract:

The presentation will highlight the fundamental soft skill sets needed to be successful in managing diverse work forces in an organization. The skill sets are invariant of the type of organizational environment. This could span R&D driven organizations to manufacturing & production to engineering, procurement and construction driven organizations and to other service industries.

To be successful, it is taken for granted for any technically driven project or organization, that to carry out his/her assigned responsibility, the manager needs to have technical skills and understanding. The factor that differentiates a truly successful manager, however, is the manager's extraordinary strength in soft skills, sometimes referred to as people skills. The necessity of these skills is becoming more important and pertinent than ever before. The reason behind this special requirement is the diversity of groups involved in any work environment. Such diversity is due to rapid globalization. Challenges associated with diversity, when managed properly, bring strength rather than weakness.

In this presentation the importance of effective communication will be emphasized. Effective communication is the most difficult skill to acquire and practice but the most rewarding when a manager/leader acquires and proficiently applies it. Through effective communication the leader brings team spirit and trust. Trust in turn brings success.

Management Careers in Consulting and Engineering

R. Benson Pair, Jr., PE, BCEE
Chief Technology Engineer - Environmental
KBR

KBR

Career Opportunities in Consulting and Engineering

Consulting companies and Engineering firms offer engineers a wide range of career opportunities in management

KBR

Management Careers in Consulting and Engineering

- Technical and Technology
- Engineering
- Project Management
- Engineering Management
- Project Controls
- Offices
- Business Units
- Corporate
- Sales
- IT/IM
- Procurement
- Construction

KBR

Consulting

- De-centralized, serving regional markets
- Smaller offices and smaller projects compared to EPC firms
- Historically has been engineering only, but that is changing
- May be focused on one market sector: Industrial or Government (Federal/State/Local)

KBR

Engineering, Procurement, Construction

- Fewer, larger offices
- Larger projects
  - Contract value
  - Schedule
  - Task force can be >100
- Often serves Industry and Government

KBR

Technical Skills

- Engineering Disciplines
- Accounting
- Cost and Scheduling
- Contracts
- HSE

KBR

People Skills

- Leadership
- Communication: Listening and talking
- Negotiating
- Respect for others
- Patience
- Humility in yourself / Pride in others
- Sense of Humor
- Delegation
- Mentoring

KBR

Career Development - The "Y" Approach

- Recognizes and values the technical specialist
- Shows transition from generalist to technology specific roles
- Identifies three broad bands of skill levels
- Continues to recognize and value the technical leaders who combine technology and management skills
- Shows transfer routes to Project Engineering / Project Management and Sales Coordinator / Sales

KBR

The "Y" Model for Career Development

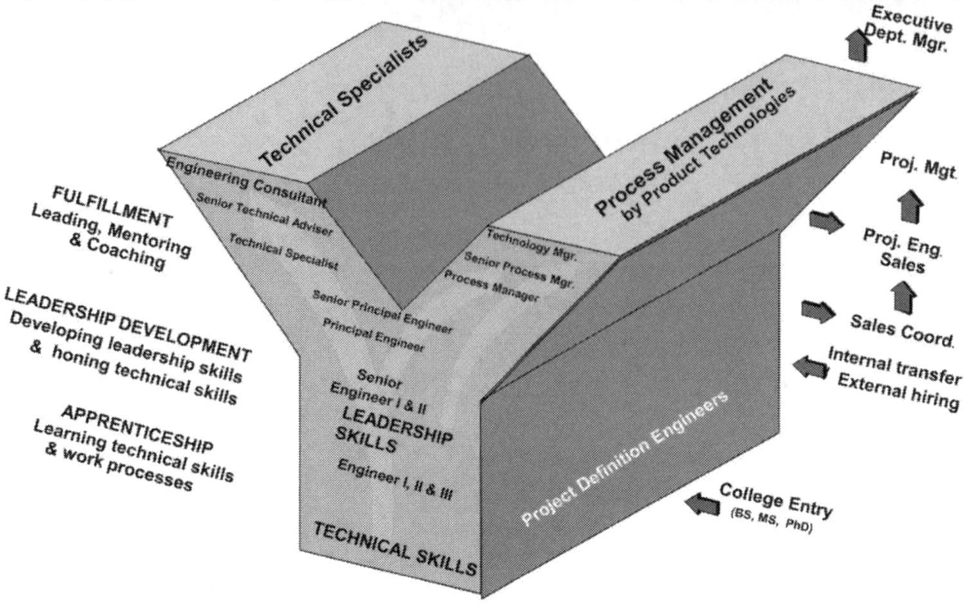

KBR

Management Is…

… a set of well-known processes, like planning, budgeting, structuring jobs, staffing jobs, measuring performance and problem-solving, which help an organization to predictably do what it knows how to do well. Management helps you to produce products and services as you have promised, of consistent quality, on budget day after day week after week.

KBR

...(Still) not Leadership

Leadership is associated with taking an organization into the future, finding opportunities that are coming at it faster and faster and successfully exploiting those opportunities. Leadership is about vision, about people buying in and about empowerment and most of all about producing useful change.

Source: "Management Is (Still) Not Leadership", John Kotter, Konosuke Matsushita Professor of Leadership, Harvard University, *Harvard Business Review*, January 9, 2013

KBR

Management and Leadership

Managers

- administer
- maintain
- focus on systems and structures
- rely on control

Leaders

- innovate
- develop
- focus on people
- inspire trust

Source: "The Wall Street Journal Guide to Management," Alan Murray

KBR

Honesty

"Your business and its employees are a reflection of yourself and if you make honest and ethical behavior a key value, your team will follow suit."

Source: "Top 10 Qualities that make a Great Leader", Tanya Prive, Forbes, 12/19/2012

KBR

Engineering Ethics

Engineer's Creed

As a professional engineer, I dedicate my professional knowledge and skill to the advancement and betterment of human welfare

I pledge:

- To give the utmost of performance;
- To participate in none but honest enterprise;
- To live and work according to the laws of man and the highest standards of professional conduct;
- To place service before profit, the honor and standing of the profession before personal advantage, and the public welfare above all other considerations.

In humility and with need for Divine Guidance, I make this pledge.

Adopted by the NSPE, June 1954

KBR

353824 Management Careers in an Industrial Setting

Abstract:

Management careers in an industrial setting offer a broad range of opportunities and challenges.

It certainly is important to gain a clear understanding of your personal goals and objectives before embarking on a career in management.

What are your motivations? Do you have the skills and aptitude? Are you willing the do the things that need done?

Learning and development are essential. Time commitments are significant.

Rewards can be amazing when you see people & operations develop beyond expectations.

The key becomes tapping into the capability of the team and help move the organization forward.

353793 Managing Through Extreme Change

Abstract:

The Department of Energy Weapons Complex, a nuclear chemical enterprise employs hundreds of thousands of scientists and engineers. Through its birth during the Manhattan project in the 1940's to the end of the Cold War, the DOE Weapons Complex was a rapidly expanding enterprise that provided career growth and service towards a national mission. After the Cold War ended, the weapons complex has been slowly decreasing in size through a series of layoffs, furloughs, reductions in benefits and contract changes. While the DOE Complex was originally built by world class industrial companies, the current contract practices and potential legal liabilities have changed the field of companies that now operate the complex for the Department of Energy. These new types of companies formed as limited liability corporations, are typically partnerships of engineering and construction firms, aerospace companies or not for profit research companies. Managing businesses in a declining industry and opportunities within the nuclear chemical industry will be presented.

353905 You Can Have it All, But Not at the Same Time

Abstract:

Women (and men) in science and engineering have struggled with the Career Life Balance. With the rise of dual income couples and more female breadwinners, the working world has a goal of becoming better at balancing work and family life for both men and women. As managers women may face additional challenges not experienced by our male counterparts. Challenges occur in all sectors of the workplace: private industry, academia and government. Numerous articles and books have been written on the subject. Recently the Huffington Post featured and article titled "Do Women Make Better Bosses?".

1

Managing Volunteers Is A Little Like Herding Cats

Joseph J. Cramer, PhD, PE

Director Emeritus, Program Development

Careers in Management Session

AIChE Spring Meeting

April 1, 2014

New Orleans, LA

2

How did I select this title?

- Clearly chosen to attract interest
- Analogy has some validity
- Both cats and volunteer Ch.E.s are:
  - Intelligent
  - Fiercely Independent
  - Can scratch when irritated
- Strict military style of management won't work
- More than 60 million volunteers in USA
- Volunteers (and their management) are very important

Background Information

3

- Have been managing volunteers for >30 years – almost entirely involving AIChE
- Three degrees in Chemical Engineering
- Didn't like life in a lab or plant
- Didn't like just doing calculations all day
- Gravitated towards project management and later department management
- Enjoyed interaction with people, like dealing with crisis situation but also liked doing planning to avoid the crisis in the first place
- Slowly transitioned from managing 'direct reports' to 'volunteers'

Types of Authority

4

- Line Authority (Traditional)
- Staff (Expertise) Authority
- Coercive/Reward Authority (penalties and rewards)
- Functional Authority (specific tasks for designated time)
- Referent Authority (based on charisma, charm, personality)

5

Developing Your Style of Managing Volunteers

- Expertise (spoke the language, been there/done that)
- Rewards (important to volunteers even if they won't admit it)
- Granting Functional Authority
- Referent Authority (personality based-honesty, credibility, trust, knowledge)
- Motivating (nurturing enthusiasm)

6

Volunteers vs Employees

- Even disinterested workers can be motivated by compensation
- Disinterested volunteers disappear (or worse, can be a negative force)
- Poor workers can be fired (not always easy)
- Poor volunteers are difficult to fire
- Must word harder to really engage volunteers
- Must match tasks with skills - carefully

Effective Volunteers

7

- Must understand the direction
- Must buy into it (engagement)
- Must have an interest (enthusiasm)
- Must have ability to comply or execute
- Good communications must be maintained
  - Clear
  - Concise
  - Consistent
  - Good listening skills

Useful Skills for Volunteer Manager

8

- Reasonable technical proficiency of knowledge
- Good Communications skills
- Enjoy interacting with people
- Goal oriented but
  - Process sensitive
  - Good at building and maintaining relationships

The Benefits

9

- The organization can accomplish so much more
- The volunteer learns about 'managing'
- Volunteers acquire management experience more quickly
- The experience is transferrable
- The skills are good for a lifetime
- Relationships can last for life

10

Thank you for your attention.

Any questions?

Comments?

Anecdotes?

354013 New Learning Curves for Industrial Biotechnology: Liquid Fuels from Bioconversion of Methane and Bioelectrosynthesis

Abstract:

Biological systems are capable of performing many useful functions with applications in energy and chemical production. However, their full potential remains unrealized due to critical gaps in our ability to engineer and control them. Two programs at the Advanced Research Projects Agency – Energy (ARPA-E) address such critical gaps by engineering microbial systems for the production of liquid fuels and will be discussed in detail in my talk. The REMOTE program, whose acronym stands for Reducing Emissions using Methanotrophic Organisms for Transportation Energy, is a first-of-its-kind program that focuses on the development of advanced bioconversion technologies to produce liquid transportation fuels from methane, the main component of natural gas. The Electrofuels program, on the other hand, focuses on the development of non-photosynthetic autotrophic organisms for the conversion of CO_2 and electricity to infrastructure-compatible liquid fuels.

353229 Scaling up of Compact Heat Exchanger Reactors

Abstract:

Chart Energy and Chemicals' Compact Heat Exchange Reactor (CHER®) is a brazed metal reactor with integrated heat exchange. CHER features a high surface area to volume ratio and short distance for heat to travel, allowing high heat transfer efficiency. This allows control of exothermic reactions in CHER that are otherwise difficult to control with conventional reactors. CHER features repeating layered structures and modular units that can promote easy scale-up. Channels can be added in parallel without altering the process conditions within the reactor.

This presentation will discuss the performance of CHER at the bench, pilot, and commercial scale. It will include recent data in pilot scale plants using CHER, along with some related reactor modeling work. Fixed-bed Fischer-Tropsch synthesis was demonstrated in a CHER with an aluminum core inside a steel vessel (based on Chart's Core-In-Kettle® patented design). Methanol synthesis was demonstrated using a stainless steel CHER.

351633 Combination of Photocatalytic and Photo-Fenton Oxidations Using Visible Light-Activated Photocatalyst for Decomposing Synthetic Chemical in Water

Abstract:

Heterogeneous photo-Fenton oxidation using CdS/multi-walled carbon nanotube-TiO_2 (CdS/MWCNT-TiO_2) under visible light relied on combination of the photocatalytic, photo-Fenton and photo-sensitizing oxidation. Heterogeneous photo-Fenton oxidation of methylene blue (MB), a model synthetic compound found in wastewater, was carried out in the CdS/MWCNT-TiO_2 aqueous suspension under visible light irradiation. The photo-Fenton reaction resulted in faster and higher removal of MB and the total organic carbon (TOC) (98% and 83 % in 30 min, respectively) than the dark Fenton oxidation and the photocatalytic degradation alone at the selected conditions. The optimum molar ratio of [MB]:[H_2O_2]:[Fe^{3+}] (1:12:3.4) in the photo-Fenton oxidation showed that the process would be a cost-effective solution to degrade aqueous methylene blue. The photocatalytic activity of the CdS/MWCNT-TiO_2 was maintained effectively even after several cyclic experiments. Besides, this study assessed the potential of using the CdS/MWCNT-TiO_2 to degrade more toxic emerging contaminants from aqueous solutions under visible light and solar light irradiation. Degradation of bisphenol A (BPA), an endocrine disruptor was evaluated using the CdS/CNT-TiO_2 under solar and visible light irradiation. Nearly 95% BPA with initial concentration of 20 ppm in a solution with pH 3, and 0.5 g/L of catalyst was removed after solar and visible light irradiation in 5 min and 4 h, respectively. Further works will include identification of intermediate formation and possible mechanisms associated with the photo-Fenton oxidation of BPA, evaluation of cytotoxicity and estrogenic activity of BPA oxidation products and process modeling of the photo-reactor.

351104 Hybrid Separations Including Organic Solvent Nanofiltration: Process Intensification for the Purification of Iso-Butylbenzene

Abstract:

The coupling of different traditional and/or new and innovative separation techniques called hybrid separations leads to process intensification at the process level within the development of new processes or within retrofitting of existing processes. Hybrid separations integrate the advantages of the individual separations techniques resulting in decreased energy consumption, increased process efficiency and/or product quality. Especially, the integration of distillation for high capacity processing and membrane for energy-efficient and selective processing offers large benefits.

Organic solvent nanofiltration (OSN) is a pressure-driven separation method in which components are separated by differences in molecular sizes or weights or by different ability to dissolve into the membrane. Pervaporation (PV) separates components by their differences in relative volatility but is not restricted by the Vapor-Liquid-Equilibrium. The main advantages of both processes are the low energy consumption compared to traditional separation methods like distillation; possibilities to recycle the separated compounds or solvents; and flexibility due to their modular design. In the case of OSN no phase transition occurs and mild processing conditions are attained by low operating temperatures [1].

In this work the catalytic alkylation of toluene by propene was investigated regarding the purification of the products: iso-butylbenzene and n-butylbenzene. The main product of the reaction is iso-butylbenzene, an intermediate to produce ibuprofen [2], while the unreacted aromatic component is recycled using distillation. Since the yield of the investigated reaction is low (<20%) recycling of toluene by distillation has a high energy demand and can also lead to polymerisation of the products [3].The conventional product purification is achieved by two distillation columns. In the first one, toluene is evaporated and recycled to the reactor. The bottom product of the first column is introduced to the second column where the separation of the two isomers is performed. Based on PV, OSN and distillation different process options for hybrid processing are generated. One of those consists of a membrane cascade for the pre-enrichment of the isomers, a distillation column for toluene recovery and recycling, and a second distillation column for the separation of butylbenzene isomers. Laboratory experiments were conducted for the membrane operations, e.g. with Puramem^{TM}280 nanofiltration membrane, at different temperatures and solute concentrations. Based on the experimental data membrane models were implemented in the simulation environment Aspen Custom Modeler^{TM}and used for the optimisation of the process options by an evolutionary optimisation method. The objective of the optimisation process was to minimise the production cost of iso-butylbenzene including investment and operational costs and to identify the best process option.

Results show that the pre-concentration of isomers by OSN significantly improves process economics by lowering the amount of toluene that has to be evaporated in the first distillation column.

References

[1] Vandezande, P., Gevers, L. E. M., Vankelecom, I.F. J., Solvent resistant nanofiltration: separating on a molecular level, Chemical Society Reviews, 2008, 37(2), 365-405.

[2] Sheldon, R. A., Organic synthesis past, present and future, Chemistry and Industry (London), (1992) 903-906.

[3] Stevens, M. G., Anderson, M. R., Foley, H. C., Side-chain alkylation of toluene with propene on caesium/nanoporous carbon catalysts, Chemical Communications, 1999, 413–414.

352041 Optimization of Floating Liquefied Natural Gas Process with Process Knowledge and Boxline Method

Abstract:

Optimization of Floating Liquefied Natural Gas (FLNG) process plant is presented. The developments in the offshore technologies and favorable market conditions led to the spot LNG market that drives the possibility of economic development of offshore stranded natural gas field and promoted the evolution of FLNG facilities in many countries. FLNG avoids the major infrastructure cost associated with onshore plants and offers a number of benefits such as elimination of the platform, pipelines and harbor jetty, and reduces the facility's environmental impact that leads to significantly reduced LNG production costs. The design criterion of FLNG is different from an onshore plant in terms of plant safety, storage, simplicity of operation and modular design. Considering the FLNG project requirements and criterion, an integrated scheme of NG liquefaction and NG liquids recovery is devised. The single mixed refrigerant NG liquefaction cycle is integrated with modified NG liquids recovery scheme. The preliminary design has several optimizing variables and to obtain the maximum benefit from the existing design, optimization was carried out. Refrigerant composition and operating pressure levels were the main variables in liquefaction plant whereas feed tray location, reflux rate, feed split rate and column number of trays are the main variables in NG liquids recovery plant. Overall plant specific energy minimization was considered as the optimization objective which in turn comprises the minimization of compression energy and reboiler duty in columns. To facilitate optimization, process knowledge of mixed refrigerant system is embedded within the search algorithm that helps in fast and robust convergence to practical solution. The optimization results in the decrease of about 70% compression energy and reboiler savings of 18% from the initial design. Furthermore the capex and opex are also calculated for the integrated process and conclusion was drawn that the annualized operating cost in FLNG plant is higher than the capital cost.

353114 Chemical Process Development - a Lean Six Sigma Perspective

Abstract:

Lean Six Sigma (LSS) is a powerful and flexible methodology that has been utilized by a range of companies for delivering significant cost and growth productivity via simultaneous reduction of waste and defects from internal processes. LSS principles have been successfully applied to a full range of both batch and continuous processes in the chemical industry.

This presentation will provide an overview of the application of both the Lean and Six Sigma Methodologies within Momentive, a global manufacturer of specialty chemicals and materials. Several case studies will be shared on how this world class framework has enhanced efforts within our Global Process Development organization.

Using Six Sigma and Experimental Design in Process Development

Mark V. M. Emonds
The Dow Chemical Company

14 AIChE
Spring Meeting, New Orleans, LA

DOW

Six Sigma in Process Development

- Six Sigma is frequently used in manufacturing to reduce defects & improve processes
- Why use Six Sigma in process development?
  - Optimize reaction conditions for yield or purity
  - Understand expected variation
  - Produce a more robust process
  - Design a process solution that fits the project needs

14 AIChE
Spring Meeting, New Orleans, LA

DOW

Design of Experiments

- Reduce the overall experimental effort
  - More experiments initially, but less overall
- Get to solution quickly, with minimal effort
- Develop an empirical model of the process
- Understand which factors have an impact on process
- Use the model to predict optimal conditions

14 AIChE Spring Meeting, New Orleans, LA

DOW®

Type of Model

- Default model is polynomial

  $Y = \beta_0 + \beta_1 x_1 + \beta_2 x_2 + \beta_{12} x_1 x_2 + \beta_{11} x_1^2 + \beta_{22} x_2^2$
- If model form is known based on scientific principles, can incorporate nonlinear relations into experimental design

14 AIChE Spring Meeting, New Orleans, LA

DOW®

Keys for Success

- Capable measurement system
  - You cannot study what you cannot measure
- Existing process is repeatable
  - DOE will provide measure of variation, but...
  - If too much variation, any signal lost in the noise
- Acceptable ranges of factors are understood
  - Importance of screening experiments
  - Understand plant/process limitations

14AIChE Spring Meeting, New Orleans, LA

DOW

Examples of Application

- Catalyst screening
  - Discovery of new catalysts for new processes
- Process optimization
  - Screen new catalysts for potential improvement
  - Determine optimum reaction conditions
  - Develop empirical model of yield
- Design new solvent system for intermediate
  - Mixture design

14AIChE Spring Meeting, New Orleans, LA

DOW

Catalyst Screening

- Use of high throughput screening to search for new catalysts/ligands
  - Initial screening of many catalysts/ligands
  - Use a screening design to detect activity
  - Depending on chemistry, may need to screen under several conditions
    - Kinetics of catalyst A may be too slow at low temp
    - Catalyst B may decompose at higher temperature

14 AIChE Spring Meeting, New Orleans, LA

DOW®

Catalyst Screening

- Most promising catalysts or ligands optimized
  - Second round of high throughput screening
  - Factors: activator, concentration, cosolvent
  - Optimize for yield and activity (reaction rate)
- Subsequently validate model in larger scale equipment
  - Verify optimal conditions

14 AIChE Spring Meeting, New Orleans, LA

DOW®

High Throughput Screening

- Easy to do large scale experimental designs
- Not always easy or possible to randomize all factors
  - Gas pressure, gas ratios, temperature
- Understand consequences of split-plot designs

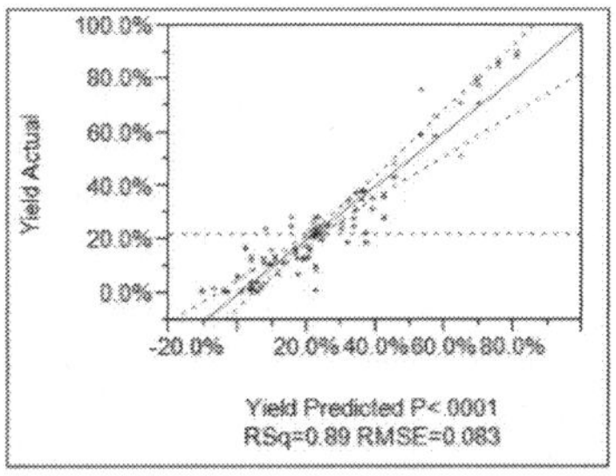

14 AIChE
Spring Meeting, New Orleans, LA

DOW

Process Optimization

- Engineering & Process Science group provides support for Amines business
- Understand how new catalysts would perform in transamination reactors
  - Seeking better product selectivity
  - Seeking better productivity
  - Potential for new products

14 AIChE
Spring Meeting, New Orleans, LA

DOW

Product Selectivity

- Complex mixture of possible products

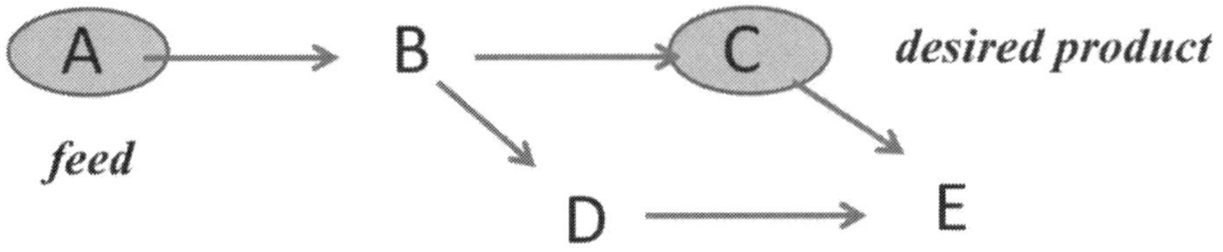

- Want to maximize selectivity and productivity

Comparing New Catalysts

- New catalysts are evaluated in continuous reactor packed bed column
- Are these really different?

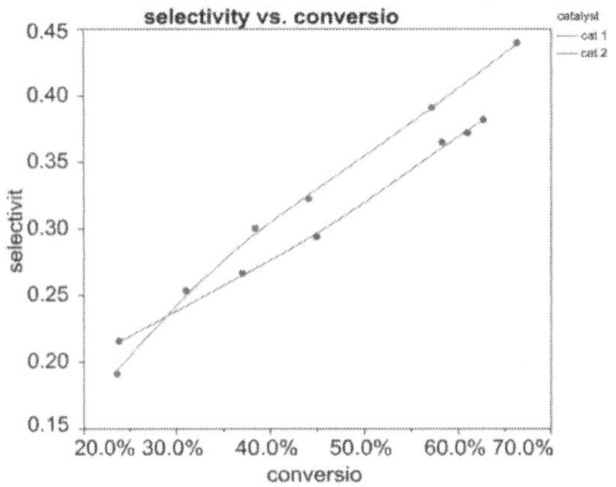

Statistical Analysis

- Fit both runs separately
  - Both fits with $R^2 > 0.99$
- Intercepts of both are same within error
- Slopes are not: outside 95% probability

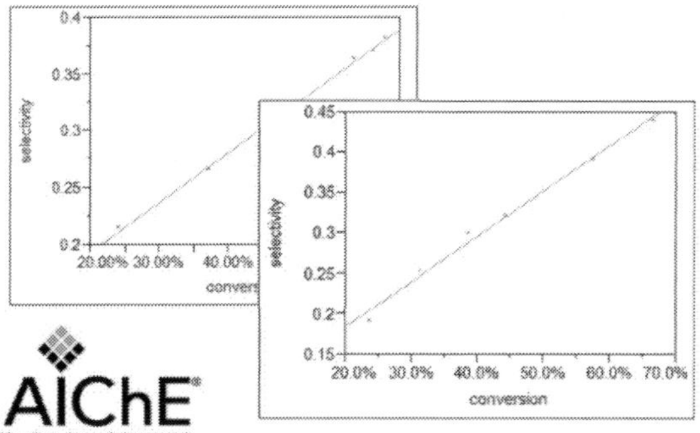

run	cat 1	cat 2
lower 95% interval	0.0848	0.0371
intercept	0.1067	0.0727
upper 95% interval	0.1286	0.1082
lower 95% interval	0.3915	0.4822
slope	0.4353	0.5597
upper 95% interval	0.4791	0.6372

14AIChE Spring Meeting, New Orleans, LA

DOW®

Process Optimization

- After catalyst is selected, further experiments define optimal reaction conditions
  - Feed rates
  - Temperatures
  - Pressures
- Result: a robust model that predicts how the catalyst will perform in the production plant

14AIChE Spring Meeting, New Orleans, LA

DOW®

New Solvent System

- Intermediate for fine chemical synthesis used as a solution to enable stoichiometric control
- Incumbent solvent system not optimal
  - Desired higher concentration of intermediate
    - Current system run at ~ 25% intermediate in solution
  - Used solvent with regulatory issues
  - Solution must be stable and intermediate soluble over a range of temperatures
    - Unheated storage tanks

14 AIChE Spring Meeting, New Orleans, LA

Dow®

First DOE

- Optimal design mixture DOE
  - Test solvents existing in process (Solvent A, Solvent B)
  - Determine effect of water levels in solvent
  - Determine if level of stabilizer affects solubility
  - Use Custom Design platform in JMP
  - 19 experiments, including three center points
  - Measured at two temperatures

14 AIChE Spring Meeting, New Orleans, LA

Dow®

First DOE Result

- Mixture of Solvent A and Solvent B provide some advantage over single solvent at dissolving intermediate
  - Statistically significant non-linear blending effects
  - Solvent A*Solvent B: Effect Test Prob > F = 0.011
- Water improved maximum solubility
  - Solvent A*Solvent B*water: Prob > F = 0.037
- Stabilizer had no effect

14 AIChE Spring Meeting, New Orleans, LA

DOW®

Model from First DOE

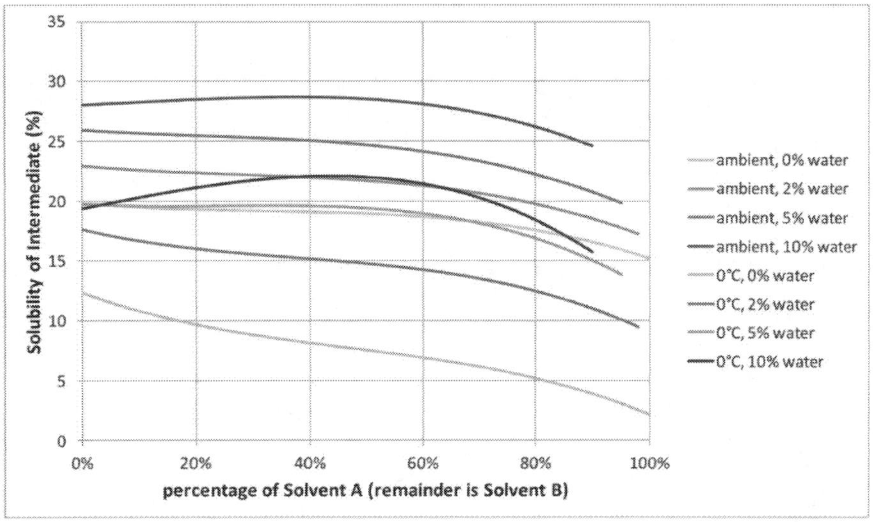

14 AIChE Spring Meeting, New Orleans, LA

DOW®

First DOE Result

- No combination of solvents reached goal
- Maximum modeled solubility at 0°C = 22%

14 AIChE Spring Meeting, New Orleans, LA

DOW

Second DOE

- After screening other solvents, decided to add third solvent to mixture DOE (Solvent C)
  - Solvent C: low cost, compatible with process
- Eliminated stabilizer from experiment
- Optimal design mixture DOE
- 17 experiments, including three center points
- Measured at two temperatures

14 AIChE Spring Meeting, New Orleans, LA

DOW

Second DOE Result

- Significant model
- $R^2 = 0.998$, adj $R^2 = 0.995$
- ANOVA prob > F: < 0.0001

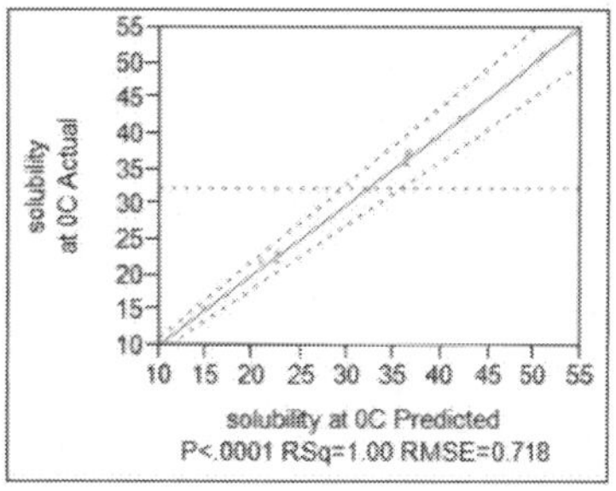

- Solubility modeled separately at each temp
- Statistically significant non-linear blending effects
  - Solvent A*Solvent C: Prob > F of < 0.0001
  - Solvent B*Solvent C: Prob > F of < 0.0001

14AIChE Spring Meeting, New Orleans, LA

DOW

Model from Second DOE

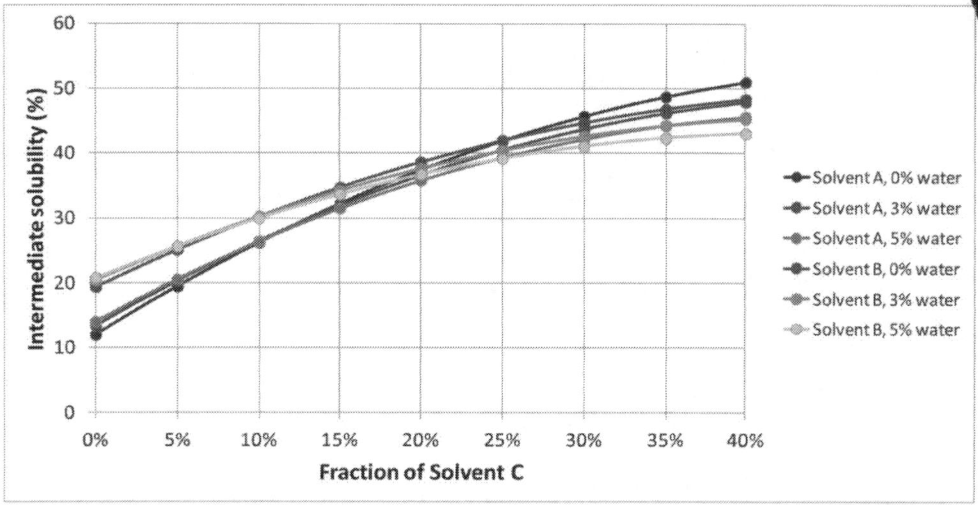

Temperature = 0°C model

14AIChE Spring Meeting, New Orleans, LA

DOW

Second DOE Result

- Either Solvent A or Solvent B with Solvent C can meet target concentration
- No advantage for three-way mixture
- Did not model with temperature as factor
  - Not required for solution to problem
  - Only collected data at 2 temperatures
  - Temperature statistically significant predictor of solubility, based on analysis of pooled data
  - Other non-linear blending effects

14 AIChE Spring Meeting, New Orleans, LA

DOW®

Solvent Optimization Results

- Met target of ≥ 40% intermediate conc.
- Solubility model validated
- Selected solutions tested down to -22°C
- Model provides framework for economic analysis
  - Cost of solvent C
  - Savings from reducing volume of solution
  - Elimination of solvent with regulatory issues

14 AIChE Spring Meeting, New Orleans, LA

DOW®

Overall Summary

- Use of Six Sigma can ensure robust, statistically validated solutions
- Optimization of process conditions
- Robust analysis of high throughput data
- Find solutions to complex problems
- Encourage use of Experimental Design in process development

14 AIChE
Spring Meeting, New Orleans, LA

DOW®

Lean Six Sigma Methods at Xerox Research Centre of Canada

Dr. Tom Enright
Manager, Process Engineering Area
Xerox Research Centre of Canada

xerox

Xerox Research Centre of Canada (RCC)

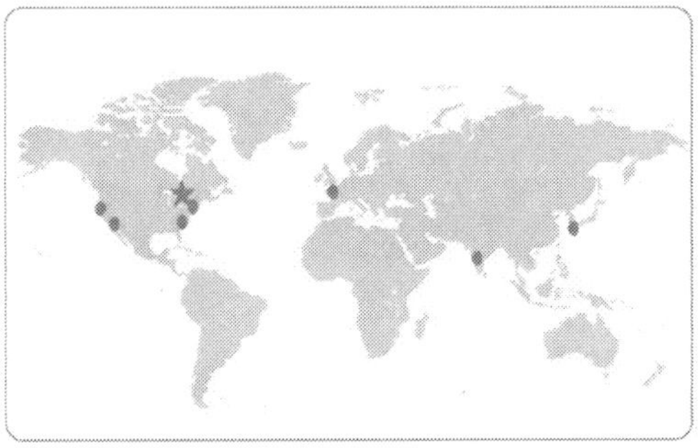

40 years Materials R&D

110 R&D + Support Staff

>150 Patents Annually

120,000 ft^2 Research Labs & Pilot Plant

Our Vision

Drive business growth by enabling commercialization of breakthrough materials, technologies, and services

2

xerox

Materials R and D at Xerox RCC

Structured Materials Films/Coatings

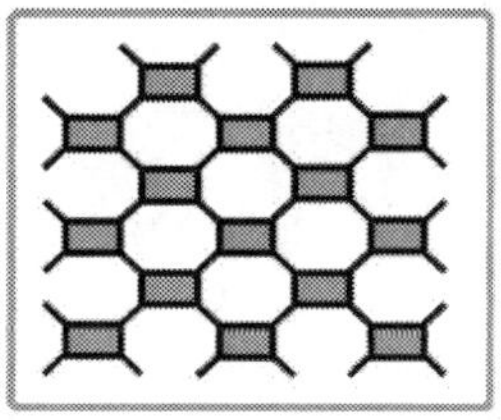

Sustainable Resins

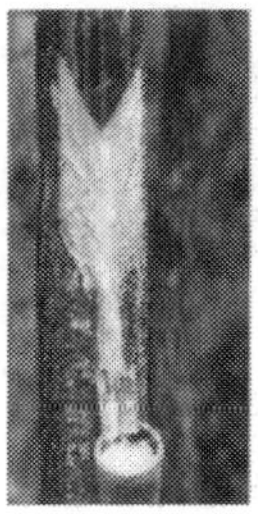

Digital Manufacturing & Electronic Materials

Composites & Nanomaterials

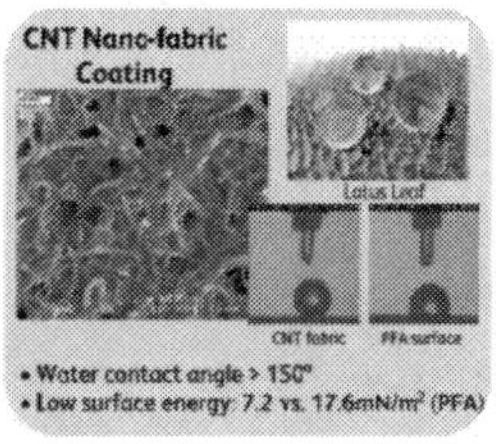

Specialty Materials

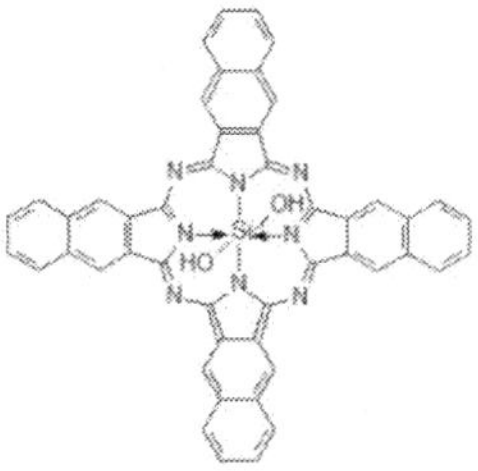

Engineering & Scale-up

From Concept to Commercialization
Xerox RCC Scale-up Engineering and Pilot Plant

I. Process Chemistry

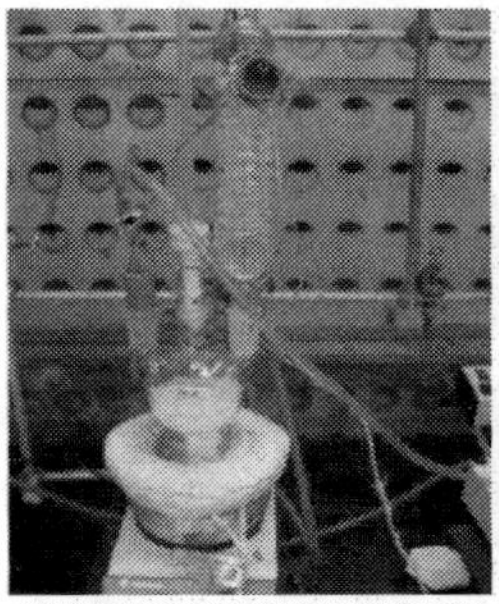

New Materials Design and Synthesis

- Process Chemistry
- QC and Characterization methods
- Performance test development and validation
- Preliminary safety assessment

II. Bench Engineering

Process Scale-Up

- Understand process fundamentals
- Identify Unit Operations
- Optimize process
- Demonstrate process reproducibility and repeatability
- Supply model material lots up to 1 kg
- Define process for piloting
- Deliver preliminary Process Operating Instructions (POI)

III. Pilot Plant

Demonstrate Process Manufacturability

- Equipment selection
- Materials handling/safety
- Process control
- Scaleability demonstration and further optimization
- Large scale sample supply up to tons
- Launch support
- Finalize POI for scale-up to manufacturing
- Technology transfer to manufacturing

IV. Manufacturing

Plant Design & Engineering

- Materials handling at scale
- Process control and automation at scale
- Cost and Quality Control
- Business Continuity assurance

Lean Six Sigma in Xerox Corporation

Lean +	**Six Sigma**
Speed + Low Cost	Culture + Quality
Lean **Speed** enables Six Sigma Quality (faster cycles of experimentation/ learning)	Six Sigma **Quality** enables Lean Speed (fewer defects means less time spent on rework)

Fast facts about Lean Six Sigma at Xerox

- 1,300 Black Belts trained or certified and active or back in operations
- 9,000 Green Belts trained or certified
- >70% Senior Executives Green Belt trained
- 50,000+ trained in QwikSolver™

For over 10 years we have employed a combination of Lean Six Sigma and Design for Lean Six Sigma methodologies, including:

- **LSS DMAIC** (Define, Measure, Analyze, Improve, Control): A standard approach for re-engineering existing processes
- **DfLSS DMEDI** (Define, Measure, Explore, Develop, Implement): Ideal for designing and optimizing new processes
- **DfLSS IDOV** (Identify, Develop, Optimize, Validate): Used, with specialized tools, for customer driven design of new technologies and services

6 April 8, 2014 xerox

LSS and DfLSS Through the Development Cycle

I. Process Chemistry	II. Bench Engineering	III. Pilot Plant	IV. Manufacturing
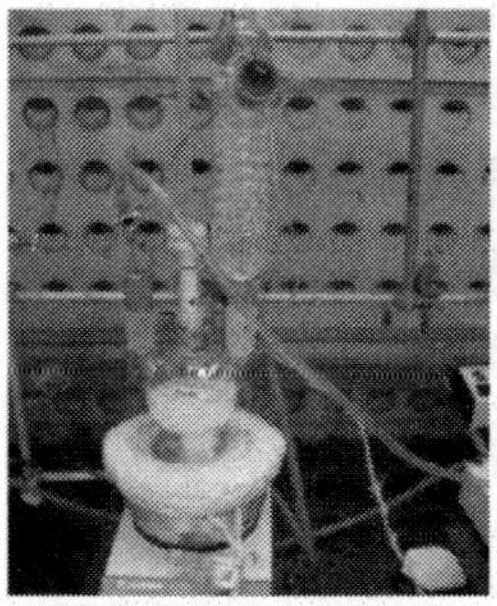			
New Materials Design and Synthesis	**Process Scale-Up**	**Demonstrate Process Manufacturability**	**Plant Design & Engineering**

DfLSS (IDOV) – New product/process design

DfLSS (DMEDI) – Design and optimize new processes

LSS (DMAIC) – Re-engineer existing processes

xerox

7 April 8, 2014

LSS and DFLSS at Xerox RCC

Design for Lean Six Sigma (DFLSS) / Lean Six Sigma (LSS)

IDOV Steps to design a new product or service

DMEDI Steps to design a new product or process

DMAIC Steps to improve an existing process

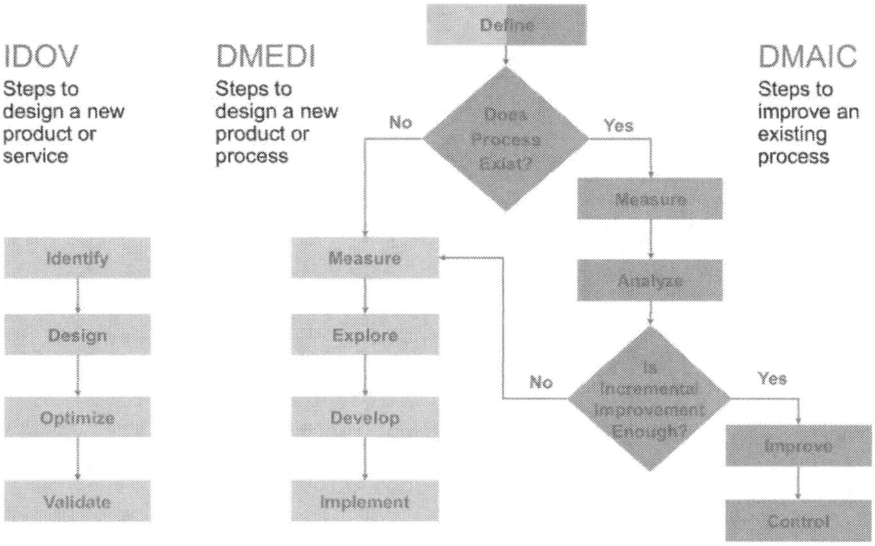

xerox

8 April 8, 2014

Basic Phase Activities

Design for Lean Six Sigma (DFLSS)

Identify		Design	Optimize	Validate
Define	**Measure**	**Explore**	**Develop**	**Implement**
• Define the opportunity from both business and customer perspectives	• Translate Voice of the Customer into Critical Customer Requirements	• Develop design concepts and high-level design	• Develop and optimize detailed design	• Validate design with pilot, develop control plan, implement

Lean Six Sigma (LSS)

Define	Measure	Analyze	Improve	Control
• Establish problem or opportunity statement • Establish success criteria	• Establish current state, identify where a process can be streamlined	• Identify root causes • Identify time intensive activities of a process • Identify how to eliminate nonvalue-added steps	• Develop, plan solutions • Run pilot • Failure analysis	• Implement solution • Monitor success of solutions • Develop process controls and mistake proofing

xerox

LSS and DFLSS Roles at Xerox RCC

- Nine Black Belt certified/trained staff
  - Mixture of DMAIC, DMEDI, DfLSS
  - Forty hours online plus five weeks in-class training
  - Lead black belt projects and demonstrate tool usage
  - Specialized skill training (e.g., Shainin, Kaizen, TRIZ)
  - Ability to select appropriate project path, lead projects, coach project leaders
  - At least one Black Belt assigned to every Xerox RCC project
- Remaining technical staff are Green Belt certified (DfLSS)
  - 40 hours online plus one week in-class training
  - Studies and tool demonstration
  - Good understanding of common tools with some specialization
- Access to Xerox corporate Master Black Belt coaches for complex projects

xerox

DfLSS (IDOV) Roadmap

Identify
- Project Charter
- Strategic Plan
- Cross-Functional Team
- Voice of the Customer
- Benchmarking
- KANO's Model
- Questionnaires
- Focus Groups
- Interviews
- Internet Search
- Historical Data Analysis
- Design of Experiments
- Quality Function Deployment
- Pairwise Comparison
- Analytical Hierarchy Process
- Performance Scorecard
- Flow Charts
- FMEA
- Visualization

Develop
- Assign Specifications to CTC's
- Customer Interviews
- Formulate Design Concepts
- Pugh Concept Generation
- TRIZ or ASIT
- FMEA
- Fault Tree Analysis
- Brainstorming
- QFD
- Scorecard
- Transfer Function
- Design of Experiments
- Deterministic Simulators
- Discrete Event Simulation
- Confidence Intervals
- Hypothesis Testing
- MSA
- Computer Aided Design
- Computer Aided Engineering

Optimize
- Histogram
- Distributional Analysis
- Empirical Data Distribution
- Expected Value Analysis (EVA)
- Adding Noise to EVA
- Non-Normal Output Distributions
- Design of Experiments
- Multiple Response Optimization
- Robust Design Development
  - Using S-hat Model
  - Using Interaction Plots
  - Using Contour Plots
- Parameter Design
- Tolerance Allocation
- Design For Manufacturability and Assembly
- Mistake Proofing
- Product Capability Prediction
- Part, Process, and SW Scorecard
- Risk Assessment
- Reliability
- Multidisciplinary Design Optimization (MDO)

Validate
- Sensitivity Analysis
- Gap Analysis
- FMEA
- Fault Tree Analysis
- Control Plan
- PF/CE/CNX/SOP
- Run/Control Charts
- Mistake Proofing
- MSA
- Reaction Plan
- High Throughput Testing

xerox

DfLSS DMEDI Roadmap

Define
- Define the Opportunity
- Complete Charter
- Develop Project Plan
- Begin Risk Management Strategy
- Develop Communication Plan
- Finalize Project Focus

Measure
- Understand the Voice of the Customer
- Translate the VOC to CCRs
- Develop SIPOC Chart
- Develop Process Map
- Analyze Existing Data
- Implement Quick Wins
- Validate Business Opportunity
- Create a Scorecard

Explore
- Generate high-level concept ideas
- Select the concept that best meets VOC & VOB
- Begin high-level design elements
- Create QFD House #2
- Analyze existing data
- Proactively manage risk

Develop
- Develop detailed design
- Create detailed design elements
- Design for Lean
- Optimize design performance
- Confirm design capability

Implement
- Plan and execute a Pilot
- Develop Process Control Plan
- Finalize Documentation
- Implement Solution
- Transition Project to Process Owner
- Identify Project Replication Opportunities
- Calculate Financial Benefits

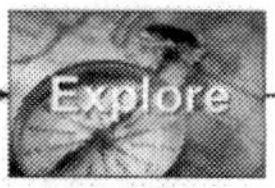

Define tools
- Project Selection Tools
- PIP Management Process
- Financial Analysis
- Project Charter
- Multi-Generation Plan
- Project Planning
- Gantt Chart
- RACI
- Risk Matrix
- Communication Plan

Measure tools
- Kano Model
- Interviews, Focus Groups, & Surveys
- VOC Plan
- Operational Definitions
- Affinity Diagram
- Structure Tree
- QFD House #1
- SIPOC, Top Down & Swim-Lane Process Maps
- Non-Value Added Analysis
- Basic Statistics/Histogram
- Pareto Chart
- Box Plot
- Kaizen Event
- Scorecard
- Target Costing

Explore tools
- Application Analysis
- Functional Analysis
- Creativity Tools
- Brainstorming
- Benchmarking
- Pugh Matrix
- Pairwise Comparisons
- Analytic Hierarchy Process
- QFD House #2
- Capability Analysis
- Statistical Tolerancing
- Queuing Theory
- Crystal Ball
- Hypothesis Testing
- FMEA
- ANOVA
- Scorecard

Develop tools
- Value-Added Analysis
- Mistake Proofing/Poka-Yoke
- 5S
- Line Balancing
- Work Control Systems
- Design of Experiments (DOE)
- Full and fractional factorial designs
- Main effects plot
- Interaction plot
- Conjoint Analysis
- FMEA
- To-Be Process Maps
- Risk Assessment Tools
- Scorecard

Implement tools
- Process Control Plan
- QFD House #3
- Control Charts
- Standard Operating Procedures
- Training Plan
- Communication Plan
- Implementation Plan
- Visual Process Control
- Process Control Plans
- Project Commissioning
- Scorecard
- Project Replication
- Financial Analysis Tools

xerox

LSS DMAIC Roadmap

	Define	Measure	Analyze	Improve	Control
Activities	• Identify Problem • Complete Charter • Develop SIPOC Map • Map Business Process • Map Value Stream • Gather Voice of the Customer & Voice of the Business • Develop CCR's & CBR's • Finalize Project Focus	• Identify Key Input, Process and Output Metrics • Develop Operational Definitions • Develop Data Collection Plan • Validate Measurement System • Collect Baseline Data • Determine Process Performance/Capability • Validate Business Opportunity	• Propose Critical X's • Prioritize Critical X's • Conduct Root Cause Analysis on Critical X's • Validate Critical X's • Estimate the Impact of Each X on Y • Quantify the Opportunity • Prioritize Root Causes	• Develop Potential Solutions • Develop Evaluation Criteria & Select Best Solutions • Evaluate Solution for Risk • Optimize Solution • Develop 'To-Be' Process Map(s) and High-Level Implementation Plan • Develop Pilot Plan & Pilot Solution	• Develop SOP's, Training Plan & Process Control System • Implement Process Changes and Controls • Monitor & Stabilize Process • Transition Project to Process Owner • Identify Project Replication Opportunities • Calculate Financial Benefits
Tools	• Pareto Charts • Project Selection Tools • PIP Management Process • Value Stream Map • Various Financial Analysis • Charter Form • Stakeholder Analysis • Communication Plan • SIPOC Map • High-Level Process Map • Non-Value Added Analysis • VOC and Kano Analysis • RACI and Quad Charts	• SIPOC Map • Operational Definitions • Data Collection Plan • Statistical Sampling • Measurement System Analysis (MSA), Gage R&R • Constraint Identification • Setup Reduction • Generic Pull • Kaizen • TPM • Control Charts • Process Capability, Cp & Cpk	• Pareto Charts • C&E Matrix • C&E/Fishbone Diagrams • Brainstorming • Detailed 'As-is' Process Maps • Basic Statistical Tools • SupplyChainAccelerator Analysis • Non Value-Added Analysis • Hypothesis Testing • FMEA • Box Plots • Interaction Plots • Simple & Multiple Regression • ANOVA	• Brainstorming • Benchmarking • Process Improvement Techniques • Line Balancing • Process Flow Improvement • Replenishment Pull • Purchasing and Sales Strategy • Poka-Yoke • FMEA • Solution Selection Matrix • 'To-Be' Process Maps • Piloting and Simulation	• Control Charts • Standard Operating Procedures (SOP's) • Training Plan • Communication Plan • Implementation Plan • Visual Process Control • Process Control Plans • Project Commissioning • Project Replication • Plan-Do-Check-Act Cycle

13 April 8, 2014 xerox

DMAIC Case Study – Silver Nanoparticle Inks

- Internal Xerox RCC research project for several years to develop silver nanoparticle inks for electronic applications
  - Designed around general material properties, not directed towards specific customer application
- Now selling inks to external clients for specialty applications
  - Customer requirements started to become more clear
  - Inks require adjustments to meet requirements for different applications
- Project path: DMAIC project to adjust material properties into desired ranges

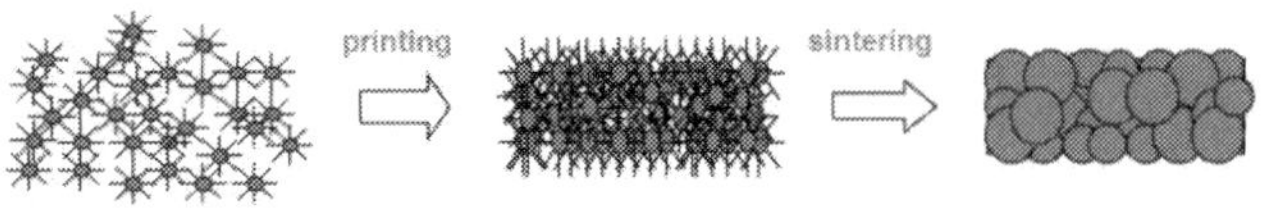

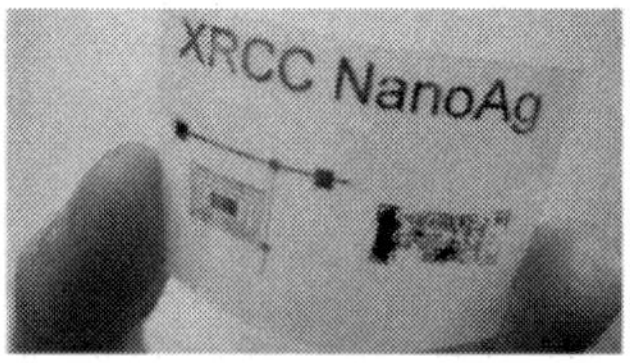

14 April 8, 2014 xerox

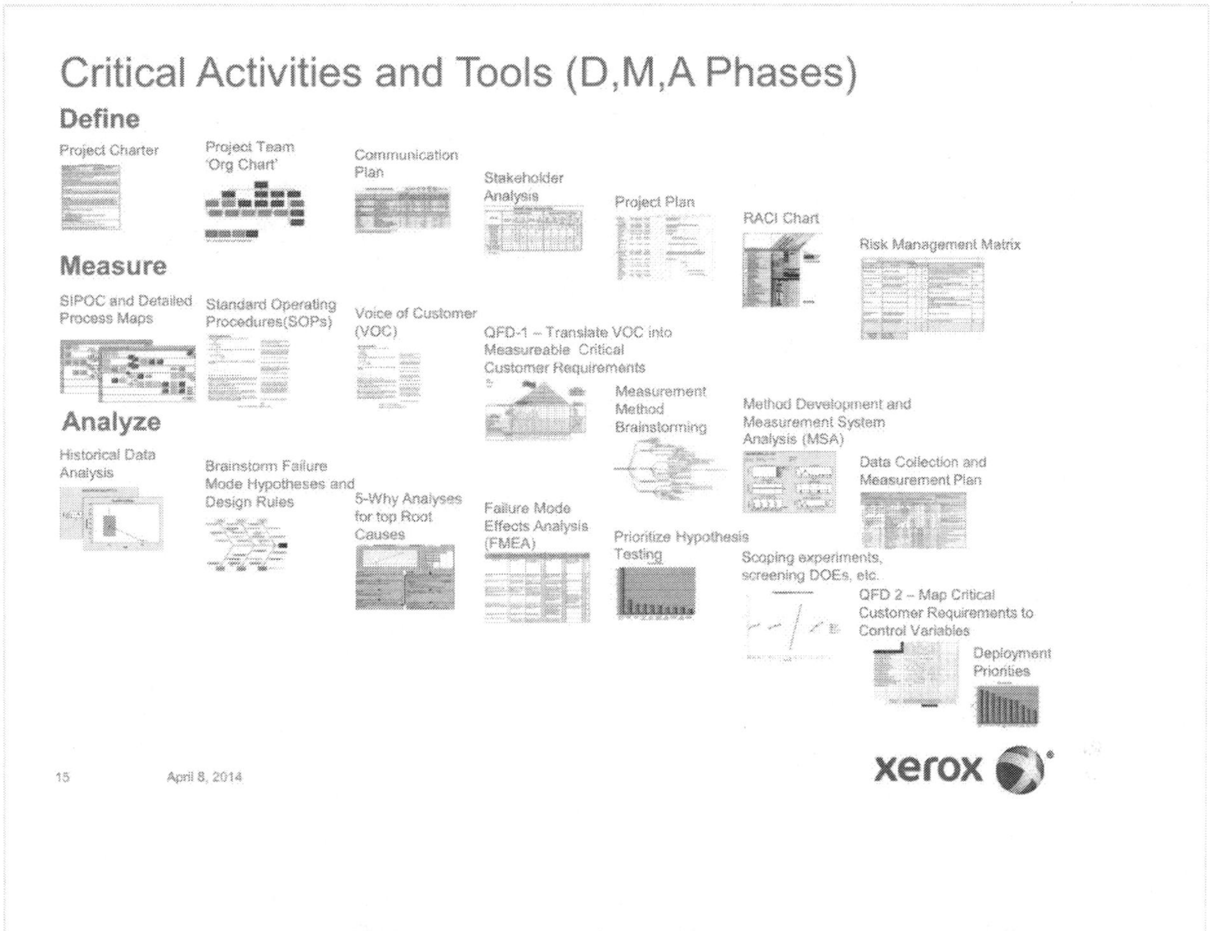
Critical Activities and Tools (D,M,A Phases)
Define
Project Charter
Project Team 'Org Chart'
Communication Plan
Stakeholder Analysis
Project Plan
RACI Chart
Risk Management Matrix
Measure
SIPOC and Detailed Process Maps
Standard Operating Procedures(SOPs)
Voice of Customer (VOC)
QFD-1 – Translate VOC into Measureable Critical Customer Requirements
Measurement Method Brainstorming
Method Development and Measurement System Analysis (MSA)
Data Collection and Measurement Plan
Analyze
Historical Data Analysis
Brainstorm Failure Mode Hypotheses and Design Rules
5-Why Analyses for top Root Causes
Failure Mode Effects Analysis (FMEA)
Prioritize Hypothesis Testing
Scoping experiments, screening DOEs, etc.
QFD 2 – Map Critical Customer Requirements to Control Variables
Deployment Priorities
15
April 8, 2014
xerox

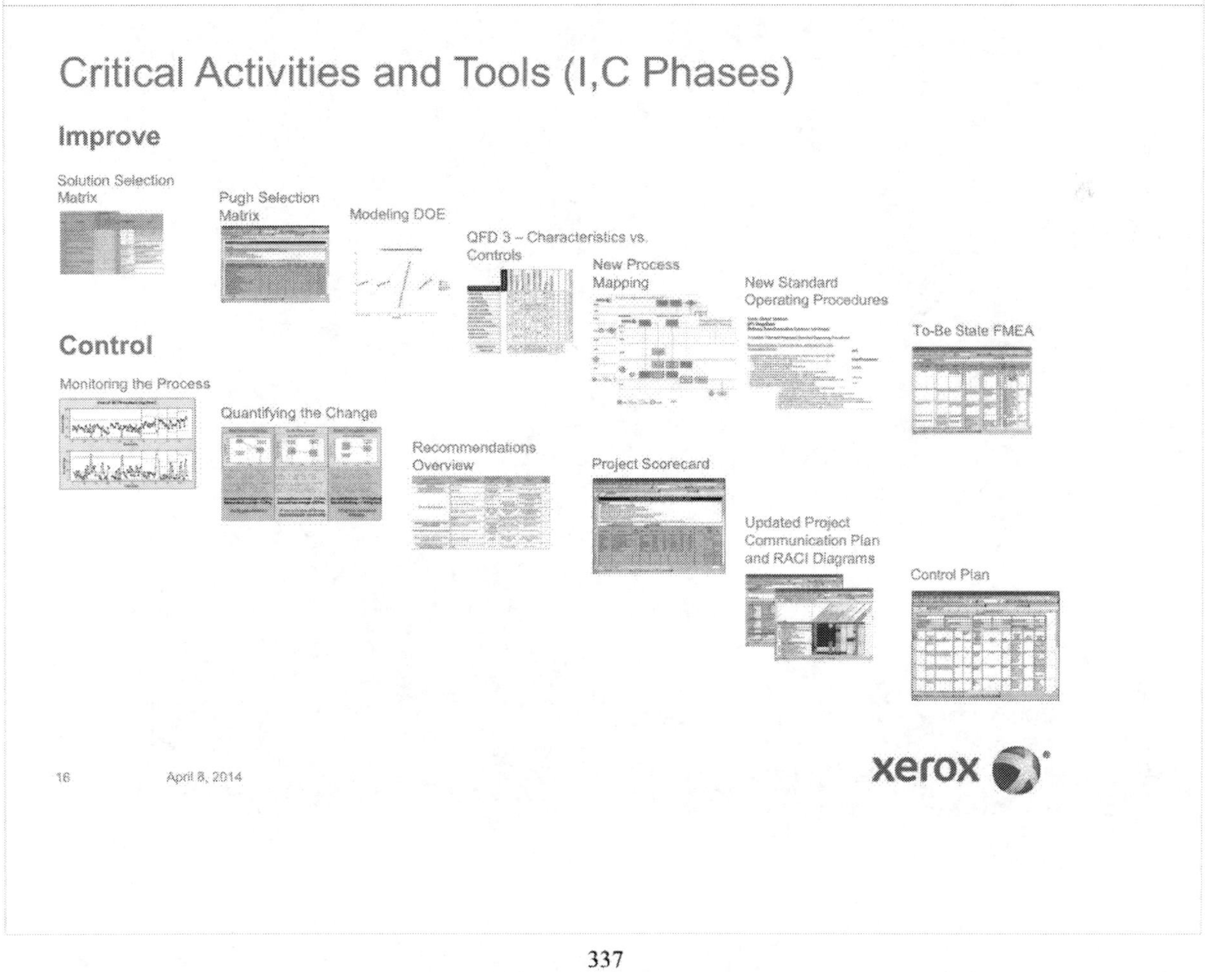
Critical Activities and Tools (I,C Phases)
Improve
Solution Selection Matrix
Pugh Selection Matrix
Modeling DOE
QFD 3 – Characteristics vs. Controls
New Process Mapping
New Standard Operating Procedures
To-Be State FMEA
Control
Monitoring the Process
Quantifying the Change
Recommendations Overview
Project Scorecard
Updated Project Communication Plan and RACI Diagrams
Control Plan
16
April 8, 2014
xerox

Summary

- Xerox RCC uses a flexible LSS/DFLSS methodology to manage projects through the development cycle:
  - IDOV for designing new technologies and processes
  - DMEDI for designing and optimizing new processes
  - DMAIC for improvement of existing processes or products
  - Series of other tools available for smaller activities
    - QuikSolver, Kaizen, etc.
- Rigorous front-end project focus with solid documentation and follow through ensures smoother project delivery cycle
- Proving to be especially effective for managing Client Services engagements

QuikSolver™

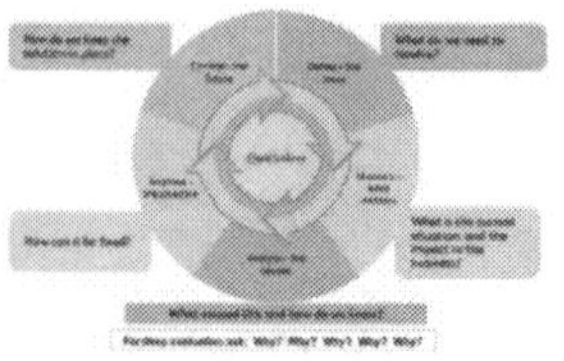

Project Delivery Cycle

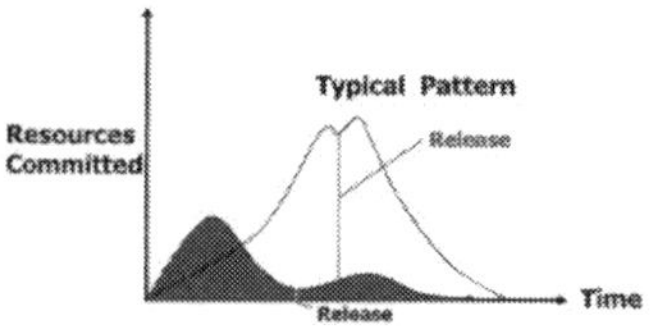

353102 Development of Continuous Improvement Network in Xerox Manufacturing

Abstract:

Xerox Corporation has five North American manufacturing sites that produce toner/developer materials for a wide variety of devices. For many years, each site worked independently and there was no formal mechanism for sharing best practices, successes, and failures. There were attempts over the years of developing some type of information sharing network, but a sustainable process never really took hold for a number of reasons.It was recently decided to make another attempt at developing a formal Continuous Improvement Network through a Lean Six Sigma Black Belt project with team members from each of the manufacturing units. This talk will give an overview of steps that were taken in the project to formulate a robust information sharing network that has checks and balances included to ensure that the network will last in the long term.

352183 Use of Six Sigma Tools in the Commercial Scale-up of a Novel PPE Copolymer

Abstract:

Poly(phenylene ether) is an engineering thermoplastic known for its excellent hydrolytic stability, dimensional stability and inherent flame retardancy. A block copolymer produced by the oxidative coupling polymerization of 2,6 xylenol and a comonomer, has been developed to deliver significant improvements in mechanical properties and flame retardancy.

Process development for new technology to culminate in production in an existing operating commercial chemical plant requires conscientious planning and risk mitigation efforts. This presentation illustrates the evolution of this commercial scale up effort as a Six Sigma project. Special emphasis is done on the application of Six Sigma tools and their impact on the planning of the project requirements, risk mitigation, and implementation analysis and control.

Achieving Environmental Compliance through Proper Destruction Efficiency of Low-Profile Flare Systems

J. D. Smith, Systems Analyses and Solutions, Inc.
Tulsa, Oklahoma

S. Smith, Zeeco, Inc.
Broken Arrow, Oklahoma

2014 Spring AIChE Meeting
New Orleans, Louisiana March 30 – April 3, 2014

SAS

ZEECO

Multipoint Flare History

Slide 2

- Originally Unstaged Candle Flares, Pipe Burners
- Staged Systems Developed Early 1970's
- Many Improvements over past 35 years in flare technology

SAS

Staging Allows Large Variation in Flowrates to Maintain Smokeless Operation

ZEECO

Slide 3

Technology for Reduction of Atmospheric Emissions in Petrochemical and Refining Industries 2014 Spring AIChE Meeting

SAS

Large Flare System Design Issues

ZEECO

Slide 4

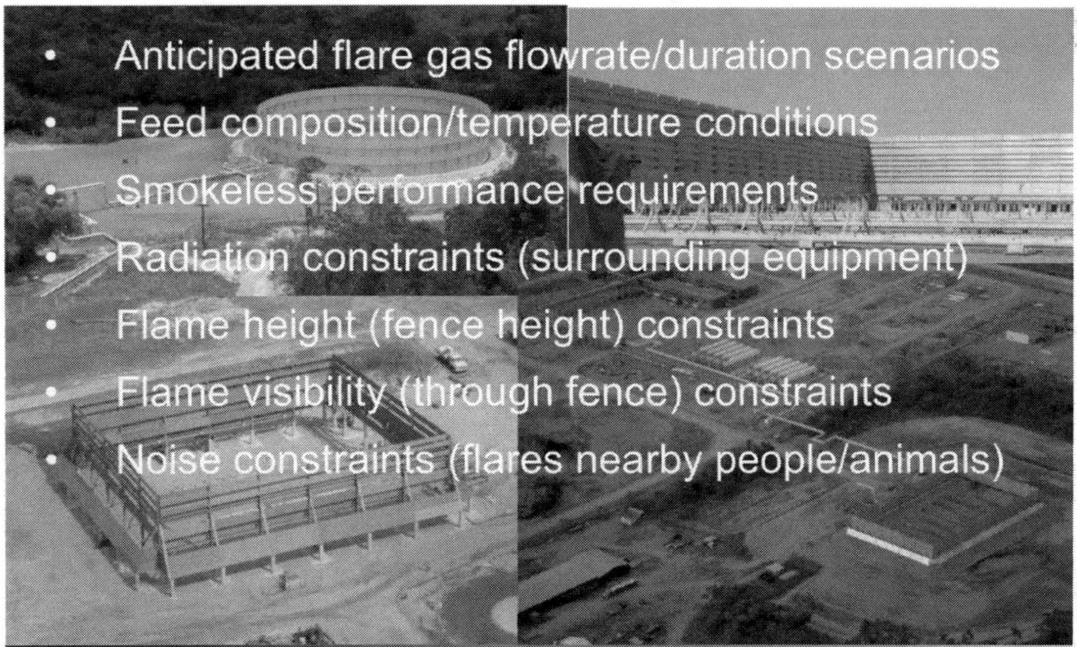

- Anticipated flare gas flowrate/duration scenarios
- Feed composition/temperature conditions
- Smokeless performance requirements
- Radiation constraints (surrounding equipment)
- Flame height (fence height) constraints
- Flame visibility (through fence) constraints
- Noise constraints (flares nearby people/animals)

Technology for Reduction of Atmospheric Emissions in Petrochemical and Refining Industries 2014 Spring AIChE Meeting

SAS

Burner Tip Design based on Multi-arm/ multi-jet configuration

ZEECO

Slide 5

- Variable Arm Area
- Investment Cast
- Pressure Tested at Factory
- 310 SS Cast Material

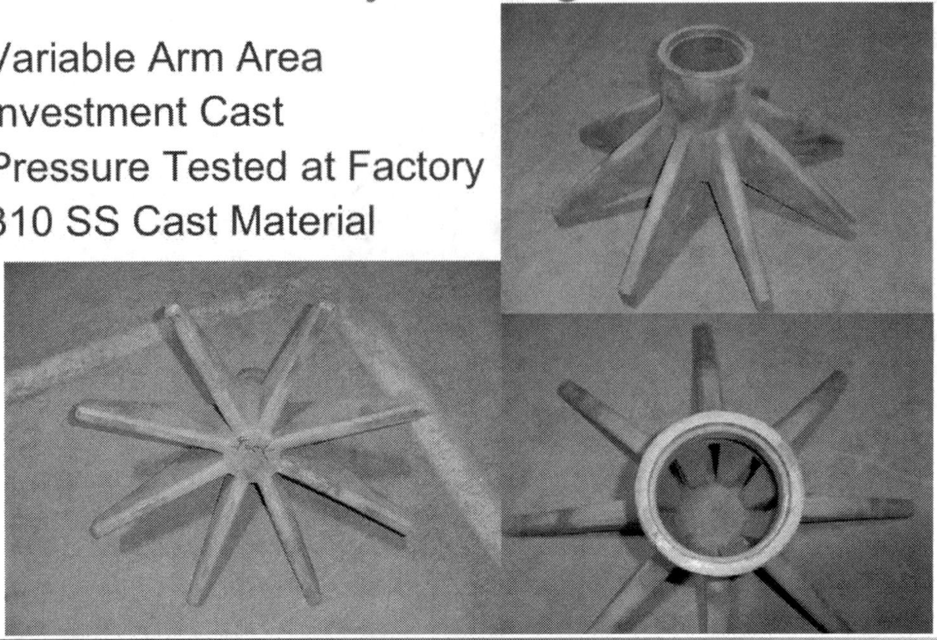

SAS

Flare Regulations and Performance Guidelines

ZEECO

Slide 6

- Flaring regulation (API guidelines) focus on utility flares and assisted flares (little attention to multi-tip ground flares)
- Flare Design/Performance guidelines established for general flares – ground flares have additional performance metrics to consider
- Flare performance difficult to quantify (large flame/high radiation flux, elevated operation)
- Multi-Tip ground flares have higher radiation inside large fenced region (larger plumes/difficult to measure emissions!)

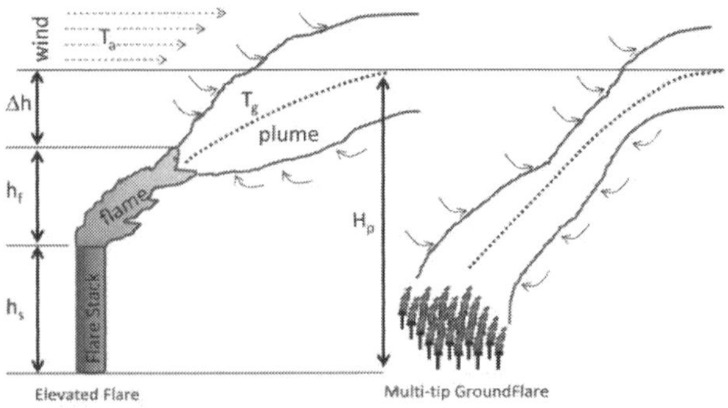

SAS

Existing Flare Regulations and Guidelines

Slide 7

- 40 CFR 60.18(b) and 63.11(b)
  - Pilot flame required
  - Sufficiently low exit velocity for stable combustion
  - Heating value high enough for flammable operation
  - Smokeless operation requires adequate mixing energy (air)
- 40 CFR 60.11(d), 60.18(d), 60.482-10(e), 63.6(e), 63.11(b)(1), 63.172(e)
  - Adhere to flare's design documents
  - Use Good Air Pollution Control Practice
- New Source Review

Federal requirements for flares are found in the New Source Performance Standards (NSPS) in § 60.18 and National Emission Standards for Hazardous Air Pollutants (NESHAP) in § 63.11. At a minimum, these rules require flares to be:

- Designed and operated with no visible emissions using EPA Method 22 (except for periods not to exceed 5 minutes in 2 hours);
- Operated with a flame present at all times, confirmed by the use of a thermocouple or equivalent device;
- Used only when the net heating value of the gas to be combusted is 300 BTU per standard cubic foot (BTU/scf) or greater (if the flare is steam- or air-assisted), or 200 BTU/scf or greater (if the flare is nonassisted); and
- Designed for and operated with an exit velocity less than 60 feet per second (ft/sec). An exit velocity of greater than 60 ft/sec but less than 400 ft/sec may be used if the net heating value of the gas being combusted is sufficiently high.

Enforcement Alert, "EPA Enforcement Targets Flaring Efficiency Violoations", EPA-325-F-012-002,August (2012).

SAS

Understanding Multi-Tip Ground Flares

Slide 8

- Understanding ground flare performance
  - Where does air come from?
  - How does wind effect flame?
  - How does tip geometry (pressure) affect flame?
- Reacting multi-phase flow in wind
  - Transient flame analysis
  - Near tip mixing
  - 100's of tips in large flare fields
- Model Validation Required
  - Calibrate Smoke and Radiation models

Tip Geometry

SAS

SAS's C3D Flare Model: Beyond Traditional Flare Analysis

ZEECO

Slide 9

- Extends standard "Steady-State" Flare Analysis:
  - Flame structure interaction with combustion chemistry and radiation flux
  - Fluid interactions with surrounding obstacles
  - Fuel specific soot formation and impact on radiation transfer
- Captures flame dynamics: Momentum vs. Buoyancy, Wind Effect on Flame Structure, Chemical Reactions, Radiation to/from flame
- Assess operating scenarios for specified conditions: wind speed, fuel flow/composition/temperature, impact on/by surroundings
- Key Assumptions:
  - Flame emissivity = F(flare gas composition, soot volume fraction, flame size/shape)
  - Optically thick flames (diffuse radiation) with time dependent view factors
  - Surroundings participate in radiation heat transfer (Re-radiation from ground)
- Overnight analysis on "standard" workstation, animations used to investigate transient flame behavior

Technology for Reduction of Atmospheric Emissions in Petrochemical and Refining Industries 2014 Spring AIChE Meeting

SAS

Multi-Tip Ground Flares: Modeling Issues

ZEECO

Slide 10

- High tip velocity increases air entrainment
  - Tip design critical to air entrainment
  - Local high velocity can translate into high sound levels
- Assist media not available to increase combustion air
  - smoke below certain tip pressure (D-stage pressure)
- Tip spacing critical
  - Flares must cross light
  - Possible Flame merge lengthens flames
  - Adjacent rows compete for air (longer flames, poor performance)

Technology for Reduction of Atmospheric Emissions in Petrochemical and Refining Industries 2014 Spring AIChE Meeting

SAS

Multipoint Ground Flare Modeling Approach

Slide 11

Single Burner Test/Model

- Detailed mesh
- Radiation/soot tests
- Single flame height/size

Multi-Burner Test/Model

- Radiation/soot tests
- Flame interactions (spacing)
- Multi-Flame shape/size

Model Full Flare Field

- Use Calibrated Models
- Predict Flare Performance (Smoke/Air Demand)
- Predict Radiation Flux

SAS

Transient Flare Analysis: Multi-step Reaction Mechanism

Slide 12

Incomplete Combustion of Flare Gas:

$$1\text{kg } \mathbf{F} + (2.87\text{-}2.6S_1)\text{ kg } \mathbf{O2} \rightarrow S_1\text{ kg } \mathbf{C} + (3.87\text{-}3.6S_1)\text{ kg } \mathbf{PC} + (50\text{-}32S_1)\text{ MJ}$$

Where: (F); Oxygen (O2); Soot (C); Products of Combustion (PC), S_1 soot parameter based on fuel type

Endothermic fuel pyrolysis (cracking reaction – soot producing):

$$1\text{kg } \mathbf{F} + 0.3\text{ MJ} \rightarrow S_2\text{ kg } \mathbf{C} + (1\text{-}S_2)\text{ kg } \mathbf{IS}$$

Where: Intermediate Species (IS), S_2 is cracking parameter based on fuel type

Soot combustion:

$$1\text{kg } \mathbf{C} + 2.6\text{ kg O2} \rightarrow 3.6\text{ kg } \mathbf{CO2} + 32\text{ MJ}$$

Where: Carbon Dioxide (CO2) is combustion product

Intermediate Species combustion:

$$1\text{ kg } IS + \frac{2.87-2.6S_2}{1-S_2}\text{kg } O_2 \rightarrow \frac{3.87-3.6S_2}{1\text{-}S_2}\text{kg } PC + \frac{50-32S_2}{1\text{-}S_2} MJ$$

Coefficients selected so complete combustion of carbon soot (C) and intermediate species (IS) produce same species and thermal energy as direct combustion of fuel; Coefficients in terms of mass weight (not moles)

SAS

Flare Burner Tests

Slide 13

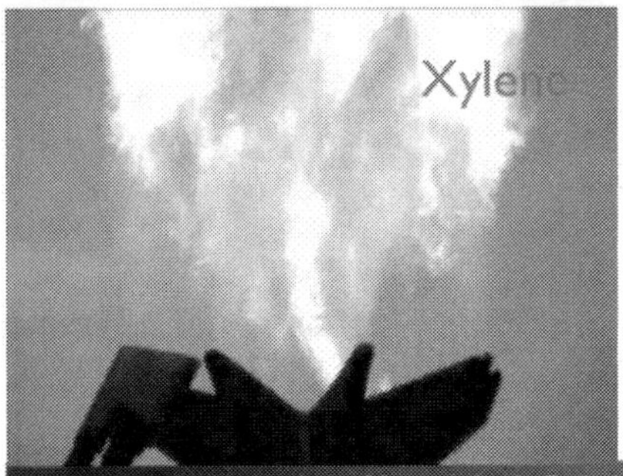

Technology for Reduction of Atmospheric Emissions in Petrochemical and Refining Industries 2014 Spring AIChE Meeting

SAS

Single Tip Burning Propane: wind vs. no-wind

Slide 14

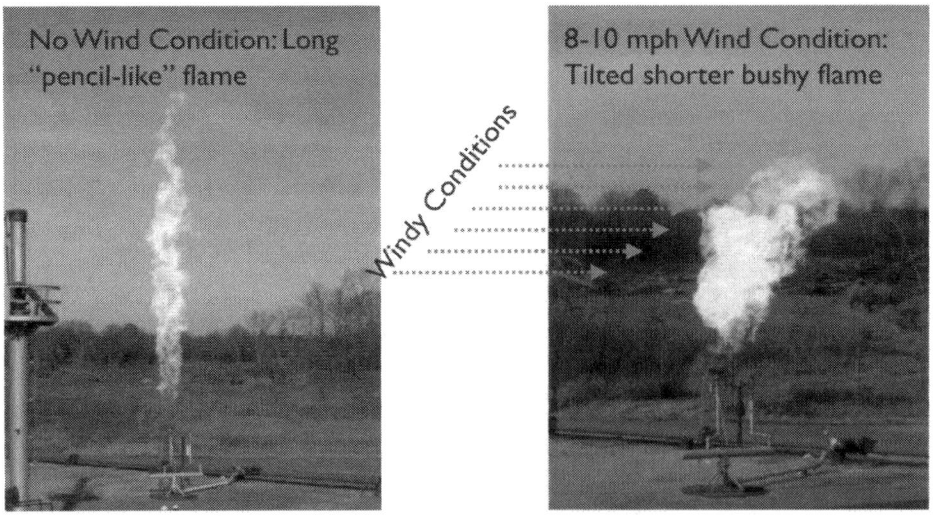

Technology for Reduction of Atmospheric Emissions in Petrochemical and Refining Industries 2014 Spring AIChE Meeting

Single Burner Flare Model

Slide 15

- 6 X 6 X 26 m physical domain
- Flare Tip located 2 m above ground level
- Turbulence and Arrhenius kinetics Included for fuel gas
  - Reaction Parameters adjusted to match observed flame characteristics
  - Soot Yield matched flame height (i.e., soot burnout)
- Flare Movies for no wind, 3m/s (7mph) wind conditions
- Predicted results for Air demand and Radiation loss from flame determined

Technology for Reduction of Atmospheric Emissions in Petrochemical and Refining Industries 2014 Spring AIChE Meeting

Combustion Models

Slide 16

Propane:

$C_3H_8 + 3.6\ O_2 \rightarrow 3\ CO_2 + 1.6\ H_2O + 0.024\ \text{Soot} + 46\ \text{MJ/kg propane}$

$\text{Soot} + 2.66\ O_2 \rightarrow 3.66\ CO_2 + 32\ \text{MJ/kg Soot}$

Ethylene:

$C_2H_4 + 0.57\ O_2 \rightarrow 0.93\ C_2H_2 + 0.64\ H_2O + 9.4\ \text{MJ/kg ethylene}$

$C_2H_2 + 2.58\ O_2 \rightarrow 2.7\ CO_2 + 0.7\ H_2O + 0.2\ \text{Soot} + 34.1\ \text{MJ/kg intermediate}$

$\text{Soot} + 2.66\ O_2 \rightarrow 3.66\ CO_2 + 32\text{MJ/kg Soot}$

Mixed Gases:

$$0.572\ C_2H_4 + 0.383\ C_2H_6 + 0.043\ H_2 + 0.982\ O_2 \rightarrow 0.53\ C_2H_2 + 0.34\ C_2H_3 + 1.1\ H_2O + 14.2\ \text{MJ/kg}$$

$$0.61\ C_2H_2 + 0.39\ C_2H_3 + 2.66\ O_2 \rightarrow 2.66\ CO_2 + 0.813\ H_2O + 0.181\ \text{Soot} + 34.4\ \text{MJ/kg}$$

$\text{Soot} + 2.66\ O_2 \rightarrow 3.66\ CO_2 + 32\ \text{MJ/kg}$

Technology for Reduction of Atmospheric Emissions in Petrochemical and Refining Industries 2014 Spring AIChE Meeting

SAS

Single-burner Mesh

ZEECO

Slide 17

- Rectangular cells
- Local refinement near burner tip
- 110,000 computational cells

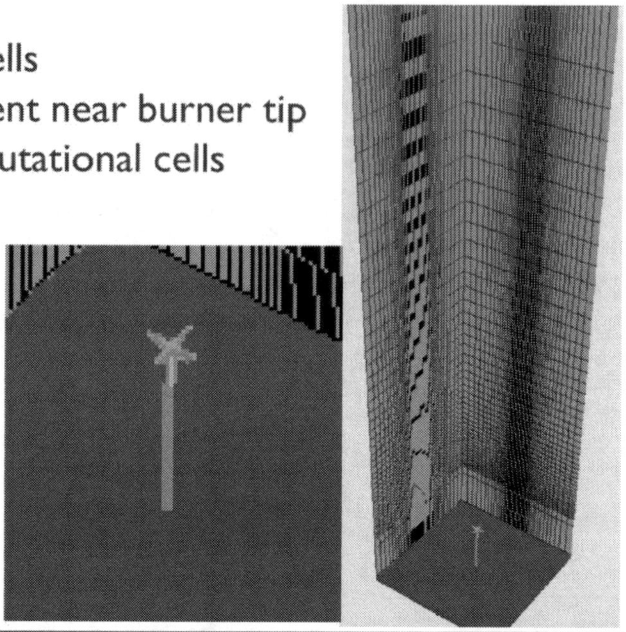

Technology for Reduction of Atmospheric Emissions in Petrochemical and Refining Industries 2014 Spring AIChE Meeting

SAS

Single Tip Burning Propane: 3.0 m/s Wind

ZEECO

Slide 18

Front View

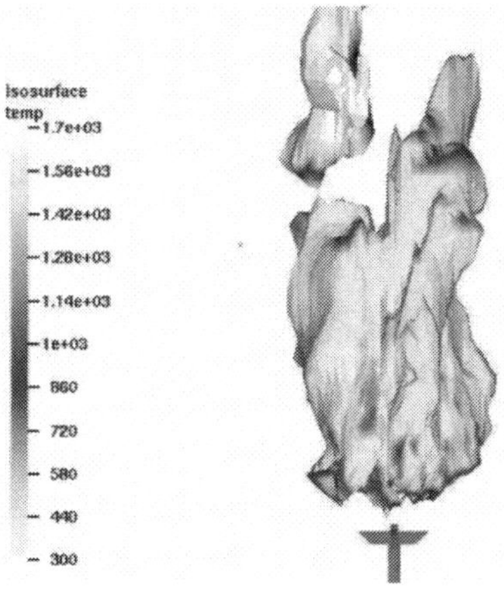

Side View

Wind produces tilted bushy, shortened flame

Technology for Reduction of Atmospheric Emissions in Petrochemical and Refining Industries 2014 Spring AIChE Meeting

SAS

Single Tip Burning Propane: wind vs. no-wind

ZEECO

Slide 19

Single Burner

No wind produces tight "pencil-like" flames

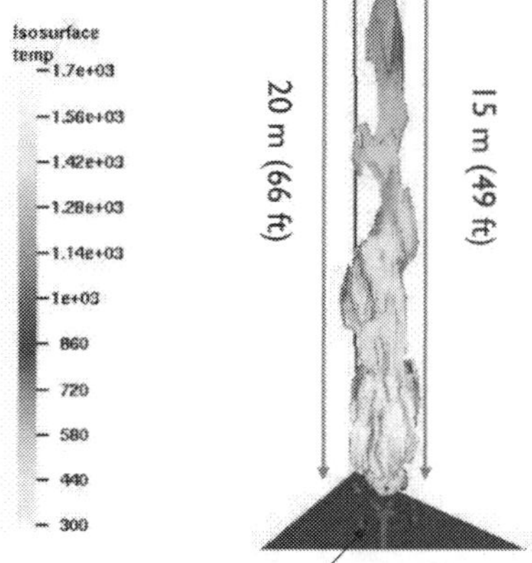

Flare tip located 2 m (~7 ft) above grade

SAS

Single-Tip Ground Flare Test Results

ZEECO

Slide 20

- 1.4" WC @ 57 °F => 7.3 psig tip pressure
- 12-16 Km/hr (8-10 mph) crosswind
- ~30% flame height reduction
- Minimum flame tilt (~8°)

Three-burner Mesh

Slide 21

- Rectangular cells
- Domain size is 30 m X 35 m X 25 m
- Local refinement near burner tips and radiation meters
- 188,000 computational cells

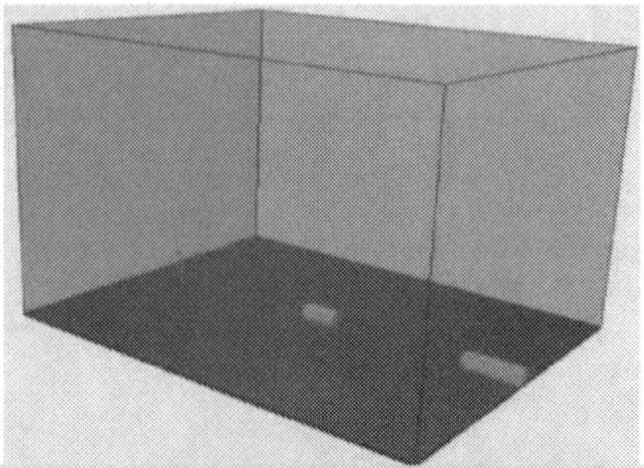

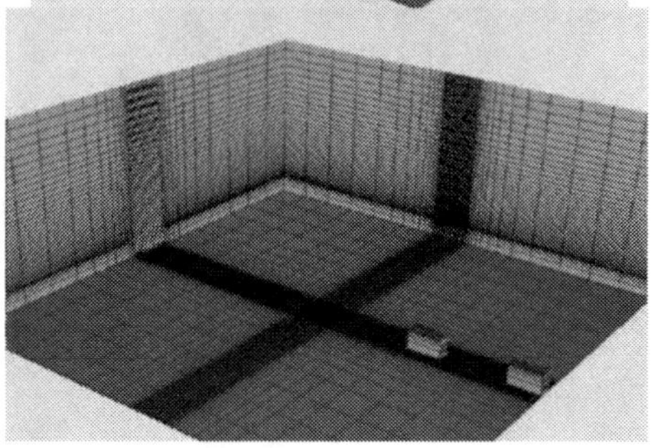

Technology for Reduction of Atmospheric Emissions in Petrochemical and Refining Industries 2014 Spring AIChE Meeting

Predicted/Measured Flame for 3-burner Ethylene test

Slide 22

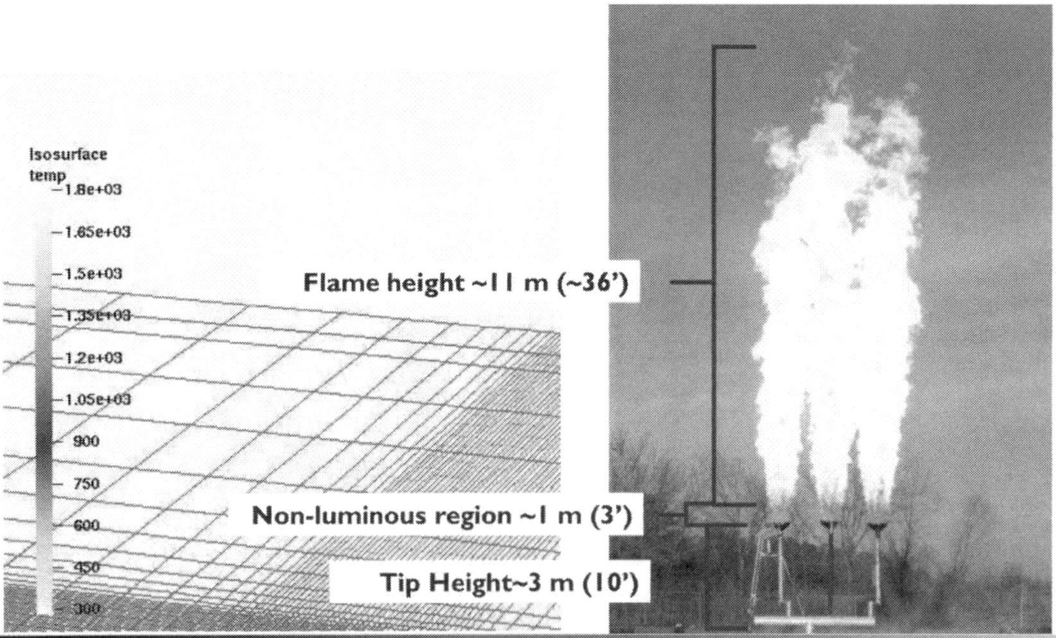

Technology for Reduction of Atmospheric Emissions in Petrochemical and Refining Industries 2014 Spring AIChE Meeting

SAS

Predicted 3-burner flare with radiation monitors

ZEECO

Slide 23

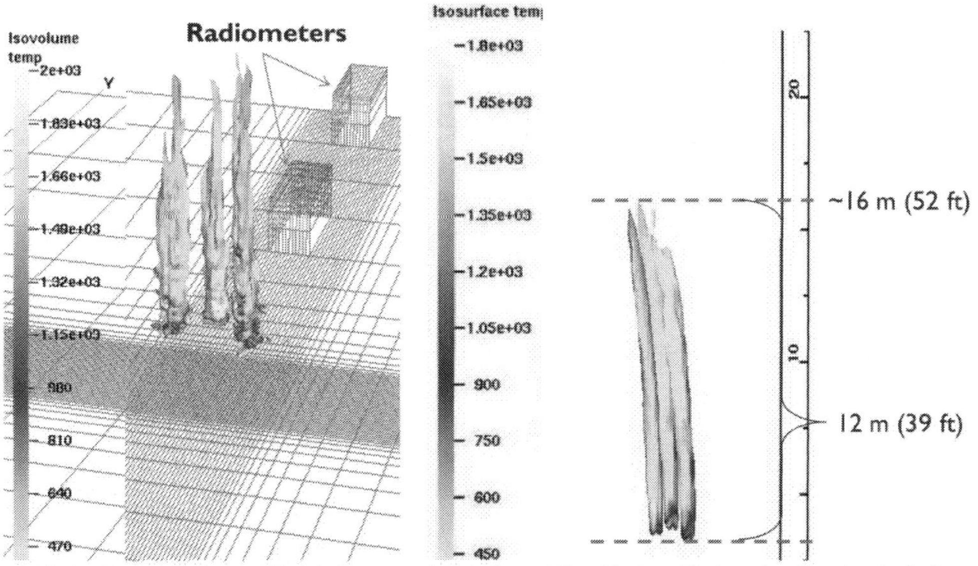

Smith, J.D., Suo-Ahttila, A., Smith, S., and Modi, J., "Estimation of the Air-Demand, Flame Height, and Radiation Load from Low-Profile Flare using ISIS-3D," *American – Japanese Flame Research Committees International Symposium, Advances in Combustion Technology: Improving the Environment and Energy Efficiency*, Marriott Waikoloa, Hawaii - Oct. 22 –24 (2007)

Technology for Reduction of Atmospheric Emissions in Petrochemical and Refining Industries 2014 Spring AIChE Meeting

SAS

Predicted Radiation from 3-Burner Flare with Modifications for Ground and Atmospheric Attenuation

ZEECO

Slide 24

Radiation Included in Prediction:

1. Ground re-radiating and reflecting incident radiation from flame
   - Assumed ground $\varepsilon = \alpha = 1$; allow ground to heat to steady state temperature
2. Atmospheric attenuation of radiation from flame
   - Model uses ambient/source temperatures with H_2O/ CO_2 absorption

WALL (plan view perspective)	Left Wall (W/m^2)	Right Wall (W/m^2)	Bottom Wall (W/m^2)	Flame Optical Thickness
PEAK FLOW				
Initial Radiation	78,000	63,000	108,000	0.275
Radiation Modification	61,000	35,000	NA	
SUSTAINED FLOW				
Initial Radiation	15,000	15,000	35,000	0.28
Radiation Modification	6,600	6,600	NA	

Technology for Reduction of Atmospheric Emissions in Petrochemical and Refining Industries 2014 Spring AIChE Meeting

SAS

Model Validation: Radiation Predictions Compared to Experimental Data

ZEECO

Slide 25

Tip Size	Position (m)	Burner Pressure (psi)	Total Predicted Radiation (W/m^2)	Measured Radiation (W/m^2)	Difference (%)
3	15	2.8	2700	3344	-20.0 %
3	15	7.3	4750	4803	-1.0 %
3	15	11.4	6150	6192	-0.7 %
3	30	2.8	650	671	-3.0 %
3	30	7.3	1350	1184	+14.0 %
3	30	11.4	1650	1532	+8.0 %
4	15	2.8	4325	6371	-32.0 %
4	15	7.3	8050	8192	-2.0 %
4	15	11.4	10000	9536	+5.0 %
4	30	2.8	1150	1513	-23.0 %
4	30	7.3	2580	2464	+5.0 %
4	30	11.4	3250	2747	+18.0 %

Smith, J.D., Suo-Ahttila, A., Smith, S., and Modi, J., "Estimation of the Air-Demand, Flame Height, and Radiation Load from Low-Profile Flare using ISIS-3D," *American – Japanese Flame Research Committees International Symposium, Advances in Combustion Technology: Improving the Environment and Energy Efficiency*, Marriott Waikoloa, Hawaii - Oct. 22 –24 (2007)

Technology for Reduction of Atmospheric Emissions in Petrochemical and Refining Industries 2014 Spring AIChE Meeting

SAS

Wind Effect on Radiative Flux

ZEECO

Slide 26

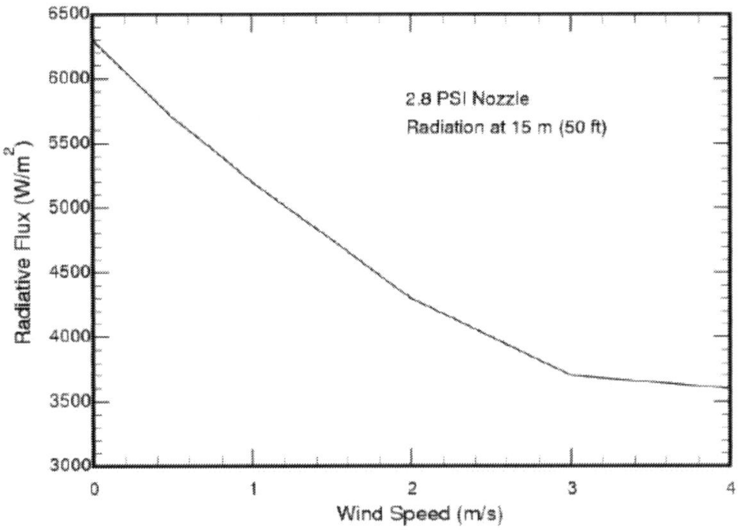

Smith, J.D., Suo-Ahttila, A., Smith, S., and Modi, J., "Estimation of the Air-Demand, Flame Height, and Radiation Load from Low-Profile Flare using ISIS-3D," American – Japanese Flame Research Committees International Symposium, Advances in Combustion Technology: Improving the Environment and Energy Efficiency, Marriott Waikoloa, Hawaii - Oct. 22 –24 (2007)

Technology for Reduction of Atmospheric Emissions in Petrochemical and Refining Industries 2014 Spring AIChE Meeting

SAS

Multipoint Ground Flare

Slide 27

Technology for Reduction of Atmospheric Emissions in Petrochemical and Refining Industries 2014 Spring AIChE Meeting

SAS

Full Field Flare Grid

Slide 28

- Domain size 10 m beyond wind fence; 25 m high
- Local refinement near burner rows/tips
- Sustained Flow Rate: 700,000 kg/hr
- Peak Flow Rate: 1,200,000 kg/hr

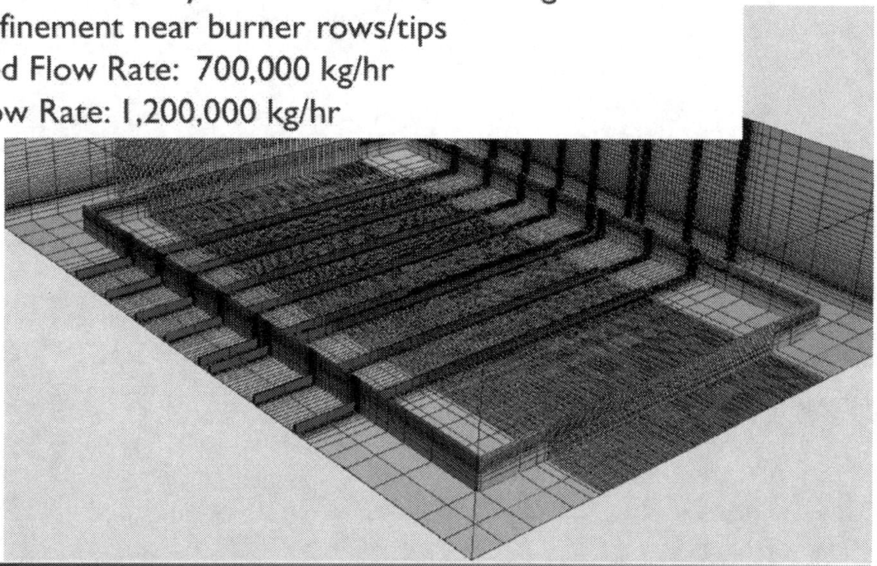

Technology for Reduction of Atmospheric Emissions in Petrochemical and Refining Industries 2014 Spring AIChE Meeting

SAS

Simulation of Flare Tests

ZEECO

Slide 29

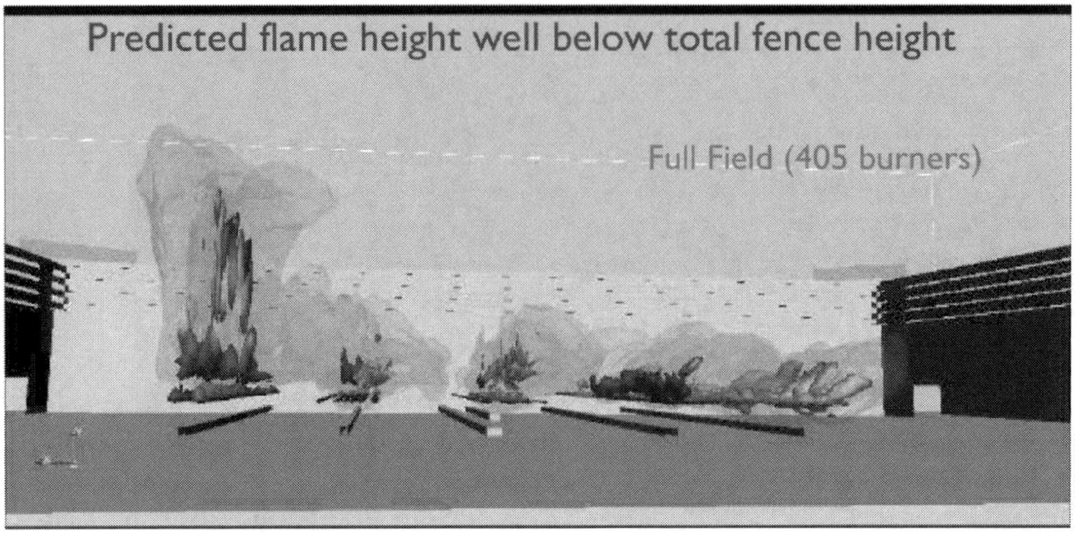

SAS

Max [C_2H_4] along line of sight for peak flow case

ZEECO

Slide 30

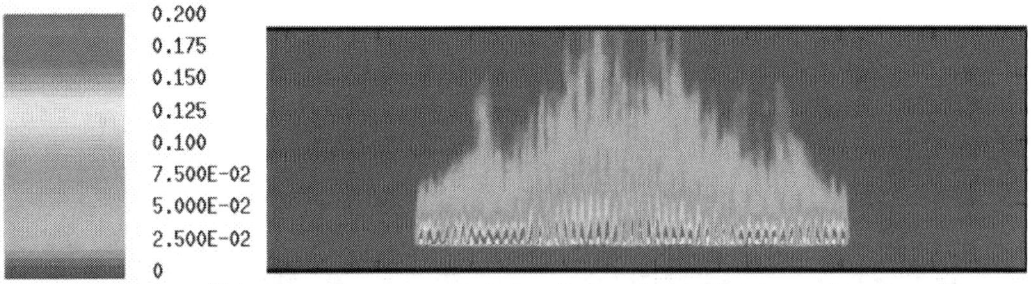

Side view along row showing flame elongation toward center of row

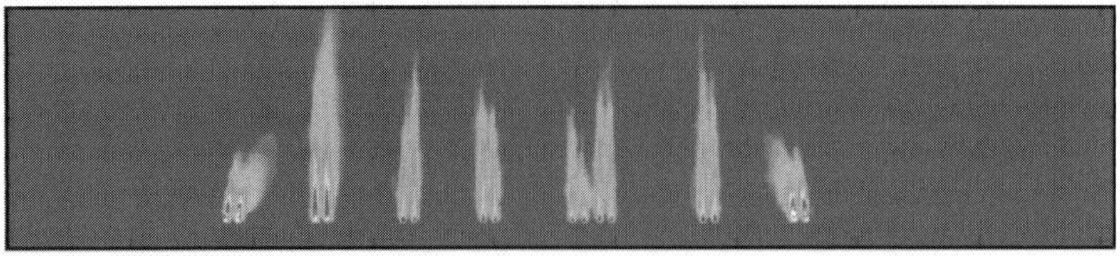

End view of flame height for each row and impact of inflow on outer rows

SAS

Predicted Air/Fuel Requirements for "no-wind" cases

ZEECO

Slide 31

Case Description	Evaluation Plane Area (m²)	Total Air Flow (kg/s)	Fuel Flow (kg/s)	Air/Fuel Mass Ratio
1 Burner Propane	14.63	60	0.46	130
1 Burner (Tip 3) Ethylene	13.26	52	0.94	55
3 Burner (Tip 3 - 7.3PSI) Ethylene	36	134	2.88	47
3 Burner (Tip 3 - 11.4PSI) Ethylene	36	144	3.84	38
3 Burner (Tip 4 - 7.3PSI) Ethylene	36	150	4.26	35
3 Burner (Tip 4 - 11.4PSI) Ethylene	36	160	5.79	28
Full Field Peak Flow Ethylene	3226	9700	262.3	37
Full Field Sustained Flow Mixed Gas	1843	4800	93.6	51.3

Technology for Reduction of Atmospheric Emissions in Petrochemical and Refining Industries 2014 Spring AIChE Meeting

SAS

Conclusions

ZEECO

Slide 32

- CFD Flare Models:
  - Single-burner model: 110,000 cells
  - Three-burner model: 188,000 cells
  - Full-field model: 700,000 cells (Sustained Flow)/1,200,000 cell (Peak Flow) (Combustion chemistry for propane, ethylene, mixed gas considered)
- Analyzed flame shape/size for 3 fuels for single and three burner tests
- Compared predictions to measurements for 12 tests (2 tip sizes, 3 operating pressures, 2 radiation sample locations)
- "*Reasonable*" estimates of radiation heat transfer and air demand for low profile flare
  - Air/fuel ratios range from 28 to 47 for 3-burner test and from 37 (Peak Flow Case) to 51 (Sustained Flow Case)
- Calibrated flare model applied to full-flare field to estimate:
  - Air demand for specified tip/row spacing
  - Radiation load on wind fence for nominal and peak flow cases
  - Expected flame height and smoke production for nominal and peak flow cases

Technology for Reduction of Atmospheric Emissions in Petrochemical and Refining Industries 2014 Spring AIChE Meeting

SAS

Questions?

Slide 33

353103 A Multi-Period Process Integration to the Management of Process Flares with Consideration of Fuel Substitution and Cogeneration Systems

Abstract:

A wide variety of process upsets can lead to flaring. Such flaring activities cause environmental emissions and result in economic losses. A possible approach to mitigating these emissions and losses is to utilize the flared gases in combustion systems (boilers and industrial furnaces) that can be used to produce utilities such as steam. Additionally, the generated steam can be further used in a cogeneration system (e.g., through steam turbines) to produce power and heating. This paper introduces a process integration approach to the effective utilization of flared gases into combustion and cogeneration systems. Since the flared gases differ in composition and characteristics from the fuel that is typically used in boilers, there is a need to make sure that the fuel quality is in the acceptable range for the boiler. Various constraints may be used to assess usability of flared gases as part of the fuel blend fed to the boiler. For instance, the Wobbe index (WI) may be used to study fuel properties sent to the boilers. A multi-period optimization formulation is developed to determine the optimal usage of flare gases in the boiler and cogeneration systems. An operational period is discretized into multiple periods corresponding to different operating modes of the process (process episodes). Economic objectives are used while including technical and environmental constraints. Additionally, the cost-benefit analysis for adding a cogeneration system with an optimal size is carried out. A case study using an ethylene plant is used to demonstrate the various aspects of the process.

PROOF OF CONCEPT TEST FOR A REAL-TIME FLARE COMBUSTION EFFICIENCY MONITOR

Presented at
2014 AIChE Spring Meeting &
10th Global Congress on Process Safety

By
Yousheng Zeng, PhD, PE and Jon Morris
Providence Photonics, LLC

Mark Dombrowski
Surface Optics Corporation

April 2, 2014

PROVIDENCE

A BRIEF REVIEW OF THE CONCEPT PRESENTED AT 2012 FALL CONFERENCE

Ref.: Zeng and Morris, "A New Method to Measure Flare Combustion Efficiency in Real-Time", presented at AWMA Louisiana Section 2012 Fall Conference, Baton Rouge, Louisiana, October 30-31, 2012

2

PROVIDENCE

Flare Combustion Efficiency (CE):

$$CE(\%) = \frac{[C]_{CO2}}{\sum_i n_i[C]_{HCi} + [C]_{CO2} + [C]_{CO}} \quad \text{Eq. (1)}$$

Flare CE – Very difficult to measure

Source: TCEQ/UT

THE CONCEPT

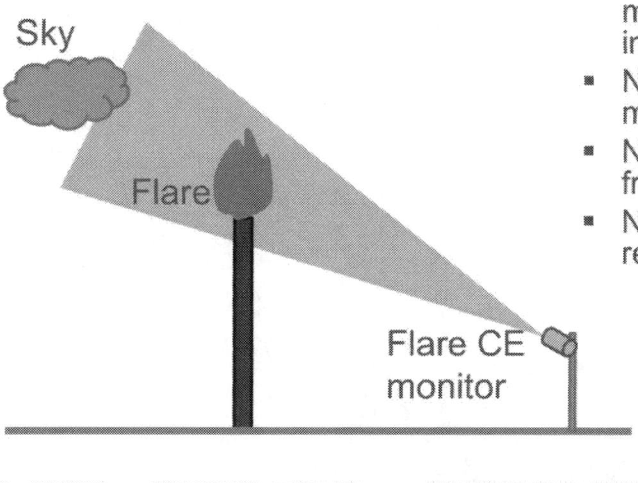

- Monitor and map flare CE in real time through a special multi-spectral IR imaging device
- Not a path measurement
- No scanning; high frame rate
- No operator required

We call it "Flare Efficiency Monitoring System" or "FEMS"

VISION

- Monitor flare CE in real-time
- Image the full flare flame; measure both overall CE and CE at a pixel level (CE mapping)
- Industrial grade device, suitable for integration with PLC or DCS
- One monitor covers multiple flares on site (step and stare)

WORKING PRINCIPLE

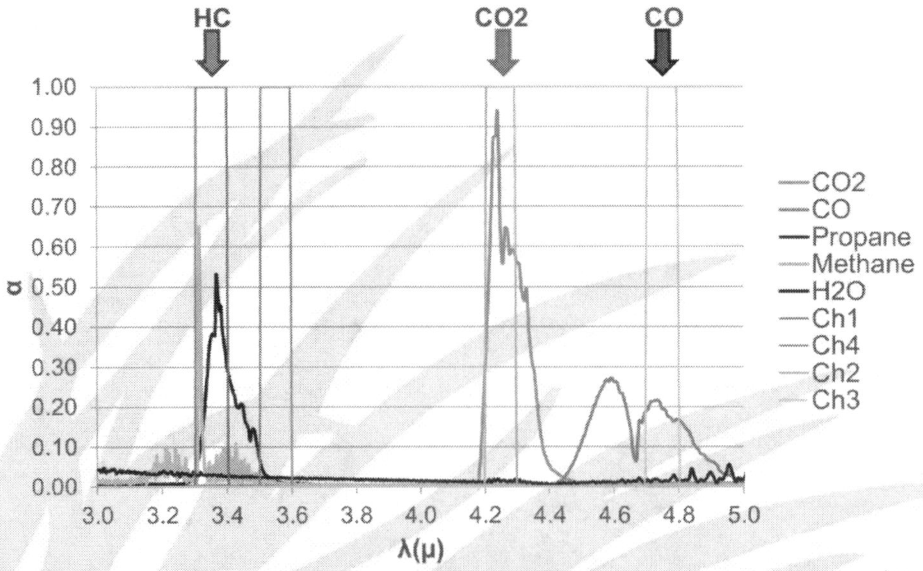

PROOF OF CONCEPT TEST

The proof of concept test was partially funded by EPA SBIR Phase I grant.

PROVIDENCE

TEST SETUP

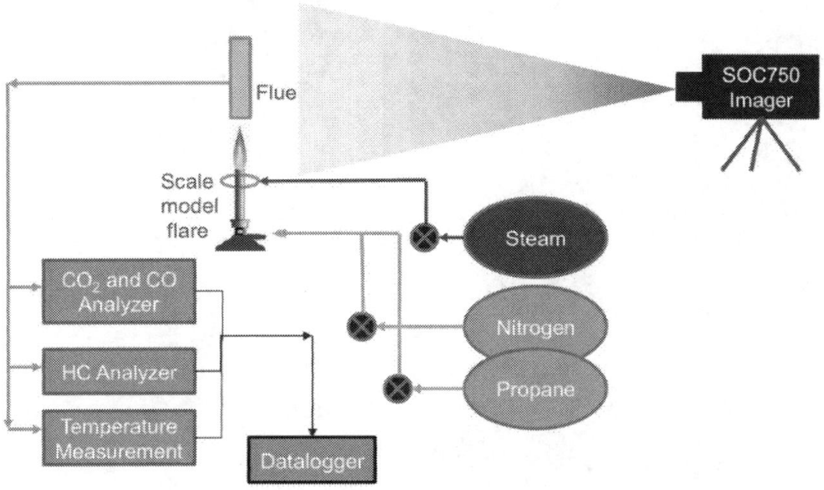

PROVIDENCE

Distance from the scale model flare to the Imager: 23 ft.

SOC750 Hyper-Spectral Imager

Scale model flare, Flue w/ Sampling Probe, and Analyzers

SCALE MODEL FLARE

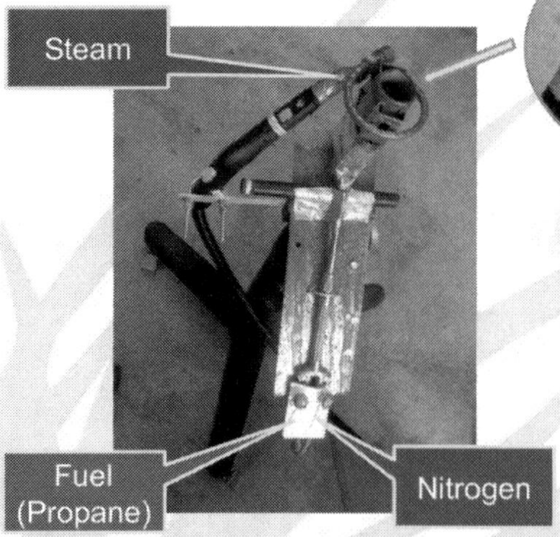

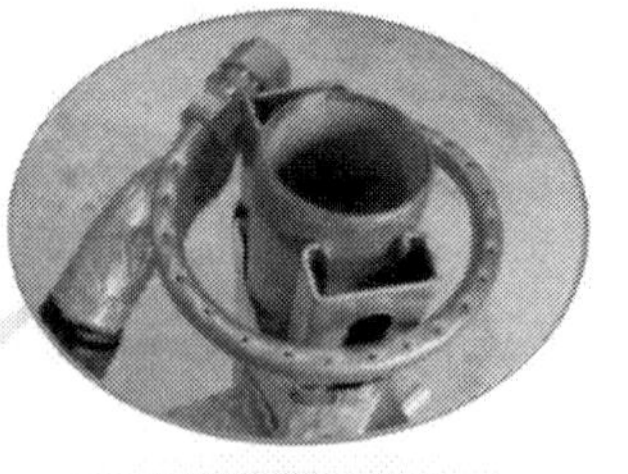

Fuel: approx. 40k Btu/hr.
Nitrogen: varied
Steam: ~ 3 lb./hr.

Flue and Sampling Probe

Analyzers and Imager

Testo 350 XL
(CO2 using NDIR, CO, O2, NOx, H2, Temp, etc.)

3010 Mini FID
Calibrated to Propane

SOC750 Hyper-spectral imager
42 spectral channels, operated at 22 cubes per sec.

TEST RUNS

PRELIMINARY RESULTS - WITHOUT CALIBRATION

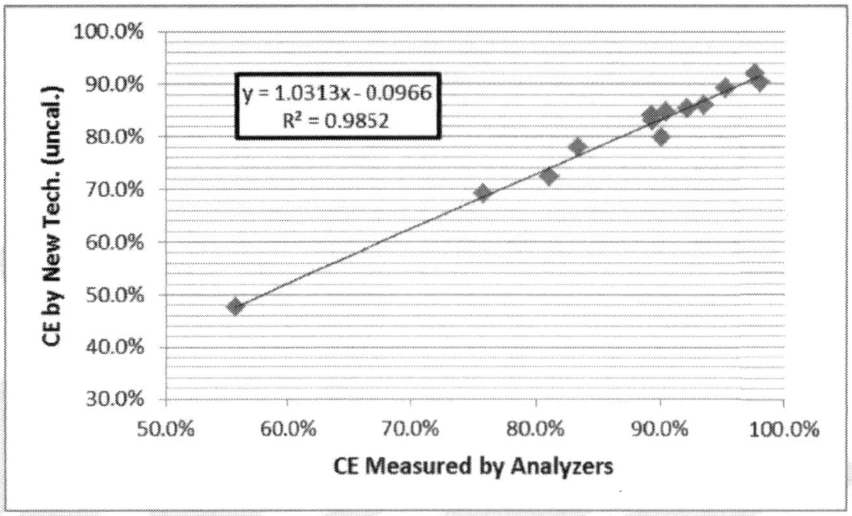

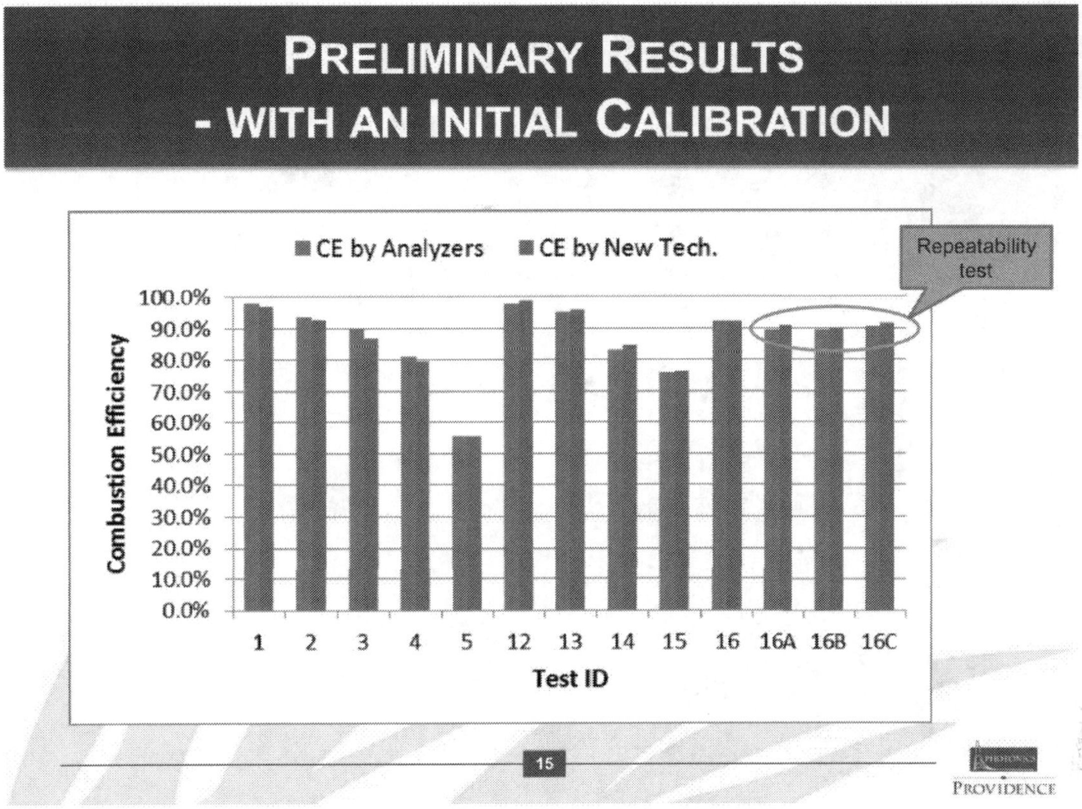
PRELIMINARY RESULTS
- WITH AN INITIAL CALIBRATION
CE by Analyzers
CE by New Tech.
Repeatability test
Combustion Efficiency
100.0%
90.0%
80.0%
70.0%
60.0%
50.0%
40.0%
30.0%
20.0%
10.0%
0.0%
1 2 3 4 5 12 13 14 15 16 16A 16B 16C
Test ID
15
PROVIDENCE

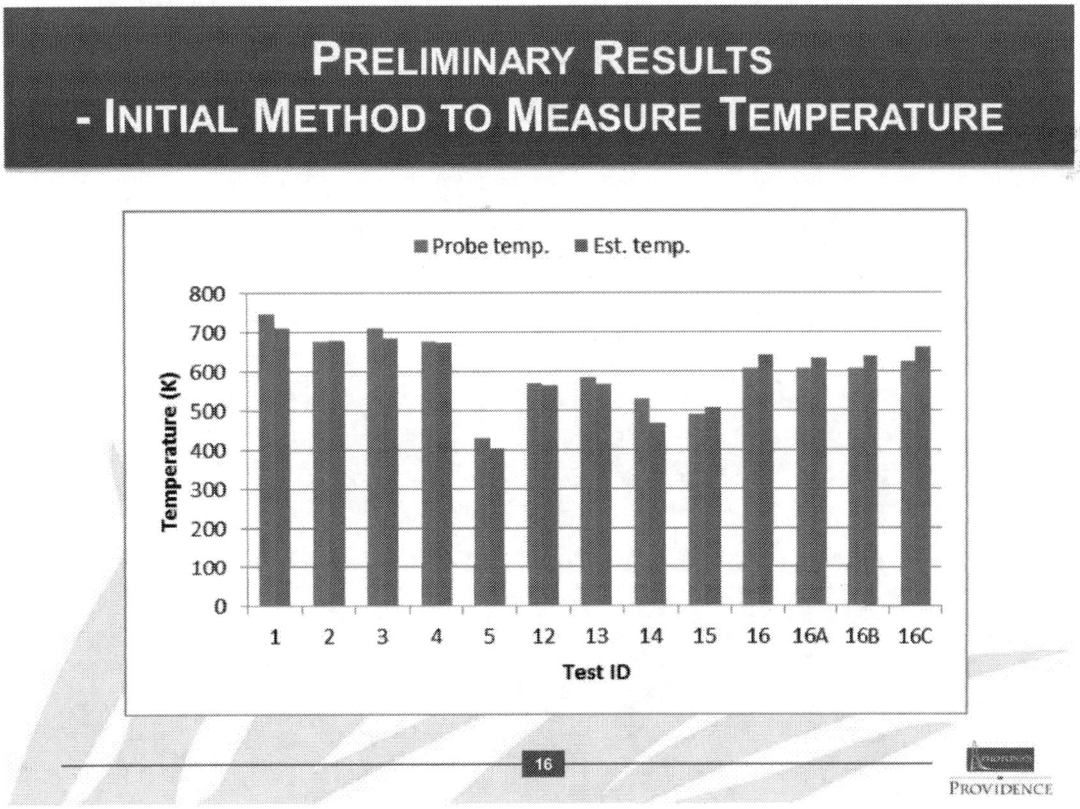
PRELIMINARY RESULTS
- INITIAL METHOD TO MEASURE TEMPERATURE
Probe temp.
Est. temp.
Temperature (K)
800
700
600
500
400
300
200
100
0
1 2 3 4 5 12 13 14 15 16 16A 16B 16C
Test ID
16
PROVIDENCE

Preliminary Results – CE Map

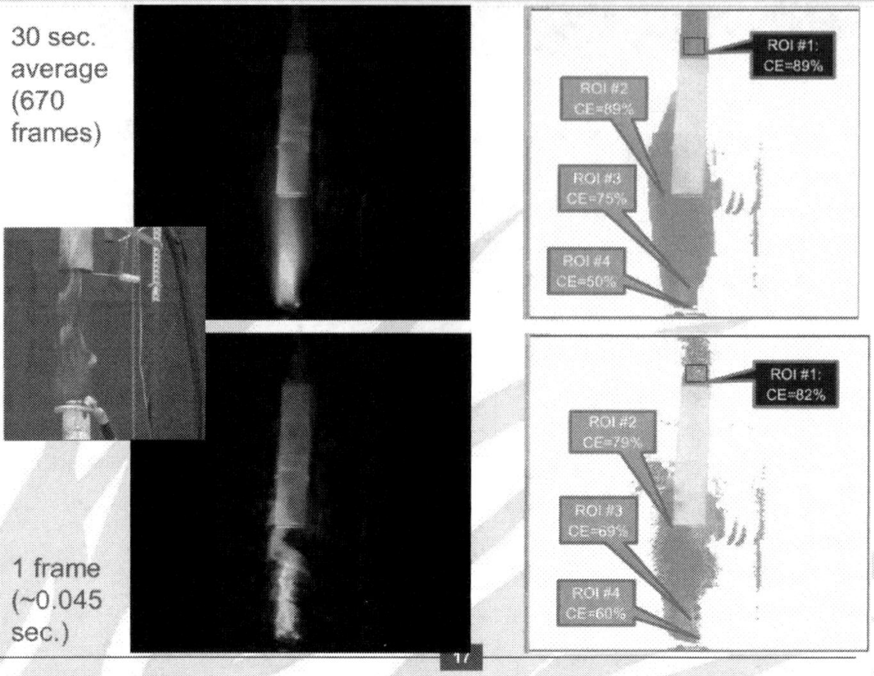

17 PROVIDENCE

Three Types of Measurement

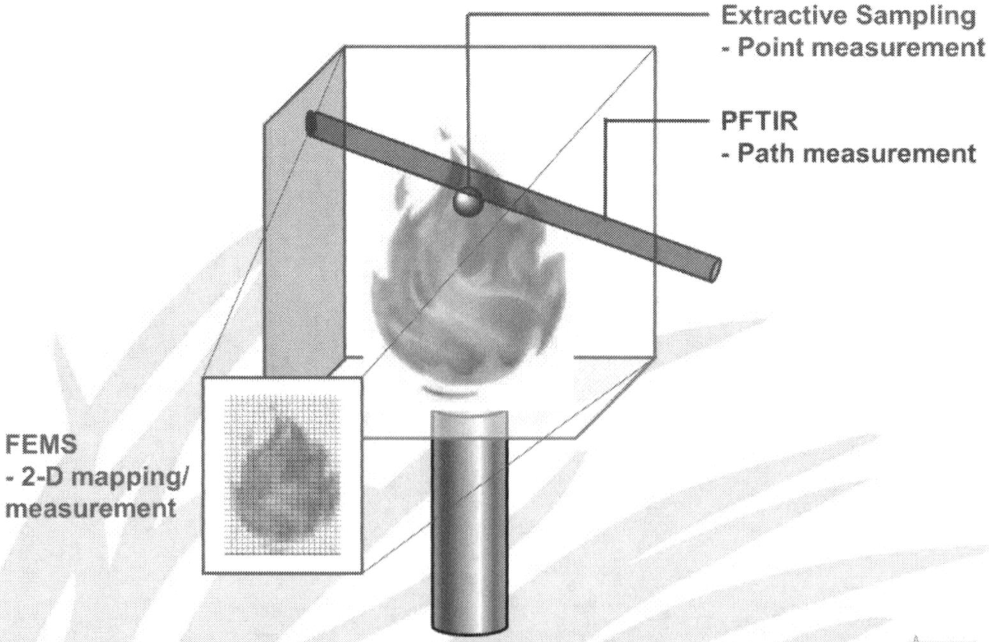

18 PROVIDENCE

COMPARISON WITH PFTIR

PFTIR

- "Scanning" - >1 sec/scan – assuming that flare is static during that time
- Path measurement – aiming required
- Human operation

FEMS

- Staring – 20-30 data cubes/sec – match the flare dynamics
- 2-D mapping of CE – no aiming required
- Automatic

19

PROVIDENCE

CONCLUSION

- The CE determined by the new technology correlate well with the CE measured by conventional analyzers
- With a further developed calibration method, real-time CE monitoring and feedback for flare optimization is feasible
- The new technology can determine CE at a pixel level, generating a CE map for the entire flare flame. No aiming issue.
- As a side benefit, it can also provide temperature mapping of the flare flame

20

PROVIDENCE

LOOKING FORWARD...

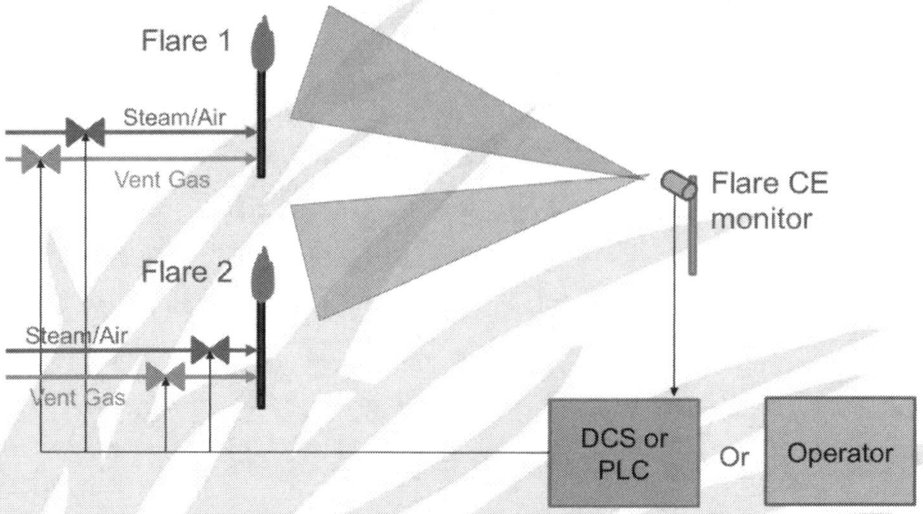

PROVIDENCE

Contact Info. youshengzeng@providenceeng.com

PROVIDENCE

348171 Flameless Thermal Oxidation

Abstract:

Linde's patented Flameless Thermal Oxidizers (FTO) technology oxidizes combustible gases and atomized liquids with air under controlled conditions in a specially designed reactor without a flame. Residence time and reaction temperatures are carefully selected to achieve 99.99 +% destruction efficiency with ultra-low NOx and dioxin emissions at very high thermal energy efficiency. These systems are commercially proven, cost effective and highly effective for the destruction of volatile organic compounds (VOC's) and hazardous air pollutants (HAP's). This paper will present an overview of the FTO technology and market applications with special emphasis on environmental performance and cost considerations.

350916 Modeling of the Dynamic Interaction in an LDPE Closed System Purge Bins-Regenerative Thermal Oxidizer (RTO) for Ethylene Emission Mitigation

Abstract:

Common designs of LDPE purge bins involve venting the purge and blend air to atmosphere containing small traces of ethylene. To mitigate ethylene emission from these bins, a new process design has been proposed to collect vent air-carrying ethylene via a closed ducting system connected to a regenerative thermal oxidizer (RTO) unit. This necessitates a blower at the downstream end to pull the purge air before feeding into the RTO to incinerate the ethylene. Concerns have been raised regarding the impact of such system on the purge bins, which are not designed to withstand any significant amount of vacuum. While conceptually steady operation of such system is feasible, operation of purge bins is never steady particularly when the purge bins are not designed to withstand any significant level of vacuum pressure. With the minimum safe operating pressure at -0.1 psig (-0.68 kPag), and the design RTO inlet pressure being -1"'WC (-0.25 kPag), the bins are forced to operate under slight vacuum conditions with a very narrow allowable margin for error. The operation is inherently dynamic as LDPE pellets are being conveyed into and out of the bins, purged and mixed in cycle by the interaction of multiple pieces of equipment. In the event of any equipment failure along the train, pressure in the bins can drop below their minimum vacuum level and cause damage to the shell. Dynamic interactions between the various components of the system including the blower with its VFD drive, purge and blend air compressors, capacitances and resistive characteristics of the pellets in the bins, as well as the response of the control system all contribute to interesting dynamic interactions that must be attended to. A model capable of simulating various dynamic phenomena has been built using Aspen HYSYS Dynamics TM v7.3 for the entire system, and used to answer what-if operational scenarios. As an example it was shown that the addition of a pull blower with its variable frequency drive (VFD), cannot maintain the bin pressure above the vacuum pressure limit of -0.1 psig even with the installation of individual pressure vacuum relief valve (PVRV) on each bin. Any interruption to air flows such as the shut-down of blend blower(s), stop/start conveying or swinging the purge air blower poses a non-negligible risk. Interaction between the PVRV's and the blower's VFD both simultaneously trying to control the bin pressure could lead to pressure fluctuations that exceed the design vacuum pressure of the bins. This paper will provide an overview of the system, state the design and operational challenges, basis of the dynamic model, and results of several what-if scenarios.

351659 Air Quality Impact of the Startup of a Single Olefin Plant

Abstract:

Ground-level ozone is one of six common pollutants defined by US EPA (Environmental Protection Agency). It can cause health effects, including lung diseases. HRVOC (Highly Reactive Volatile Organic Compounds) emissions from industries are highly responsible for ground-level ozone pollution, especially around industrial district. Flare emissions from MSS (maintenance, startup and shutdown) & events are major sources of HRVOC emissions.

Most of air quality modeling treats industrial flares as constant emissions, which means the emissions keep constant for the whole scenario days, and the total amount of emissions equal to the total emissions reported to environmental authorities. However, real flare emissions from MSS and events are time-various. At certain time, flare emissions could be very huge, and at other time, flare emissions could be very small. During emission peaks, the ozone formation could be much higher than a flat emission curve.

In this paper, time-various flare emissions of startup procedures are considered for air quality modeling. Emission data are obtained from dynamic simulation of a typical ethylene plant. The dynamic simulation results were validated with real plant data and emission data of startup events. The time-various flare emission data were added into Houston-Galveston-Brazoria (HGB) 2006 episode base case modeling as a point emission source. Three startup scenarios were simulated, which include a baseline scenario and two flare reduction scenarios. For each scenario, 120 cases were built to simulate different starting time covering the whole episode days.

Simulation results show that for the baseline scenario, the maximum 8-hour average ozone increment could be 15 ppb at certain starting time. It means starting a single plant could temporary increase the regional ozone concentration by 15 ppb, which is 20% of current ozone standard (75 ppb). For the two flare reduction scenarios, the maximum 8-hour average ozone increment could also reach 10 ppb. Simulation results also show that ozone increment is high when starting time is before 8 am. Ozone increment is low when starting time is after 12 pm. For each scenario, air quality simulation gives least impact period, which means ozone impact could be low when starting time is within that period. For the baseline scenario, the least impact period is only 2 hours. For the two flare reduction scenarios, the least impact periods are 4 hours and 8 hours.

In conclusion, dynamic simulation and air quality simulation show that the flare emissions from MSS and events are highly time-various, and may not be treated as flat emissions. Air quality simulations also give least impact periods of startup time, which could help chemical plants to arrange their startup schedule to reduce ozone formation.

352632 A Qualitative in Field Test to Characterize Oil-Shale Reservoirs

Abstract:

Shale oil recovery, despite its extraction costs and low yield per shale matrix is becoming a more feasible solution to the rapidly emerging global fuel demands. However, many oil-shale reserves are located several thousand feet below the earth's surface, which makes it extremely difficult and expensive to extract. In addition, the quality of the shale matrix may have a broad range such that the potential yield through and other methods may vary significantly across neighboring reservoirs. And determining the quality of the oil-shale matrix and its potential yield of synthetic/tight oil using laboratory analytical methods can take several weeks. This work presents a rapid and simple qualitative "in-field" test-tube test method that categorizes and prioritizes a given shale matrix based upon its composite. This test method can be a significant cost benefit and a proof positive indicator of where to focus exploration and recovery interest.

Pressure Regulation and Kick Attenuation with Wired Drill Pipe Technology in Managed Pressure Drilling

Hector Perez
Reza Asgharzadeh Shishavan
John Hedengren
Brigham Young University

Outline

- Overview
- Motivation for Pressure Control
- Method and Structure of the Control System
- Results
- Results of Automated BHA Pressure

Motivation for Pressure Control

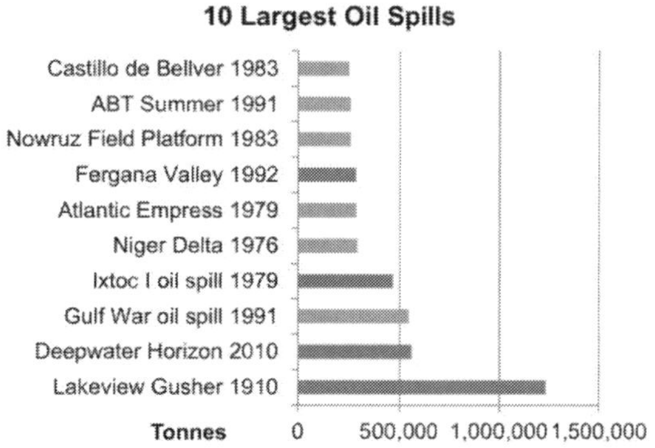

Motivation for Drilling Control

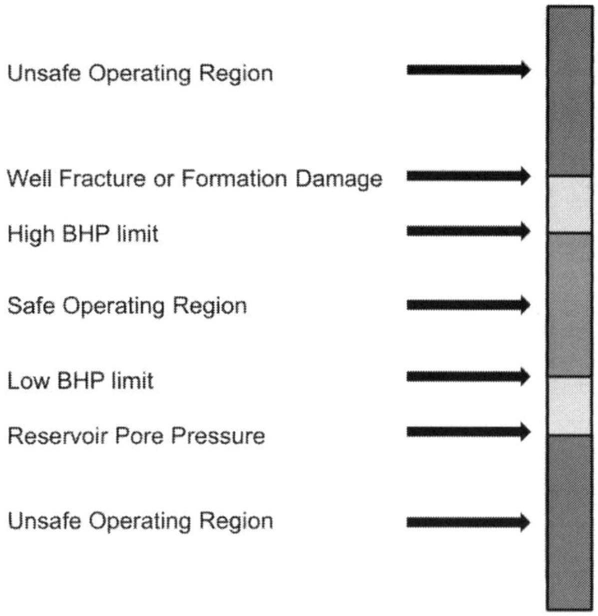

BHP Control Objectives

- Pressure Regulation
- Kick Attenuation

Managed Pressure Drilling System

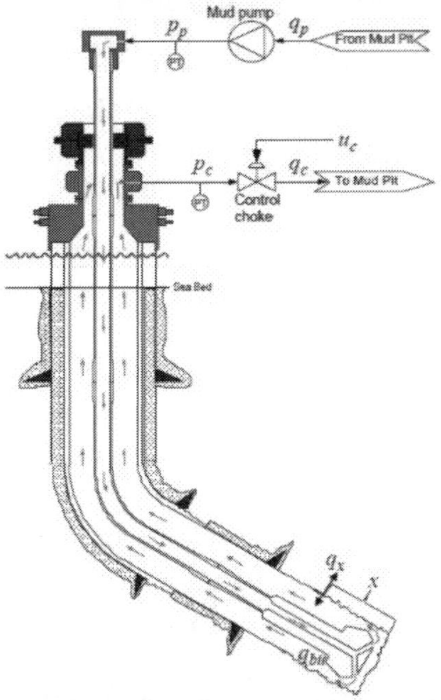

- ❑ Overbalanced Drilling
- ❑ Managed Pressure Drilling (MPD)
- ❑ Underbalanced Drilling

MPD offers:

- ❑ More precisely control of borehole pressure
- ❑ Decrease NPD and mitigate drilling hazards

Manual Control versus Automated Control

Challenges

- Large Time Delays
- Complex Behavior of Multiphase flow
- Sudden Change in Formation Properties

Benefits of BHP control Automation

- Improve Reliability of Pressure control
- Reduce the possibility of human errors

Objective

Designing a comprehensive controller which Simultaneously

- Regulates the BHA Pressure within specified range
- Attenuates Unwanted gas influx when it happens
- Uses latest automation technologies
- Has improved performance in comparison with manual and conventional controllers

Wired Pipe Technology to be used in Control Scheme

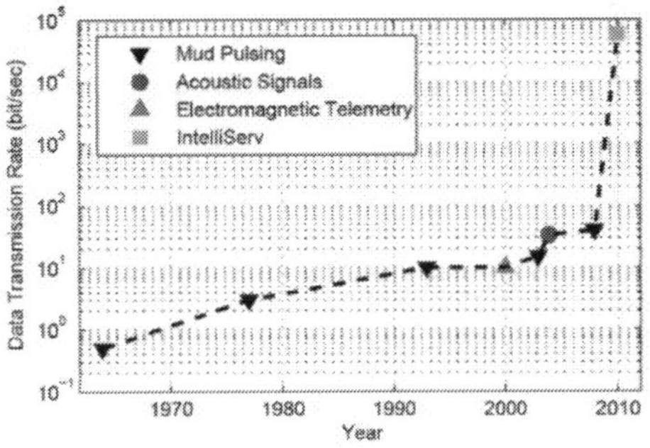

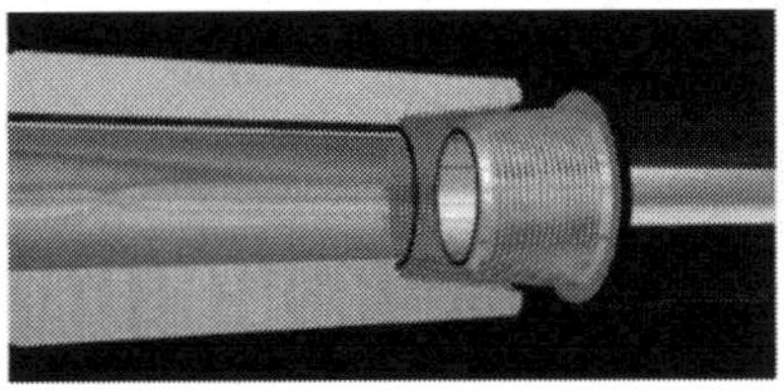

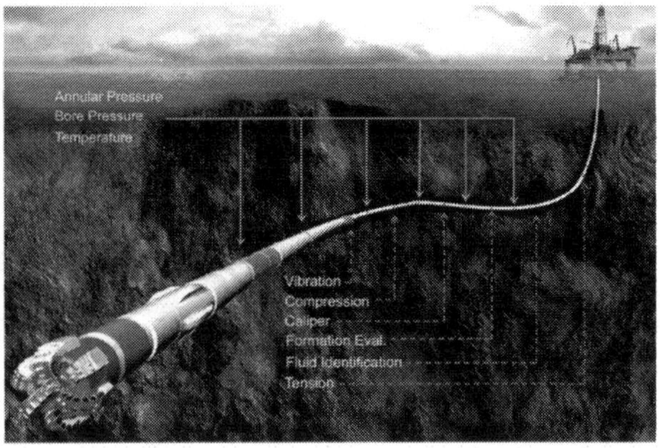

Challenges in the designed controllers

Mud Pulse Telemetry Based

- ❑ Indirect control of BHP
- ❑ High uncertainty in estimated variables values
- ❑ Still Large time delay in kick detection

Wired Pipe Telemetry Based

- ❑ Directly control of BHP and annular pressures ✔
- ❑ Increased precision in estimating the unmeasured process variables ✔
- ❑ Improved kick detection ✔

Schematic of the MPD Controller

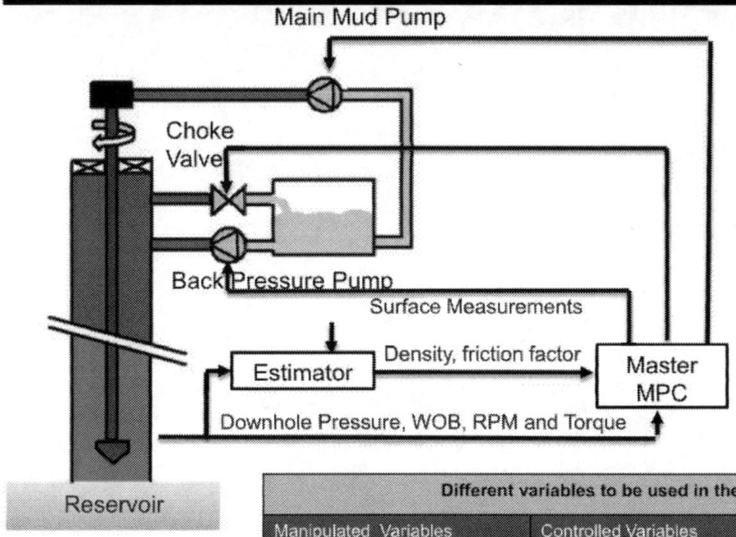

Different variables to be used in the current research		
Manipulated Variables	Controlled Variables	Estimated Variables
Rig Pump Flow Rate	BHP	Annulus Drill Fluid Density
Choke Valve Opening	Annular Pressures	Annulus Friction Factor
		Reservoir Pore Pressure

Controller Structure

Multivariate Nonlinear Model Predictive Controller (MPC)

What are advantages over conventional controllers?

Nonlinear Feature of the Controller

Multivariate Nonlinear Model Predictive Controller (MPC)

- Drilling is a non linear process
- Current controllers use linear (simplified) models
- A nonlinear controller can capture the dynamics better
- Designing nonlinear controllers are more challenging

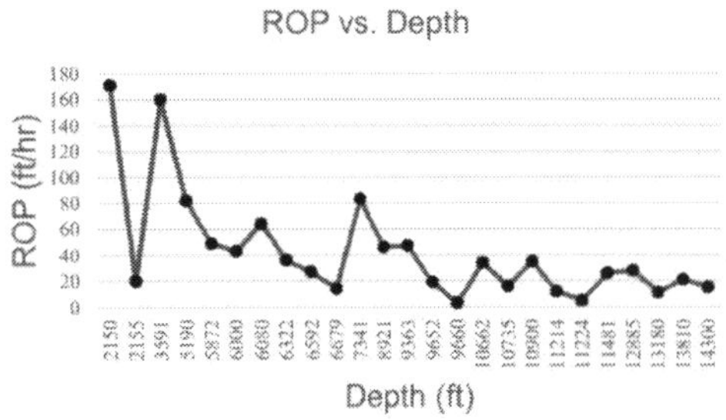

Figure 1. Change of ROP with drilling depth. Field data from (Kinabalu East-1, 1990)

Model Predictive Control and Moving Horizon Estimator

Controller

Multivariate Nonlinear Model Predictive Controller (MPC)

- Future prediction capability – Improved control performance
- MPC vs. Conventional Controllers

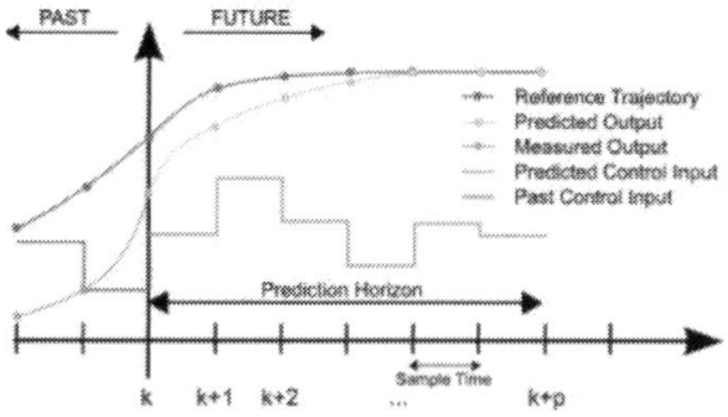

Estimator

- Moving Horizon Estimator
- Uses both downhole and surface measurements
- More precise and faster parameter and state estimation

Software

- APMonitor Optimization Suite
- Designed Especially for Dynamic Nonlinear Processes
- Using the Most Efficient Solvers Available
- Solution Time: Fraction of a Second

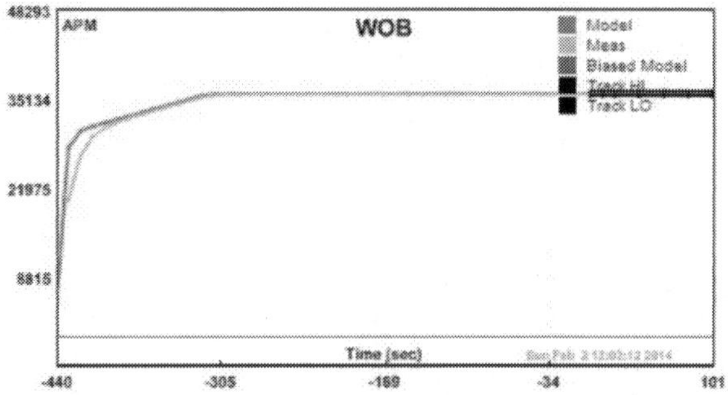

Lower Order Model

$$\dot{p}_p = \frac{\beta_d}{V_d}(q_{pump} - q_{bit})$$

$$\dot{p}_c = \frac{\beta_a}{V_a}(q_{bit} + q_{back} - q_{choke} + q_{res})$$

$$\dot{q}_{bit} = \frac{1}{M}(p_p - F_d|q_{bit}|q_{bit} + \rho_d g h_{bit} - p_{bit})$$

$$\dot{\mathbf{h}}_{bit} = \boldsymbol{ROP}$$

Balance Equations

$$M_a = \rho_a \int_0^{l_w} \frac{1}{A_a(x)}dx$$

$$M_d = \rho_d \int_0^{L_{dN}} \frac{1}{A_d(x)}dx$$

$$M = M_a + M_d$$

$$p_{bit} = p_c + \rho_a \times F_a|q_{bit} + q_{res}|(q_{bit} + q_{res})h + \rho_a g(h_{bit} - (n-1)h)$$

$$q_{choke} = K_c z_{choke}\sqrt{\rho_{an}(p_c - p_0)}$$

Closure Relations

Zhou, Kaasa, et al. (2008

Kick Detection and Attenuation

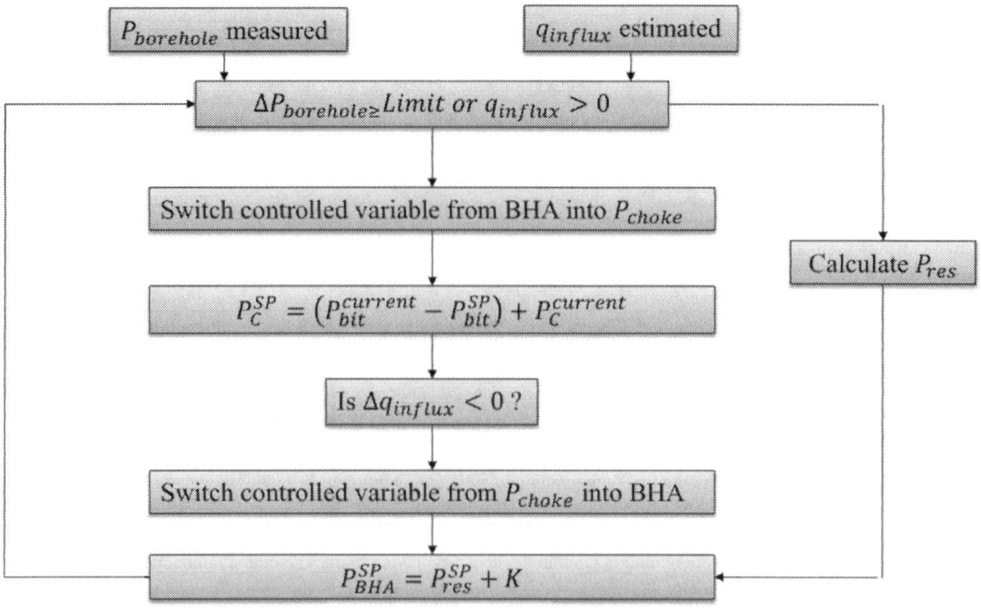

Results

Automated BHA pressure control

Table 1. Initial well conditions used for simulations

Variable	Value (SI units)	Value (US units)
Distance between two successive pressure sensors h_a	150 m	492.1 ft
ROP	0.01 m/s	0.033 ft/s
Fluid density in the drill string - ρ_d	0.012×10^{-5} kg/m^3	74.9 lb/ft^3
Fluid density in the annulus - ρ_a	$0.014^a \times 10^{-5}$ kg/m^3	87.4 lb/ft^3
Bulk modulus in the annulus β_d	14000 bar	2.03×10^5 psi
Bulk modulus in the annulus β_a	14000 bar	2.03×10^5 psi
Annulus cross sectional area	0.021 m^2	0.226 ft^2
Friction factor in the drill string F_d	10000 bar m^{-6} s^2	116.3 psi ft^{-6} s^2
Friction factor in the annulus F_a	1250^a m^{-5}	3.29 ft^{-5}
Lumped density per length parameter for the drill string - M_d	5700×10^{-5} kg/m^4	108.5×10^{-5} lb/ft^4
Lumped density per length parameter for the annulus – M_a	800×10^{-5} kg/m^4	15.22×10^{-5} lb/ft^4
Well depth (h_{bit})	2000 m	6562 ft
[a]For connection procedures, the values used for simulation were 0.01 (density) and 1000 (friction factor)		

MHE Estimator

Within 30 seconds of operation, the estimator has accurately determined the friction factor and has determined the annulus density within 1.5% of the actual value.

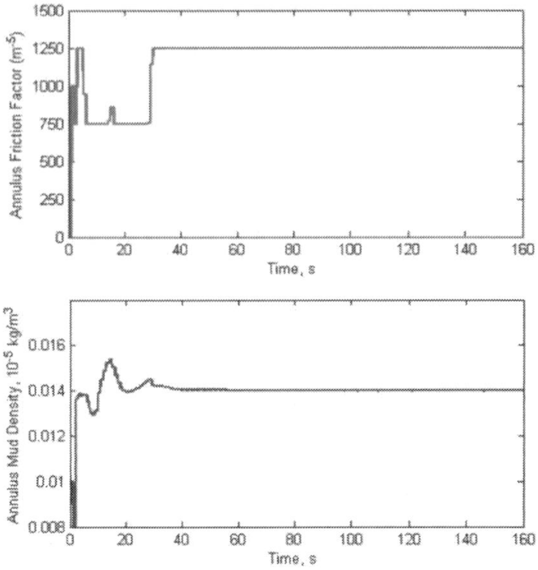

Case 1: Normal Operation

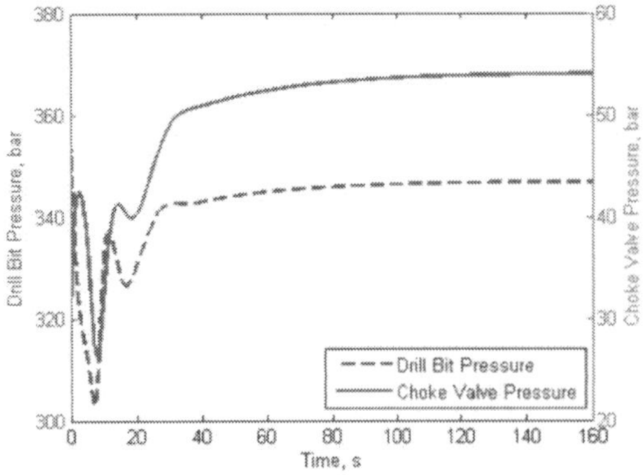

Within 30 seconds, the drill bit pressure has been controlled within 0.72% of its set point of 345 bar

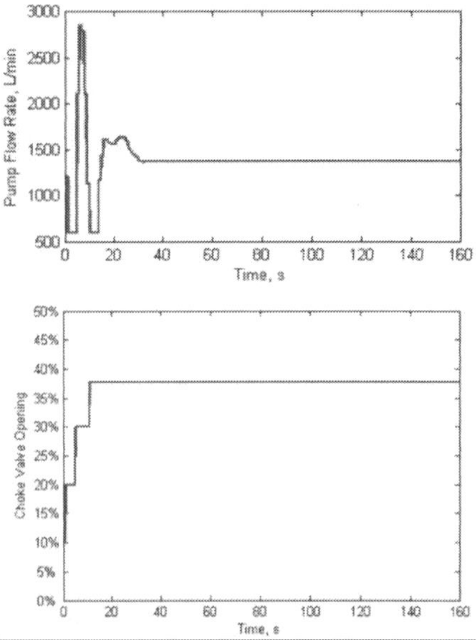

Case 2: Connection Procedure

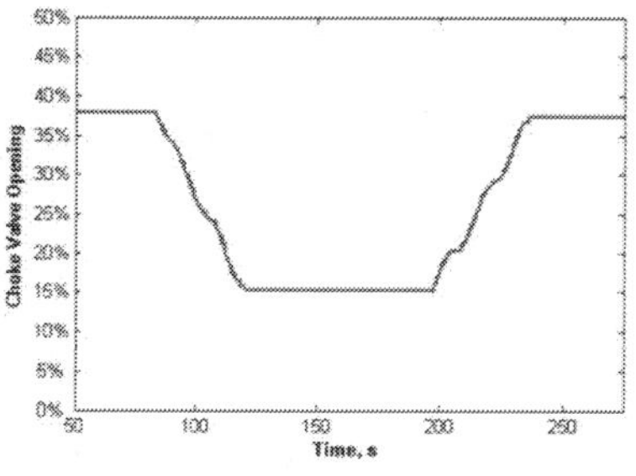

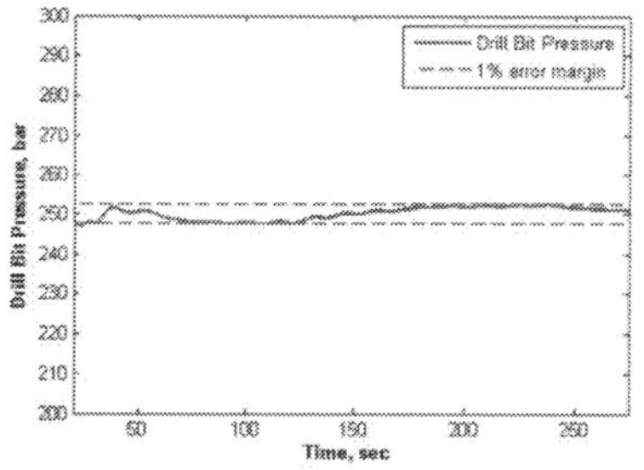

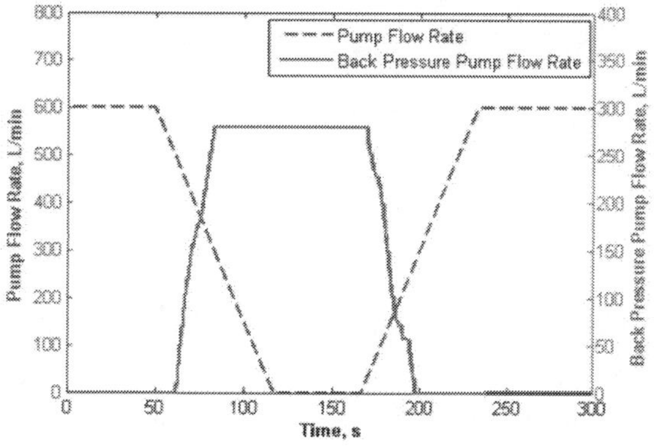

Apr 2, 2014 Pressure Regulation and Kick Attenuation with Wired Drillpipe in MPD

Case 3: Unwanted Gas Influx

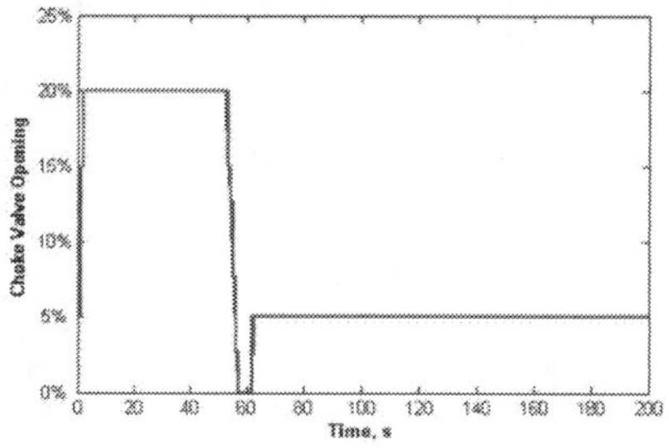

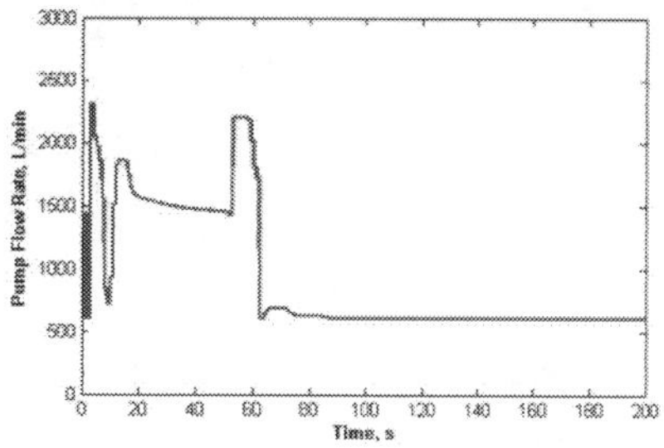

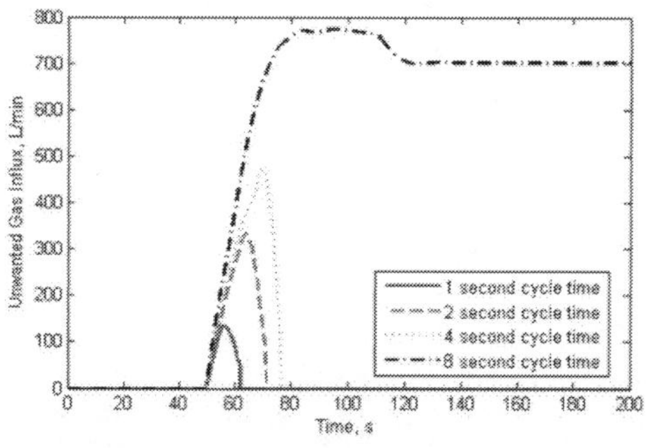

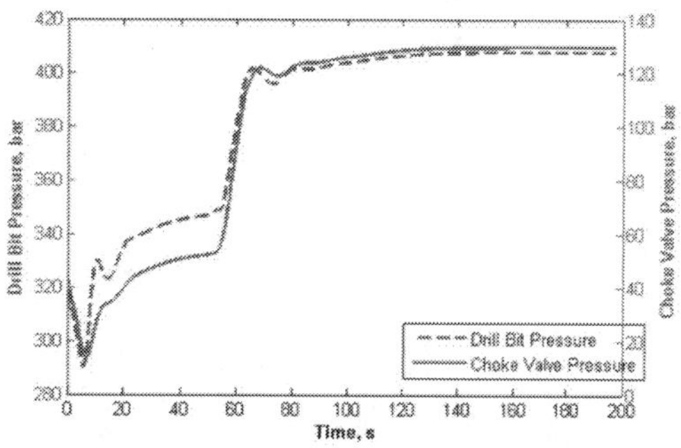

Apr 2, 2014 Pressure Regulation and Kick Attenuation with Wired Drillpipe in MPD

Thank You for Your Attention

Apr 2, 2014 | Pressure Regulation and Kick Attenuation with Wired Drillpipe in MPD

Risks from Unconventional Wells Gasification

Dennis B. Butts
Americas Risk Practice
Environmental Resources Management (ERM)
CityCentre Four
840 West Sam Houston Parkway North, Suite 600, Houston, TX 77024
dennis.butts@erm.com

Prepared for Presentation at
American Institute of Chemical Engineers
2014 Spring Meeting
Topical 2: Shale Gas and Tight Oil
New Orleans, LA
March 30 – April 2, 2014

UNPUBLISHED

Risks from Unconventional Wells Gasification

Dennis B. Butts
Americas Risk Practice
Environmental Resources Management (ERM)
CityCentre Four
840 West Sam Houston Parkway North, Suite 600, Houston, TX 77024
dennis.butts@erm.com

Keywords: LNG, Risk Management, Process Safety, Unconventional Wells, Shale

Abstract

With the continued differential between diesel and natural gas costs, onshore well operators have been working to convert both drilling rigs and hydraulic fracturing pumps to utilize natural gas as fuel. Associated with this conversion is a change in the risk profile that may not be fully recognized or understood in the field. Risk management philosophies and assumptions need to adjust to the difference in hazard characteristics between the fuels, as well as the variety of delivery methods.

Specific topics will include:

- Motivations for fuel conversion and the differences in scale for rig power, natural gas substitution on diesel hydraulic fracturing pumps, and dedicated natural gas powered hydraulic fracturing pumps. Risk mitigation and field training requirements may not fully accounted for in investment decisions.
- Scale of conversion and location specific availability drives natural gas sourcing, which can then drive personnel risk and well pad layout issues depending on whether the gas is sourced from local wellheads, pipeline, Compressed Natural Gas (CNG), or Liquefied Natural Gas (LNG). Separate Simultaneous Operations (SIMOPS) issues can arise based on the specifics of each sourcing method.
- Associated risks need to be appropriately managed in a manner that involves the personnel directly exposed. And there is a need to ensure that well sites which did not previously present a threat to surrounding activities do not become a threat due to either direct risk from the fuel conversion itself or from escalation risks not previously of concern.

1. Introduction

Over the past few years there has been a significant increase in efforts to substitute natural gas for diesel to provide fuel for drilling and fracturing onshore wells. This is an expansion of the conversion efforts that were already underway to convert transportation fleets away from diesel fuel. While the reasons for this may vary, and do include environmental benefits, the most significant reason is cost savings.

Natural gas has been averaging about $4.50/mmbtu [1] over the past 6 months, which is about $0.59 per diesel gallon equivalent (DGE) [1]. Even with added distribution costs, if pipeline gas is available, the economics are very attractive. Wellhead gas is even more attractive. And if LNG is the only alternative at a well site, at $4.50/mmbtu, the natural gas component of LNG is about $0.34 per gallon [1], with a markup of about $0.15 per gallon for liquefaction and storage [3] plus transport costs of about $0.15 per gallon (up to 100 miles) [4], giving a breakeven price of $0.64 per gallon or $1.10 per DGE. With typical supplier pricing to bulk customers as a discount to diesel typically around $1.50 per gallon, and more substantial discounts for large scale contracts, there are significant financial benefits for both LNG suppliers and drillers when diesel fuel is in the $4 per gallon range.

And as a bonus to the economic justification, drillers can tout emissions reductions of as much as 20 – 30% relative to diesel [5]. Rarely in these discussions is the subject of added costs or effort for process and technical safety brought up.

Substitution of natural gas for diesel has been underway for some time in heavy vehicle transportation fleets. With added competition in engine conversion markets and among storage tank suppliers, conversion costs have been reduced to the point where heavy vehicles can recover costs in less than 2 years [6]. Efforts to convert drill rigs originated in the Marcellus region, due to environmental concerns, but the economics have driven natural gas rigs into most of the major markets in the US, as well as into western Canada. Since hydraulic fracturing pumps use even more diesel than rigs, recent efforts have been in this direction, though technical challenges regarding power delivery have resulted in two types of conversion. One is the substitution of gas for diesel in existing engine drivers to ensure sufficient horsepower. The other is to use high powered turbine pumps, which are very well suited to gas power. Large scale conversion of fuel sources is not unprecedented, as the US heavy truck fleet went from less than 10% diesel powered to nearly 100% diesel powered from 1950 to 1980 [6].

So what does this mean in terms of risk management? Aside from the business and financial risk of converting equipment to a fuel in natural gas that is historically more volatile in pricing than diesel fuel at a time of historically high oil to gas price differentials, which is a subject for another paper, there are practical technical safety considerations that come into play as drilling sites convert from diesel to natural gas. These can result in increases to the risk profile of a well pad that is not generally accounted for, as well as changes in the risk profile within the drilling operation itself.

This paper will discuss some of the issues and options regarding conversion and highlight specific risk issues associated with each.

2. Sources of Natural Gas

2.1 Typical Volume Requirements

A heavy truck may use 20,000 gallons of diesel per year [4], with significant potential savings being the result of a conversion to natural gas fuel. Conventional fluid drill rigs use an average of 425,000 gallons of diesel per year [5], creating a fuel cost savings equivalent to converting a fleet of over 20 heavy trucks, but at an investment cost that is much lower. But even these volumes pale in comparison to the amounts of diesel used in hydraulic fracturing. Fracking crews can use 5,000 to 10,000 gallons of diesel per day [7], and it is common for the crews to be in operation for as many as 300 days per year, accounting for mobilization and demobilization time. That means 1.5 to 3 million gallons of diesel per year can easily be used by a single crew, often going well to well for the same operator. An unfortunate outcome of this is that conversions become a foregone conclusion, without due consideration to the resulting process safety risks.

2.2 Options for Sourcing and Delivery

Natural gas for a well site can be sourced as gas directly from the wellhead or gathering system, or it can be piped in from a commercial distribution line, when those are available. In more remote locations, distribution lines may not be available, and if the gas is too wet, wellhead or gathering gas may also be unsuitable for use. In those situations the options are to utilize either CNG or LNG. CNG would require considerably more transportation than LNG due to the lesser amount of gas that can be moved and comes with higher costs for storage and handling equipment. Therefore, CNG has limited utility for well sites, though its lower fuel cost makes it useful in some narrow applications. The use of LNG can be cost effective when other options are not available, as it is more available than in the past and although containing less energy than diesel per gallon, an onsite tank containing 15,000 gallons of LNG is still equivalent to over 8600 gallons of diesel [2] and is about 2.4 times the energy density of CNG [9].

2.3 Issues Associated with Each Option

Use of wellhead gas or gas from gathering lines requires either very dry gas or extensive treatment, as most engines converted to utilize natural gas are designed for gas approximating domestic gas. This is in contrast to offshore facilities where engines are often optimized to take advantage of fuel gas diverted from separation trains. Pipeline gas can be the best option, for those sites with ready access to a distribution line, and will provide sufficient flow and pressure for most natural gas drilling rigs. However, pipeline gas may have to have compression added for certain types of fracking pumps, and available flow may not be sufficient for some fracking operations. CNG can be useful in areas that do not have dry gas or ready pipeline access and may not have easy access to LNG. In such areas, CNG facilities can be set up to provide the product to users in the surrounding region much more readily than LNG facilities. To utilize CNG does require more transportation of the gas and more onsite storage than using LNG. With LNG being the last supply option, but becoming very common in the field, this is an option that carries significant process and technical safety risks. LNG is a material than can bring many benefits, but needs to be treated with considerable respect, often more than is witnessed at well sites.

3. Associated Risks

Associated process and technical safety risks arise in several stages of the conversion from diesel to natural gas. In most cases there is an increase in those risk levels relative to diesel fuel, though this is not always fully contemplated by decision makers or well communicated to the field level staff. Whereas diesel fuel is classified as a combustible liquid, NFPA Class II [8], is difficult to ignite if spilled and primarily a localized threat, natural gas is highly flammable vapor and depending on delivery method could have the potential to disperse considerable distances if released.

In use in equipment, natural gas can create the need for hazardous area class or hot work permit zones that were not required around diesel equipment. This issue has been missed on well sites and often is not included in discussions about conversion costs until after being identified in PHA studies or other risk assessments. The same consideration extends to gas monitoring and shutdown systems. System risks are manageable, but need to be given due consideration and treated with the proper respect. Some expansion of this thought is provided below.

3.1 Source of Gas

Wellhead gas primarily introduces risk through the separation and dehydration process. Even relatively dry gas needs to be treated prior to use in most of the engine drivers used for onshore production. To utilize it during drilling usually means that at least one well has already been put into operation on a multi-well pad. The risks associated with production while drilling and completing adjacent wells have to be assessed. Controls that may need to be in place during certain phases of drilling could interfere with the supply of gas. An alternative that may be available within a field of well sites under one operator is to use gas from the gathering system. While treatment would also be required for that gas, the potential SIMOPS issues of a producing well on a pad with drilling activity. In either case the gas being treated will be a relatively high pressure and is likely to result in some amount of condensate or natural gas liquids (NGL) production and handling. This can increase the fire and explosion risks above those normally found during drilling activity.

Pipeline gas is only an option when a distribution line is nearby. But if available, this can provide a supply of natural gas that is already treated and can be adjusted to the necessary pressure at the takeoff location from the distribution network. In analysis of fire and explosion risks, pipeline gas supply most often has the smallest consequence zones and the least amount of equipment susceptible to failures. This is even the case if high horsepower fracking pumps are in use that requires some degree of compression of the natural gas to reach necessary operating pressures.

CNG introduces risk due to its high pressure, typically near or above 3000 psig [9]. This can both drive the spread of a gas cloud in the event of a release as well as provide the means through depressurization to cool the gas to a low enough temperature that it will disperse for some distance along the ground, further lengthening dispersion distances. Storage under pressure also introduces a finite risk of boiling liquid expanding vapor explosion (BLEVE) that could spread debris.

LNG usage at well sites is capable of imposing a larger potential impact zone than a well blowout scenario. And this is excluding a catastrophic failure of the storage medium, which is low likelihood event due to the dual tank construction of trailers that handle LNG. When released, LNG will initially disperse in a manner very similar to liquefied petroleum gas (LPG), but as the edges of the dispersing cloud warm, flammable vapors will loft upward. Assumptions regarding layout and spacing that are based on LPG can address the flammable breadth of a possible LNG cloud, an example of which is shown in Figure 1 from PHAST [10], going out about 850 feet (260 meters) to the 50% lower flammable limit (LFL). Maximum distance to a 1 psi overpressure, good for window breakage, trailer damage, and upsetting neighbors, would extend another 300+ feet (100 meters) beyond the 50% LFL distance. In general, the height of the 50% LFL section of the cloud will remain under 15 feet (5 meters), but this also needs to be confirmed relative to the drill rig flare height. And the failure of a LNG loading hose would impose an even larger impact zone. Though many operators end up using gas detection, this is only effective if appropriate emergency response procedures are in place.

Figure 1: 50% LFL Dispersion Distance for 1-inch LNG Leak [10]

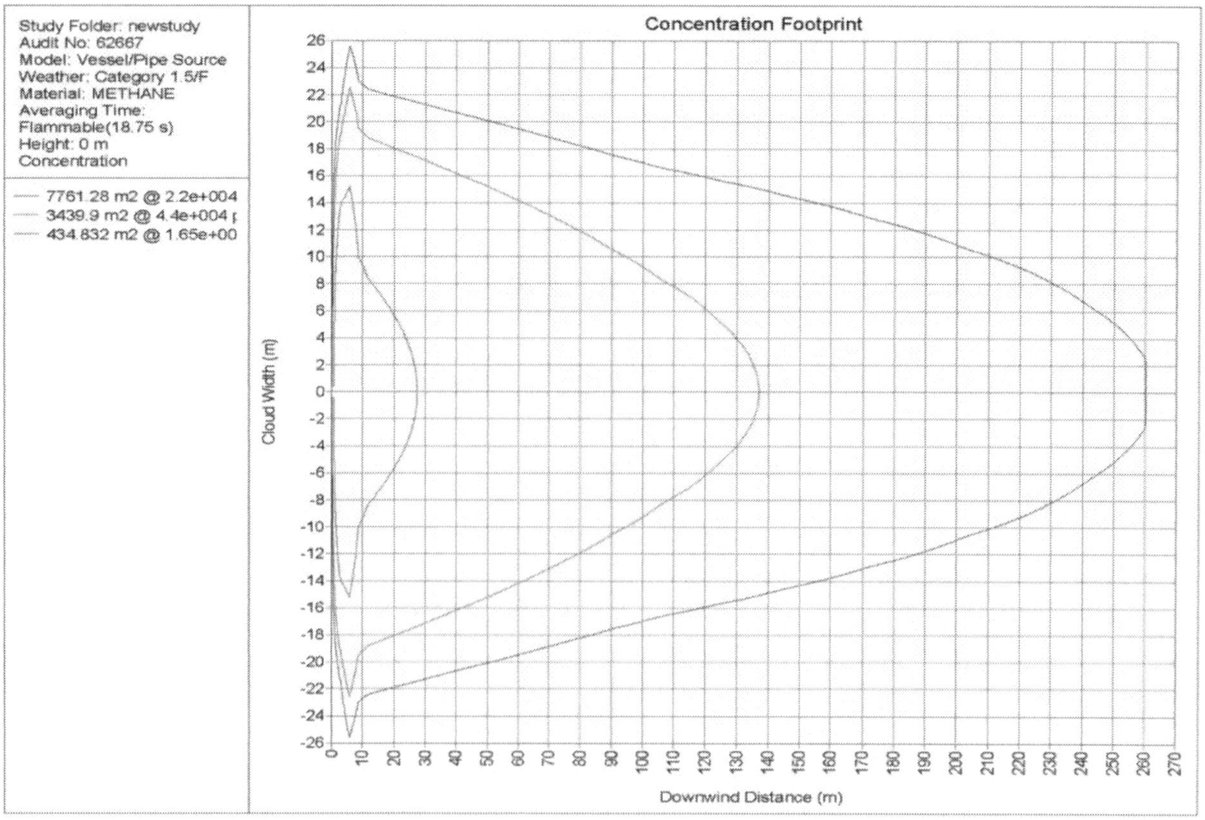

For LNG most releases behave as impinged releases, due to the low pressure environment. For CNG, the best case is for a free field release, which provide dispersion distances of less than half that of LNG. However, horizontal releases of CNG are capable of dispersion distances beyond those of LNG, in particular if they impact other equipment and become partially impinged, as shown for dispersion to 50% LFL for a 1-inch CNG release in **Figure 2**. An interesting phenomenon is how the LNG dispersion slumps outward from the beginning, due to release at temperatures low enough for the methane to be denser than air, but the CNG is initially released at warmer temperatures, then settles into a much lower temperature, resulting in a slumping spread.

Figure 2: 50% LFL Dispersion Distance for 1-inch CNG Horizontal Leak [10]

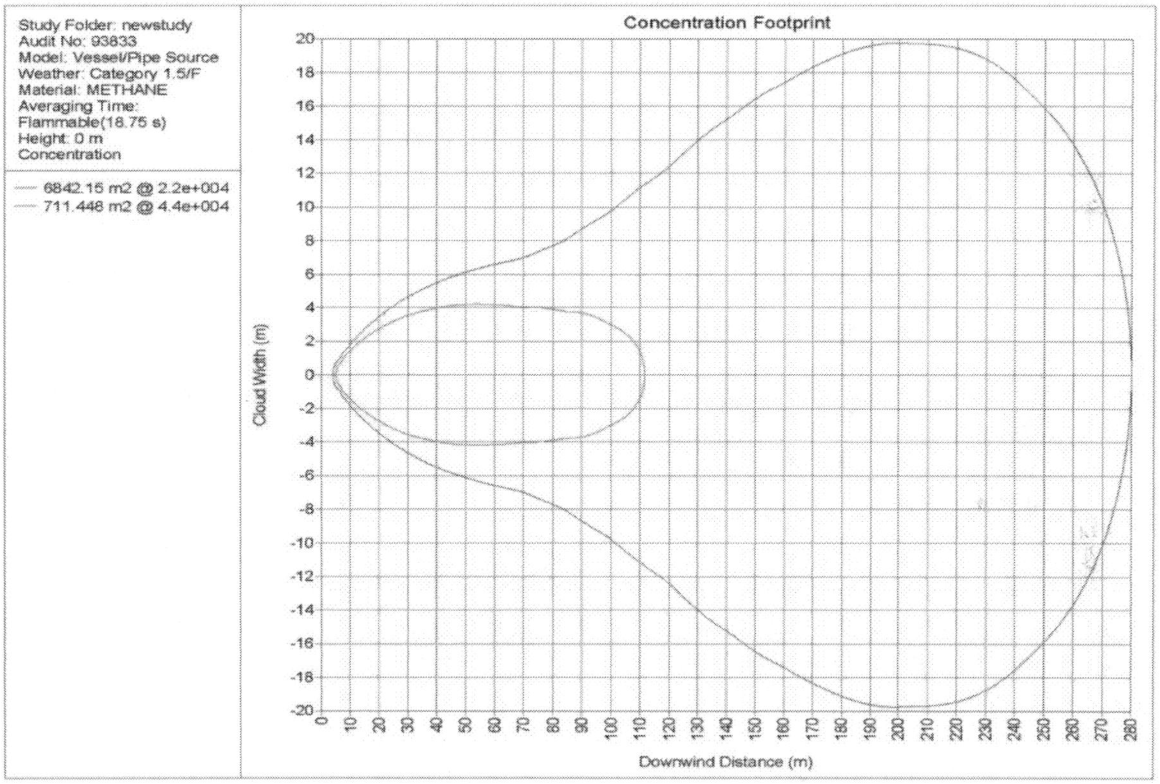

As a reference, an ignited well blowout, traditionally considered a worst case event at a well pad, will typically have an effect distance to 1500 Btu/hr-ft^2 (4.7 kW/m^2) of about 500 feet (150 meters).

3.2 Refueling at Site

When utilizing CNG or LNG as the source of gas, there are additional considerations for storage at the site and for refueling deliveries. Both require delivery by truck, just like diesel fuel, but the release of diesel from an overturned truck is not the same as a release of compressed or liquefied natural gas. This is why there are stringent design requirements on CNG and LNG transport vehicles. However, when assessing risks it is wise to keep in mind that those standards were developed for paved highway transportation, not gravel access roads to well sites. In addition to possible visibility issues, there are the impacts such roads could have on the expected behavior of safety features on the vehicles.

3.3 Compression and Delivery of Gas to Equipment

Further issues arise in situations where either pipeline gas or LNG is used to supply equipment such as hydraulic fracturing pumps that require higher pressure feed gas. In these circumstances, arrangements have to be made to provide compression at the site. A site that is typically already very crowded with equipment. In terms of risk from the equipment it is normally a localized threat, however the extra equipment adds a level of complexity that can complicate operations. The added complexity can add a human factors risk dimension that is rarely assessed in drilling situations.

4. Conclusion

Considerable financial and environmental benefits can be had by converting drilling operations from diesel fuel to natural gas, as long as the current differential in crude oil versus natural gas pricing exists. To achieve those benefits does entail some risk, with this paper having focused on the process and technical safety risks. In some instances, the benefits and risks can pull in opposing directions creating a challenge to find equilibrium between the two.

Depending on the source of natural gas for the drilling and completions operations, due consideration needs to be given to:

- Training needs for field staff that are used to dealing with combustible diesel fuel and not highly flammable natural gas;
- Flammable gas dispersion distances as well as resulting maximum potential zones for 1 psi overpressures, in particular for LNG or CNG applications;
- Review of drill rig flare location relative to natural gas supply point(s) to account for potential height of dispersion;
- Road conditions and their potential impact on transportation of fuels;
- Hazardous area classification zones and resulting equipment alterations, layout changes, or work permit requirements;
- Gas detection needs and the emergency response planning to accompany the detection; and
- Hazard assessment requirements and when they are activated to address changes made to drilling operations or well pad layouts once in service.

As we further explore this question, it is necessary to keep in mind not only the benefits of the ongoing process of converting away from diesel fuel to natural gas at well sites, but also the risk management challenges that can arise.

5. References

[1] US Energy Information Administration, 2014

[2] US Department of Energy, Fuel Energy Spreadsheet, 2014

[3] Quine, Tom, "Fundamentals in LNG", ASGMT White Papers, 2010

[4] Center for Energy Economics, University of Texas at Austin

[5] Oil & Gas Technology, "Gas Powers Fracking Plays", April 10, 2013

[6] Blomerus, Paul and Oulette, Patric, "LNG As a Fuel for Demanding High Horsepower Engine Applications: Technology and Approaches",

[7] Schutt, Bryan, "Diesel to Gas Switching Trend Takes Off in Drilling Sector", SNL Financial, September 3, 2013

[8] NFPA 30, National Fire Protection Association, 2012

[9] Envocare Ltd., "Liquefied Petroleum Gas (LPG), Liquefied Natural Gas (LNG), and Compressed Natural Gas (CNG)", 2007

[10] PHAST 6.7, DNV GL AS

353108 Supercritical CO_2 / Brine / Hydrocarbon Interactions in Shale Rock Fracture: A Geo-Microfluidics Approach

Abstract:

Shale oil and gas is a critical unconventional resource for the energy sector in the US and worldwide. The booming of its production in the recent decade, due to the hydraulic fracturing technique, has led to serious environmental concerns such as huge water demands and toxic additives into fracking water which may pollute groundwater. A proposed solution is to use supercritical CO_2 ($scCO_2$) as a more environment-friendly alternative fracking fluid to replace water for enhanced oil recovery (EOR) in oil shale. Many challenges lie in the way of this scheme such as to understand multiphase flow of $scCO_2$ at injection sites, its interactions with brines, hydrocarbons, and surrounding geologic matrix. We present here a microfluidic approach to investigate the micro-scale multiphase fluid flow and reactive transport within low-permeability shale rock. Unlike the conventional microfluidic techniques, the geo-microfluidic chip is directly fabricated from a thin-section of shale rock, hence provides the genuine surface properties. Simple fracture-like patterns are created on the shale rock surface by machining and femto-second laser etching to simulate the natural structures. Flow experiments in the geo-microfluidic chip are conducted within a high-pressure, high-temperature apparatus to simulate the in situ reservoir conditions. The interactions between $scCO_2$/brine/hydrocarbon will be quantified and the hydrocarbon extraction efficiency of using $scCO_2$ and using fracking water will be compared. The displacement patterns, the dissolution rate of $scCO_2$ in brine/hydrocarbon, and the surface wettability of $scCO_2$ and shale rock will also be investigated. These micro-scale experiments results will help to reveal the underlying mechanisms of multi-phase fluid flow under in situ temperatures and pressures.

352468 Model Shale Gas Properties from a Statistical Mechanical Perturbed Hard-Sphere Chain Cubic Equation of State

Abstract:

Shale gas is an increasingly important energy source in North America. Hence, an accurate description of the physical, transport and thermodynamic properties of shale gas mixtures are a key prerequisite for process design. A particularly advantageous feature is that multiple properties can be obtained from a single equation of state (EOS) for a model system. Traditional EOS's are cubic in molar volume and are derivatives of the basic van der Waals (1873) EOS. Redlich and Kwong (1949), Soave (1973) and Peng and Robinson (1976) are attempts to improve the accuracy of the basic vdW EOS; however, as is shown early in this talk, these equations of state overcorrect the attractive term to compensate for deficiencies in the repulsive term. On the other hand these EOS's have the advantages of a computationally simple form and *three well-defined molar volume roots*. Nonetheless, while these EOS's often give a reasonably accurate representation of the phase composition of simple binary and ternary alkane mixtures (model shale gas systems); poor results are obtained for other properties including liquid density, mixture second virial coefficient and diffusion coefficient.

On the other hand, a class of accurate molecularly-based but complex and computationally intense equations of state: the statistical associating fluid theory (SAFT) equations (Huang and Radosz, 1990; Chapman et. al., 1990), which account for individual molecular level effects (i.e., hard-sphere repulsion, London dispersion, covalent bond chain formation, multipolar forces and hydrogen bonding) have successfully emerged in the past few decades to account for multiple effects in pure fluids and mixtures including associating and/or polydisperse components. These EOS's capture the major molecular level effects; however, convergence difficulties and the existence of several molar volume roots complicate implementation of these EOS routines.

Consequently, considering the above difficulties, we have developed an approach which incorporates the rich molecular description of the SAFT equations into a cubic in molar volume form. Specifically, molecular simulation data for hard-sphere compressibility factor is nearly exactly fit to a simple empirical function in packing fraction, which ensures an overall cubic in molar volume equation of state and from which two universal parameters (applicable to all fluids and fluid mixtures) are obtained. Furthermore, London dispersion attractive forces are described by an extremely simple polynomial refit of data generated for Chen and Kreglewski's (1977) *thirty-six term* double sum in reciprocal absolute temperature and packing fraction. Finally, chain formation (covalent bonding) is described by a modified form of Wertheim's (1984) thermodynamic first-order perturbation theory (TPT1) for segment covalent bonding, in which a diimer reference radial distribution function is employed to obtain more accurate results. Conventional forms of the hard-dimer radial distribution function (Chiew, 1991; Yethiraj and Hall, 1992; Ghonasgi and Chapman, 1994; Sadus, 1999) were refit to a novel empirical functional form which results in an equation of state which is overall cubic in molar volume.

The resulting theoretically-based EOS was applied to predict several model shale gas properties including compressibility factor versus packing fraction isotherms, the vapor-liquid coexistence curves, the second-virial coefficient versus temperature and the self-diffusion coefficient versus pressure isotherms for the constituent pure fluids; and vapor-liquid equilibria and binary diffusion coefficients versus pressure for binary systems. The observed performance of the molecularly-based *cubic* EOS presented in this paper is generally better than the performance of conventional cubic equations of state and about equal to the performance of the more complex non-cubic SAFT equations of state for the same fluids but with much less computational effort. Ongoing work involves development of *cubic* EOS terms which account for multipolar and hydrogen bonding effect to include water and other species in the shale gas mixtures.

REFERENCES

Chapman, W. G., K. E. Gubbins, G. Jackson and M. Radosz, *New Reference Equation of State for Associating Liquids*, Ind. Eng. Chem. Res. **29**, 1709-1721 (1990).

Chen, S. S. and A. Kreglewski, *Applications of the Augment van der Waals Theory of Fluids: I. Pure Fluids*, Berichte der Bunsengesellschaft fur Physikalische Chemie, **81**, 1048-1052 (1977).

Chiew, Y. C., *Percus-Yevick Integral Equation Theory for Athermal Hard-Sphere Chains. II. Average Intermolecular Correlation Functions*. Molec. Phys. **73**, 359-373 (1991).

Ghonasgi, D. and W. G. Chapman, *A New Equation of State for Hard-Chain Molecules*, J. Chem. Phys. **100**, 6633-6639 (1994).

Huang, S. H. and M. Radosz, *Equation of State for Small, Large, Polydisperse and Associating Molecules*, Ind. Eng. Chem. Res., **29**, 2284-2294 (1990).

Peng, D.-Y. and D. B. Robinson, *A New Two-Constant Equation of State*, Ind. Eng. Chem., **15**, 59-64 (1976).

Redlich, O. and J. N. S. Kwong, *On the Thermodynamics of Solutions. V. An Equation of State. Fugacities of Gaseous Solutions*, Chem. Rev. **44**, 233-244 (1949).

Sadus, R. J., *A Simple Equation of State for Hard-Sphere Chains*, AIChE J., **45**, 2454-2457 (1999).

Soave, G. *Equilibrium Constants from a Modified Redlich-Kwong Equation of State*. Chem. Eng. Sci., **27**, 1197-1203 (1972).

van der Waals, J. D., *Over de Continuiteit van den Gas- en Vloeistoftoestand (On the Continuity of the Gas and Liquid State)*. Ph.D. thesis, Leiden, The Netherlands, (1873).

Wertheim, M. S., *Fluids with Highly Directional Attractive Forces I. Statistical Thermodynamics*, J. Stat. Phys., **35**, 19-34 (1984).

Yethiraj, A. and C. K. Hall, *Monte Carlo Simulations and Integral Equation Theory for Microscopic Correlations in Polymeric Fluids*, J. Chem. Phys., **96**, 797-809 (1992).

SOLID FUEL TO NATURAL GAS CONVERSIONS FOR EXISTING BOILER APPLICATIONS

William Gurski
Director Power Sales
Zeeco Inc.

John Guarco
Technical Director Boiler Burners
Zeeco Inc.

Nando Nunziante
Manager Power Controls
Zeeco Inc.

Prepared for Presentation at the Year 2014 Spring National Meeting
New Orleans, LA, March 30 – April 3, 2014

Abstract

Recent discoveries of vast natural gas reserves in the United States have resulted in increased production of Natural Gas, resulting in lower prices. Petrochemical facilities are performing solid fuel conversions on their boilers to natural gas as a cost-effective and efficient fuel solution. Natural Gas is not only economically beneficial but also environmentally efficient with cheaper prices and reduced SO_2, NO_x, and CO_2 emissions. As more facilities consider boiler fuel conversions, it is important to understand all facets of the conversion, from the thermal evaluation of the boiler, to the complete design, supply and installation of the new firing system.

We will provide specific details and recommended practices from a recent Circulating Fluidized Bed (CFB) Boiler fuel conversion application designed for 1.4 billion Btu/hr of heat input for the maximum continuous steam rating. We will also review thermal performance results, design and supply of the complete new gas firing system, and installation conversion assistance for the boiler modifications and firing system installation details.

Introduction

Zeeco was contacted by a large industrial plant owner to assist them with the fuel changeover of a utility designed solid fuel CFB. Due to current contract requirements and the competitive nature of the industry, the plant owner will remain confidential.

The primary goals of the project are as follows:

- Full Conversion from Petcoke firing to Natural Gas
- Maintain boiler steam design capacity (~1200 kpph)
- Meet permitted NO_x requirements – 0.07 lbs/MMBtu
    - 25-100% heat input for NO_x emissions
- Minimal or no impact on the boiler waterside design

The owner provided the original boiler design information for Zeeco's review of potential boiler burner retrofitting options. This was the first of two boilers that would require retrofitting Two sister units were built/commissioned on solid fuels approximately three years previously. The units were originally designed with natural gas start-up burners and commissioned for main fuel support via the petcoke. The petcoke was to be fired and combusted via a circulating fluid bed system. Upon initial advisement from the plant owner, they noted that their petcoke fuel contract was no longer cost effective, as this fuel feedstock was significantly higher in price than originally estimated in the plant system's budget. The boilers were only in operation for less than two years when this decision to switch to natural gas was determined.

Zeeco will review the major facets of this boiler conversion—from the initial proposal phase to the commissioning and startup of the boiler and firing system. We will also discuss the combustion parameters of the technology behind the project's success– Zeeco's DT Ultra-Low NO_x Free-Jet Burner technology.

Combustion Parameters

In order to understand why the ZEECO® Free-Jet technology was chosen, formation of thermal NO_x emissions must first be examined. For gaseous fuels with no fuel-bound nitrogen (N_2), thermal NO_x is the primary contributor to overall NO_x production. Thermal NO_x is produced when flame temperatures reach a high enough level to "break" the covalent N_2 bond apart, allowing the "free" nitrogen atoms to bond with oxygen to form NO_x.

Stoichiometric Equation describing typical combustion in a natural gas fired burner:

Methane & Air with Excess Air

$$2CH_4 + 4\ (XA)\ O_2 + 15\ (XR)\ N_2 \text{ ---> } 2CO_2 + 4H_2O + (XA)\ 15N_2 + (XR)\ O_2$$

Natural air is comprised of 21% O_2 and 79% N_2. Combustion occurs when O_2 reacts and combines with fuel (typically hydrocarbon). However, the temperature of combustion is not normally high enough to break all of the N_2 bonds, so a majority of nitrogen in the air stream passes through the combustion process and remains diatomic nitrogen (N_2) in the inert combustion products. Very little N_2 is able to reach high enough temperatures in the high intensity regions of the flame to break apart and form "free" nitrogen. Once the covalent nitrogen bond is broken, the "free" nitrogen is available to bond with other atoms. Basic chemistry dictates that free nitrogen, or nitrogen radicals will react to other atoms or molecules that can accept them to create a more stable atom. Of the possible reactions with the products of combustion, free nitrogen will most likely bond with other free nitrogen to form N_2. However, if a free nitrogen atom is not available, the free nitrogen will react with the oxygen atoms to form thermal NO_x. As the flame temperature increases, the stability of the N_2 covalent bond decreases, allowing the formation of free nitrogen and subsequently increasing thermal NO_x. Burner designers can reduce overall NO_x emissions by decreasing the peak flame temperature, which can reduce the formation of free nitrogen available to form thermal NO_x.

The varied requirements of Power and Industrial plants entail the use of numerous types and configurations of burners. The method utilized to lower NO_x emissions can differ by application. However, thermal NO_x reduction is generally achieved by delaying the rate of combustion. Since the combustion process is a reaction between oxygen and fuel, the objective of delayed combustion is to reduce the rate at which the fuel and oxygen mix together and burn. The faster the oxygen and the fuel gas mix, the faster the rate of combustion and the higher the peak flame temperature.

Figure 1 plots Peak Flame Temperature against Thermal NO_x created. NO_x emissions increase as the adiabatic flame temperature increases. Slowing the combustion reaction reduces the flame temperature, which results in lower thermal NO_x emissions. The challenge in achieving lower thermal NO_x emissions is not the theory; however, it is in retrofitting the latest burner technologies into older existing boilers without adding expensive external components or processes.

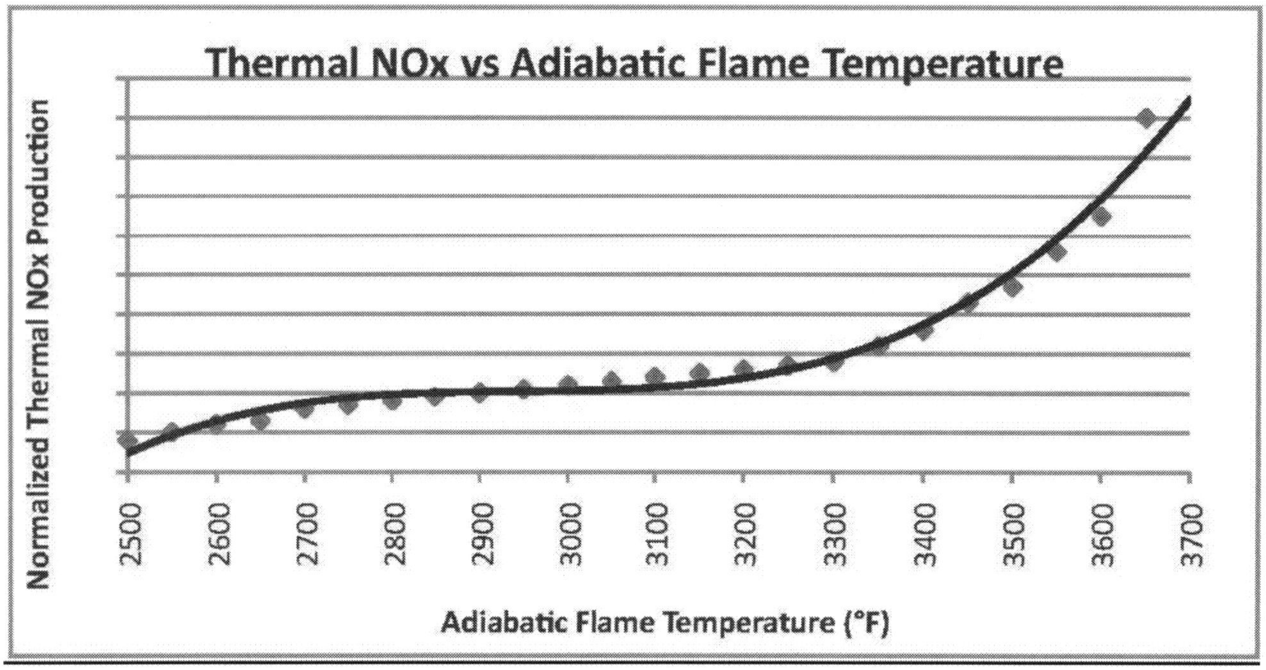

Figure 1: Calculated Peak Flame Temperature vs. Thermal NO_x Production

The industry's standard method to reduce thermal NO_x is to mix the fuel gas together with the inert products of combustion to recondition the fuel before combustion occurs. Since the reconditioned fuel is mainly comprised of inert components, the resulting composition burns at a lower peak temperature. To best utilize the inert products of combustion (flue gas) within the boiler, the fuel gas is introduced along the outside perimeter of the burner tile in an area where flue gas is present while the boiler is in operation. As the fuel gas passes through the inert products of combustion, mixing naturally occurs. This changes the composition of the fuel, and stabilization occurs at the tile exit. Since the reconditioned fuel mixture is 80 to 90% inert in most cases, the resulting flame burns at a lower peak temperature and generates less thermal NO_x.

The mixing of the fuel gas with flue gas prior to combustion is called Internal Flue Gas Recirculation (IFGR). When IFGR is too aggressive, it can result in an increased blower power usage, decreased burner turndown, and increased flame destabilization. Through Free-Jet Theory, maximizing IFGR while maintaining flame stability and flame length can become a challenge.

Flame Stabilization

In order to fully utilize the Free-Jet Theory, Zeeco devised a method to stabilize the burner flame with a highly inert fuel gas/flue gas mixture. This type of combustion is achieved when the flame is stabilized in a low-pressure area created on a series of specially designed hot refractory ledges. As combustion occurs, the refractory ledge retains heat and flame stability is enhanced. Thus, to achieve improved stability and extreme Thermal NO_x reduction, the Free-Jet Technology:

1. Mixes inert flue gas through free-jet methods with all of the fuel gas before combustion occurs, lowering flame temperature.
2. Stabilizes the flame on a refractory ledge, improving flame characteristics

Before combustion is initiated, a furnace is typically filled with normal air, which contains 21% oxygen. Once the burner is ignited, the oxygen content inside the furnace decreases until the burner achieves maximum duty. At this point, the oxygen content in the furnace is normally 2 to 3%. To keep the burner stable throughout the transition from start-up with 21% oxygen to maximum duty with 2 to 3% oxygen, Zeeco developed a series of stabilization ledges as shown in **Figure 2**. These ledges are a design feature of the Zeeco® Free-Jet Burner chosen for the boiler retrofit.

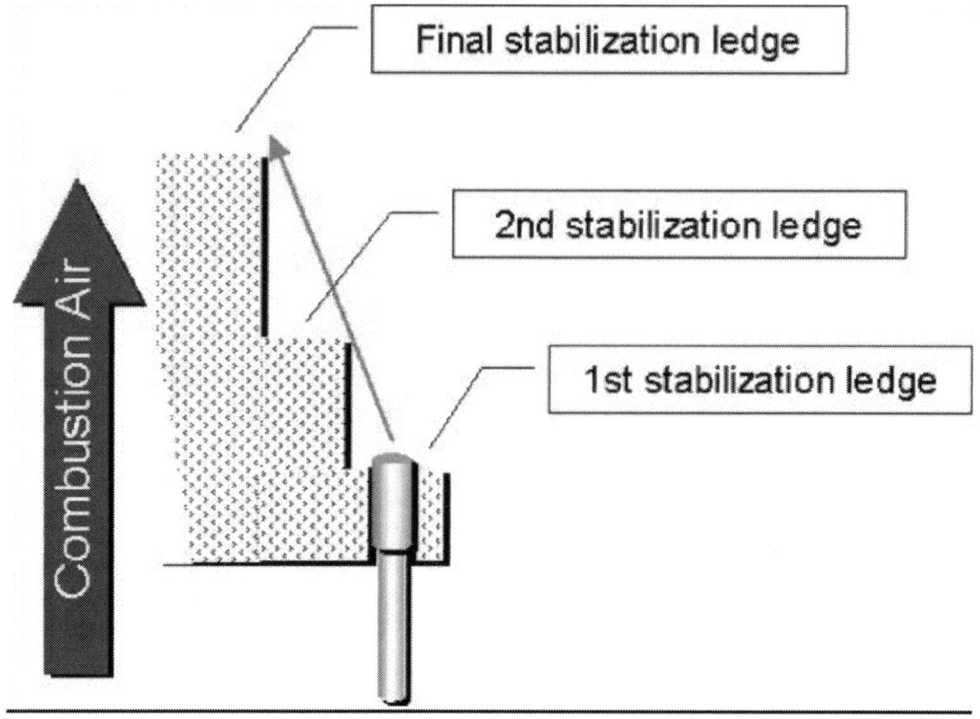

Figure 2: Free-Jet Flame Stabilization Method Illustration

Burner Technology Description

Introduction to Free-Jet Technology

The Zeeco Free-Jet burner series was designed with the specific purpose of maximizing the amount of IFGR to reduce thermal NO_x emissions without sacrificing burner performance with respect to flame length, turndown, and stability. The maximization of IFGR means many of the problems associated with using high levels of EFGR to achieve low emissions can be reduced or eliminated. Specifically, Zeeco's Free-Jet design dramatically reduces or eliminates the need for EFGR by reducing blower power usage, increasing turndown, reducing maintenance and improving flame quality.

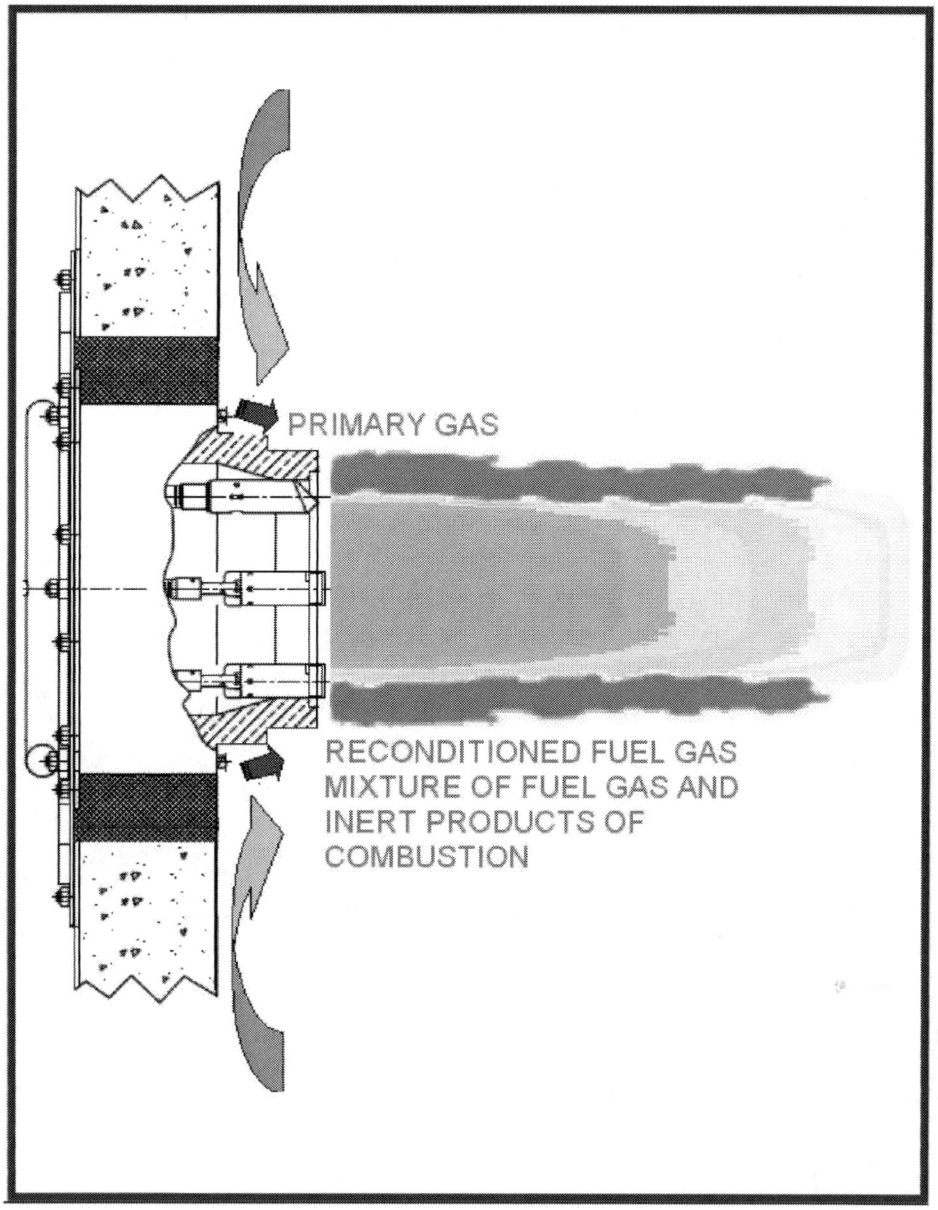

Figure 3: Inert Products of Combustion (Flue Gas)

Pre-Award Engineering

Since the boiler was only in operation for a short period of time, the plant needed assistance with the selection of the new gas firing system and options that would allow them to operate the boiler at the same steam flow capacity and pressure as originally designed for petcoke firing. Zeeco designed three possible boiler conversion options for the operator to consider. Every option required removing the existing start-up burners and closing off the tube openings or retrofitting with new larger burners:

1) Remove the fluid bed bottom and vertically fire the boiler. This option was feasible for a full 100% conversion as the entire fluid bed and nozzle/air system would be removed in favor of six new burners with a corresponding windbox and air system (1500 MMBtu/hr).
2) Removing the four start-up burners and increasing the firing capacity to the necessary levels for full load steam conditions (1500 MMBtu/hr).
3.) Removing the four start-up burners with new burners, plus adding a second level of four burners above to achieve the necessary heat input for full load steam conditions (1500 MMbtu/hr).

Option One was not selected, as it would have required an entire new lower section of the boiler, plus major reconstruction of the fluid bed in order return to solid fuel firing. Option Two was not selected, as the heat release/absorption area within the boiler would have been too great for the water/steam production side.

Option Three was selected, as the owner determined that they wanted the flexibility to return to petcoke firing should the price for natural gas rise to a level higher than petcoke fuel. This allows ~80% of the system's equipment to remain available for future petcoke firing. This solution also offered the least cost to retrofit back to solid fuel firing. The plant owner also had the challenge of the water/steam side of the project, as the boiler system would now be required to operate on 100% natural gas firing instead of a high pressure recirculating fluid bed system. Zeeco assisted the plant owner and the boiler's Original Equipment Manufacturer (OEM) with the system design and selection. Several iterations were submitted by the Boiler OEM on the redesign, and Zeeco worked as a team with the three parties to select the best option for firing the boiler.

The challenge with the boiler redesign interactions was to ensure that the following conditions were met, as required by the plant owner:

- Maintain main steam flow per original design
- NO_x emissions at permit levels
- Boiler able to follow steam needs per electrical generation requirements
- Ability to ramp per the original design conditions
- Turndown of 10-1
- No design impact or changes on existing water or steam circuit metallurgy (tube material changes)

Two design challenges were discovered throughout the Pre-Award Engineering Phase. The first included the air system redesign due to the fluid bed operating on primary and secondary air systems. This is further detailed in the Contract Execution Section below. Another challenge was the integration of the natural gas firing Burner Management System (BMS) and controls into the existing Distributed Control System (DCS). This is also further explained in the Contract Execution Section.

Project Engineering

Once awarded, Zeeco immediately began the burner and BMS systems design. The plant owner required less than 26-week delivery for all components, as the need to fire natural gas was critical to the financial status of the project.

Zeeco began the contract with a full-site visit with the entire team, including the plant owner and boiler OEM. The technical requirements and time schedule were reviewed to ensure all those involved were informed of the critical nature of the project. The technical details of the project varied, as the unit was being fully converted from a solid fuel CFB to a natural gas fired utility boiler. Many items required detailed analysis during the contract phase, including:

- Burner integration into the existing waterwall and boiler structure
- Addition of one new elevation of burners and effects of flame in this furnace region
- Complete redesign of the burner combustion air system and air heater outlet to a single pressure and delivery system to the new burners
- Integration of the new BMS panel and controls logic into the existing DCS system
- Design and location of all new gas fired fuel valving and rack systems, plus local termination cabinets
- Refractory removal and modification to support new natural gas firing (lower boiler refractory was no longer needed due to the elimination of petcoke and recirculating solids from the cyclone)
- Structural modifications to support all equipment and ancillaries for the natural gas firing (combustion air ducting, burner supports, valve racks, etc.)

Burner integration posed several challenges. The new burners were required to fit into the existing waterwall opening and modified from a downfired arrangement to a horizontal firing arrangement. Zeeco advised the plant owner that the new Zeeco Free-Jet DT burner would be able to fit within the existing opening, and would allow the exact tube opening design to be used for the new upper burner elevations. Zeeco also assisted with the design of the support systems and relocation of all structural steel for the new burner locations (maintenance platforms, etc.).

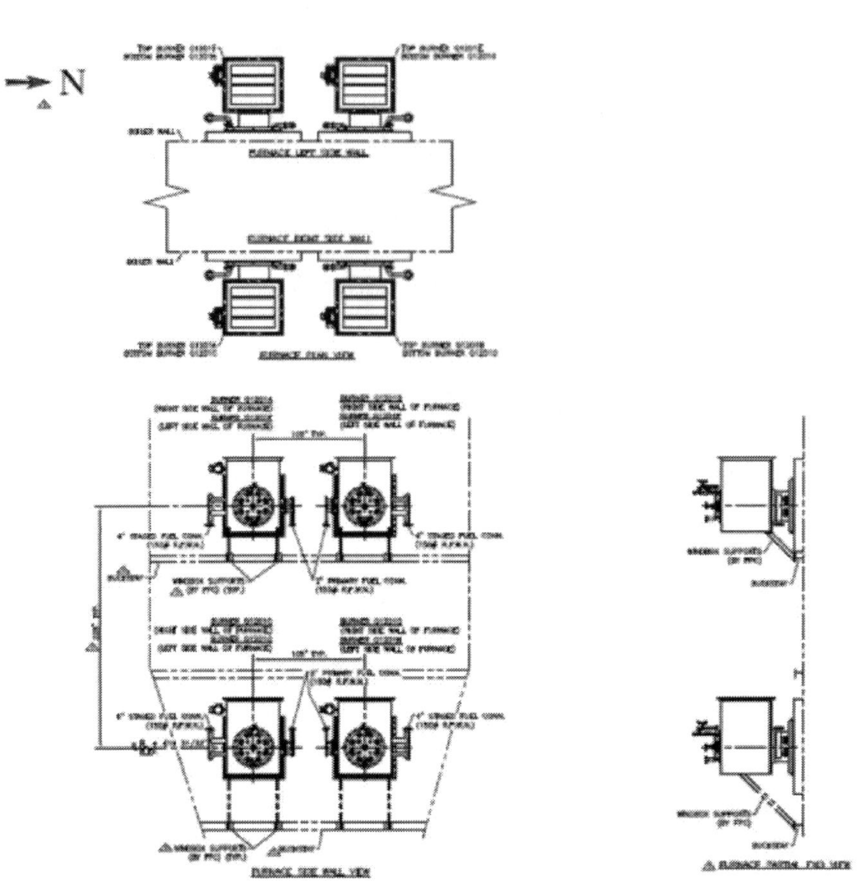

Figure 4: Boiler Design Diagram

It was important to ensure the new upper burner elevation had a proper flame fit within the boiler furnace and would not adversely affect heat transfer to the waterwalls or radiant furnace super heat. An elevation of ~19 feet was selected for the new burners above the existing burners as the ideal location, for not only proper flame fit but also ease the installation of the burners without major boiler structural modifications.

Figure 5: Zeeco's DT Free-Jet Burner

The combustion air system was completely redesigned from a two-system design (primary for the fluid bed nozzles and secondary air system for bed support), to a single pressure design. Zeeco was contracted to assist the plant owner with the entire system redesign, as well as routing to the new burner elevations. Zeeco also assisted with a new design for the air heater outlet and fan control due to the single pressure design needed for the new burners. Zeeco's Model Simulation Engineering Team spent significant time on-site with the Plant Owner's personnel designing the new system. Zeeco's engineers designed and constructed a scaled Plexiglas model of the entire combustion system and ran the Physical Flow Simulation model in our Connecticut facility. The completed model ensured +/- 3% distribution to each burner with less than 10% peripheral distribution around each burner's air entry.

Figure 6: Free-Jet Physical Flow Simulation

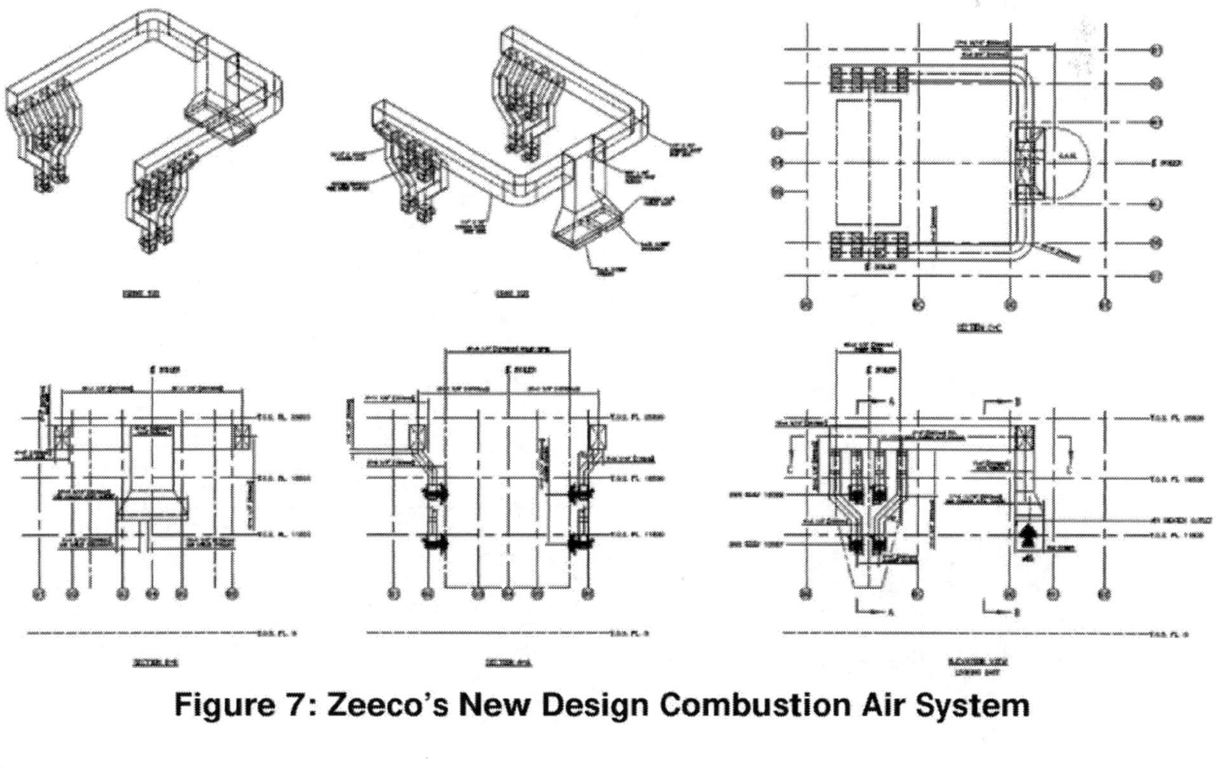

Figure 7: Zeeco's New Design Combustion Air System

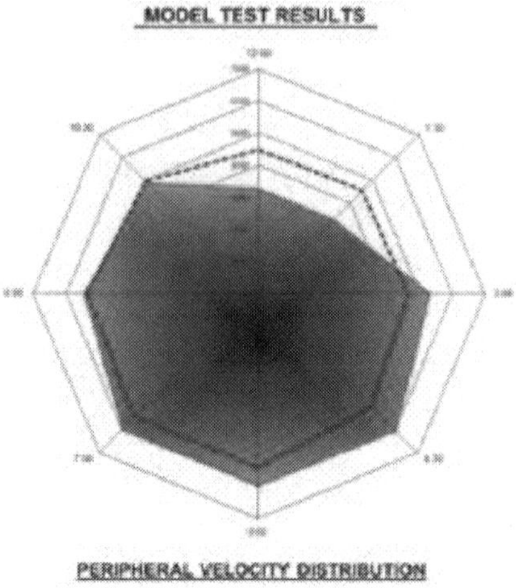

Figure 8: Example of Model Study Peripheral Distribution Chart

Due to the boiler's original controls and Burner Management System's design for main fuel petcoke firing and natural gas start up burners, they were not designed for full load natural gas firing. During the initial design proposal phase, it was agreed that a separate Redundant PLC BMS system would be necessary to integrate into the main DCS system. This would allow the main system to remain "as-is" in the event the unit was to return to petcoke firing in the future. Essentially, it would be a "fuel selector" switch to return the boiler's operation back to petcoke firing. This system was designed in working with the plant owner's original DCS and BMS supplier to ensure a smooth integration into the DCS system.

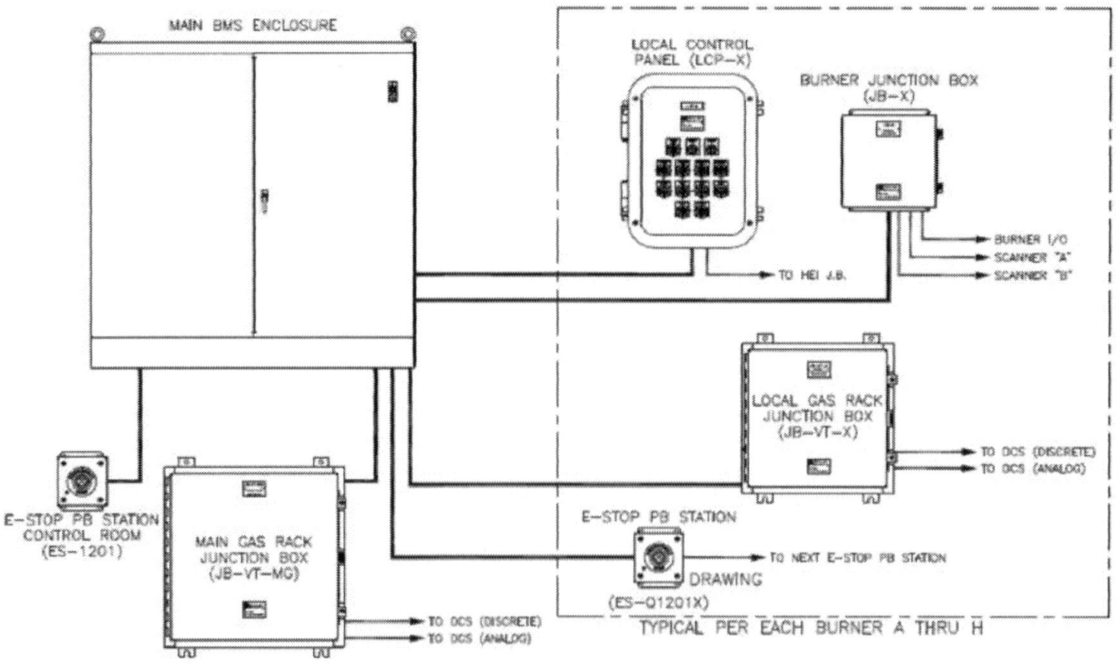

Figure 9: Boiler's Burner Management System

In order for the new burner system to be modified to fire on natural gas, several modifications were required for successful full load natural gas firing, including:

- Removal of furnace's lower slope refractory- During petcoke firing, the turbulent nature of the fluid bed, including the solids return from the cyclone, required the lower slope tubing to be fully protected from potential erosion. When the solid fuel is removed, this lower slope no longer required the refractory coverage to protect the tubing from solids within the cyclone, promoting further heat absorption in the furnace for natural gas firing, and to assist with lower emissions.

- As noted above, the cyclone design in a CFB is to remove solids from the furnace exit flue gas and return them to the fluid bed to promote high efficiency firing and lower carbon loss. With petcoke firing removed, the cyclone was no longer needed for this operation. The inlet to the cyclone plated off, effectively "closing" the inlet so no flue gas would pass through the cyclone. The loop seal exit was "plugged" with refractory to protect it from furnace gasses. This sealing can be deemed 'temporary", as the plate and refractory plug can be removed in the event petcoke firing is deemed cost effective over natural gas firing.

- All back-end clean up equipment are placed out of service due to natural gas firing.

Upon all final design "freeze" and plant owner approvals, fabrication and procurement proceeded and all equipment was delivered to the site by the end of March 2013. Due to the critical schedule, the plant owner was required to take the boiler out of service in late January 2013 to begin the modification process for the natural gas firing. The equipment, upon on-site arrival, was immediately transported to the boiler and the installation process began.

Due to the diligence of the entire engineering team, the integration and installation of the burners, valving and BMS proceeded very smoothly throughout the outage. All the upfront engineering and time ensured the design was cost effective and the outage was completed on time.

Commissioning and Startup

The plant owner also contracted Zeeco's Service Engineers to provide full commissioning and installation assistance, to ensure that all mechanical and electrical equipment was installed per Zeeco's requirements. Zeeco's Service Engineering team was on-site to review and check out the electrical/mechanical/BMS/DCS installation. This proceeded very smoothly during checkout, and all minor issues were addressed in the time allotted for commissioning. Actual equipment start-up proceeded immediately after the final commissioning. Burner performance and full steam boiler operation was achieved in less than two weeks.

Final Conditions are as noted:

- Final emissions performance met the contract and permit requirements of 0.07lbs/MMBtu
- Full load steam boiler requirements were met within the two week timeframe

Conclusion

Boiler steam load and emissions were met in a very timely fashion, and the plant owner was able to generate plant power within less than a month after the initial boiler start-up. The plant owner operated the boiler and turbine/generator near design capacity for over three months after final commissioning. The effort placed on the entire project, pre and post-award, led to the overall success of the boiler's conversion. The integration by the plant owner and the three primary engineering parties led to a fully integrated design that was reviewed by all to ensure the timely success of the project. Based upon the success of the first boiler, Zeeco was awarded the contract for the second CFB boiler conversion, and will begin with installation and commissioning to be in mid-2014.

351797 Vapor Intrusion - Models and Their Value

Abstract:

Vapor Intrusion- Models and Their Value

The problem of vapor intrusion into structures built atop contaminated properties has received increasing attention over the last decade. On the national level, new guidance from the US EPA is imminent, but in the meantime different states have adopted a wide range of policies to address the problem. Many approaches to characterization are empirical, based upon measurements at impacted locations. Since these can involve alarming and intrusive measures involving many residents, some emphasis has been placed on screening, using simple mathematical descriptions of the problem. With the vastly increased availability of powerful fluid dynamics codes, these models can guide more of the site investigation work, as well as evaluate possible possible mitigation strategies. Examples will be given of how such tools can be used to assess both petroleum-based and chlorinated solvent-based situations, based upon recent research carried out in Brown University's Superfund Research Program.

The approach involves solving the equations of vapor transport in the subsurface below the structure of interest, which is typically the soil of the vadose zone. The transport is dominated by diffusion, though entry into the structure may be strongly influenced by local advective transport near the building foundation, as a result of small pressure differences between ambient and the indoors. The indoor environment is typically modeled using a CSTR model. Subtle factors can have a very large influence on predicted indoor air exposures. These include effects related to the resistance of a capillary zone in the case of a chlorinated solvent source in the groundwater, or the form of the assumed biodegradation rate law in the case of petroleum contaminants. The models point to what might represent reliable or problematic field measurement strategies. Also, once a reliable site model is developed, the efficacy of mitigation methods such as subslab depressurization, may be more reliably examined.

(137c) Synthesis, Fine Structural Characterization, and CO2 Adsorption Capacity of Cobalt and Nickel-Based Metal Organic Framework-74

Adhikari, A. K., Yuan Ze University

Lin, K. S.,

Originally presented on: 4/2/2014 14:30:00 - 15:00:00

Abstract: Two metal organic frameworks of MOF-74 group (cobalt and nickel-based) were synthesized, characterized, and evaluated for CO_2 adsorption. The BET specific surface areas of MOF-74(Co) and MOF-74(Ni) MOFs were 1,406 and 1,418 m^2g^{-1}, respectively with the identical pore volume of 0.84 cm^3g^{-1}. CO_2 adsorption capacities of MOF-74(Co) and MOF-74(Ni) were 3.35 and 4.07 $mmol{\cdot}g^{-1}$, respectively measured at 273 K and 1.1 bar. The oxidation state of central atoms in MOF-74(Co) was composed of Co(II) and Co(III) confirmed by XANES spectra while MOF-74(Ni) was Ni(II) central atoms. The bond distances of Co??O and Ni??O were 1.95 and 1.97 Å, respectively.

1. Introduction

Global population and the industrialization are increasing and hence, the energy consumption is also increasing swiftly. Till now, the main source of energy is the burning of fossil fuels. However, burning of fossil fuels releases a vast amount of CO_2 greenhouse gas into atmosphere which causes the raise of average surface temperature of the earth. As the present developments in CCS technology are not sufficiently effective by means of cost and energy consumption, it has prompted to improve the existing technologies and find some new alternatives [1]. Metal organic frameworks (MOFs) are a new class of materials that have attracted substantial attention in past few years [2]. MOFs are synthesized using organic linkers and metal joints that self-assemble to form different dimensional networks. It can generate unusually large diameter channels and cavities which is particularly attractive for CO_2 gas storage application. Here we report the CO_2 gas adsorption capacity on same type of two MOFs; such as MOF-74(Co) and MOF-74(Ni).

2. Experimental

In a typical experiment, 0.75 g of 2,5-dihydroxyterephthalic acid (DHTA) and 3.0 g of $Co(NO_3)_2{\cdot}6H_2O$ or $Ni(NO_3)_2{\cdot}6H_2O$ were dissolved in 150 mL of DMF with sonication in a 250 mL Duran bottle. The bottle was capped tightly and placed in a 110°C oven for 20 h. After cooling to room temperature, the mother liquor was decanted and the crystals were washed with DMF. The product was immerged into methanol for six days. The products were then evacuated and heated under vacuum to150°C over a period of 12 h.

3. Results and discussion

The FE-SEM images of MOF-74(Co) and MOF-74(Ni) samples are shown in Fig.1a and 1b, respectively. It shows that MOF-74(Co) has cauliflower type structure with the particle size of about 4-6 mm and MOF-74(Ni) has uniformly distributed spherical shape with the average diameter of 2-4 mm, although some particles are aggregated. The XRD patterns (Fig.1c) shows that both samples have the unique crystalline phase indexed on the basis of a face-centered hexagonal unit cell of MOF-74. Both samples indicate almost similar diffraction patterns at identical 2?Æ values ascribed the formation of similar crystallographic phases.

Figure 1. -SEM images of (a) MOF-74(Co), (b) MOF-74(Ni), and (c) XRD patterns of synthesized MOF-74 samples.

MOF-74(Co) was thermally stable up to 258°C while in case of MOF-74(Ni) the value was 344°C confirmed from the TGA analysis shown in Fig. 2. The thermogravimetric curves of both samples are consistent with prominent weight-loss steps. It can be seen that the thermal stability is not similar for these two samples although they follow the same trend of loosing weight. In case of MOF-74(Co), a fully hydrated molecular sieve contains about 27 wt.% of water and other guest molecules. and in case of MOF-74(Ni) this value is almost similar, but there is one additional step of weight loss of about 13 wt%.

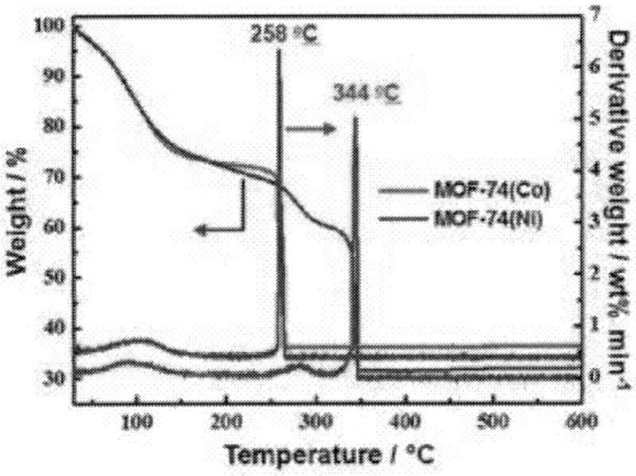

Figure 2. - TG/DTA curves of as-synthesized MOF-74(Co) and MOF-74(Ni)

From the N_2 adsorption isotherms it has been evaluated that both sample follow type-I isotherms with almost no hysteresis loop which corresponding to microporous material. The BET specific surface areas of MOF-74(Co) and MOF-74(Ni) MOFs were 1,406 and 1,418 m^2g^{-1}, respectively with the identical pore volume of 0.84 cm^3g^{-1}. DFT pore size distribution of MOF-74(Zn) and MOF-74(Cu) are shown in Fig. 4a and Fig. 4b, respectively while the values of average pore size and volume are presented in Table 1. These results suggest that these samples are microporous materials.

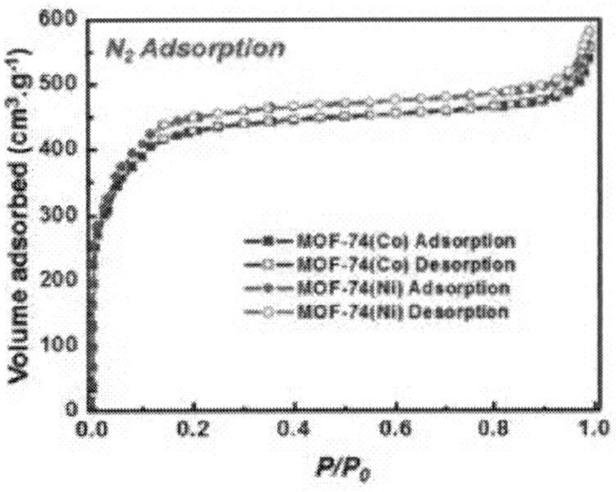

Figure 3. - N_2 adsorption/desorption isotherms of as-synthesized MOF-74(Co) and MOF-74(Ni). Filled and open symbols represent the adsorption and desorption of nitrogen gases, respectively.

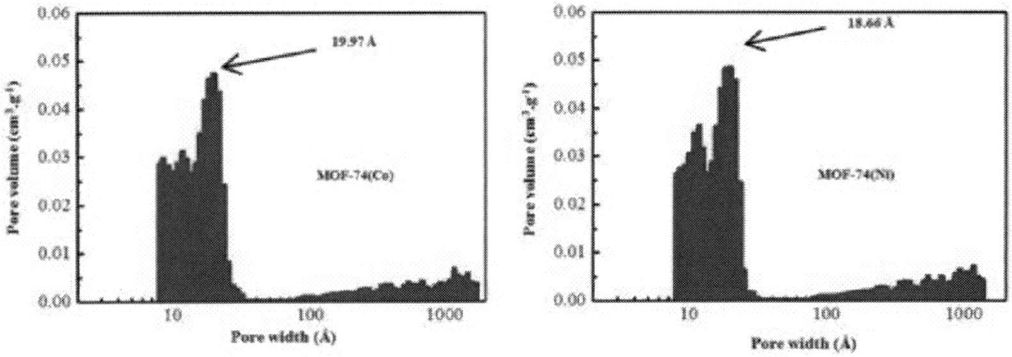

Figure 4. DFT pore size distribution of (a) MOF-74(Co) and (b) MOF-74(Ni)

Table 1. Fine structural parameters of MOF-74 samples analyzed by Co and Ni K-edge EXAFS.

Sample	BET SSA $(m^2g^{-1})^a$	Langmuir SSA $(m^2\ g^{-1})$	Pore volume $(cm^3g^{-1})^b$
MOF-74(Co)	1,406	2,120	0.87
MOF-74(Ni)	1,418	2,123	0.84

[a] *The specific surface area was measured by BET method; and*
[b] *Pore volume was measured by density functional theory (DFT) method.*

Fine structure and oxidation state of Zn and Cu species in MOF-74 were further investigated through X-ray absorption spectroscopy. The data were collected several times and standard deviation was also calculated. The XANES region of the X-ray absorption spectrum is very informative on both the oxidation and coordination states of metal ions. XANES spectra of MOF-74(Co) and MOF-74(Ni) samples are shown in Fig. 5a and 5b, respectively along with the standard materials. The comparison is designed to determine the oxidation states of center atoms in MOFs and understanding the effects of DHTA linkers and guest molecules on the crystal structures. It was found that the edge position (7709.46 eV) of as-synthesized MOF-74(Co) sample is very close to the Co(II). However, compared with the standard Co(II) the offset was +0.46 eV, revealed that Co ions were not entirely surrounded by oxygen atoms and hence formed hydroxides (OH^-) through covalent bond. This result reveals that the as-synthesized sample was not totally moisture or guest molecule free. The source of the OH^- ions is either adsorbed moisture or the organic linkers that cause the little distortion in the crystal structure. In case of MOF-74(Ni) the offset was +0.57 eV. Fine structural parameters for both samples are shown in Table 2 where it can be seen that the bond distances of Co??O and Ni??O were 1.95 and 1.97Å, with the coordination number of 4.27 and 4.23, respectively.

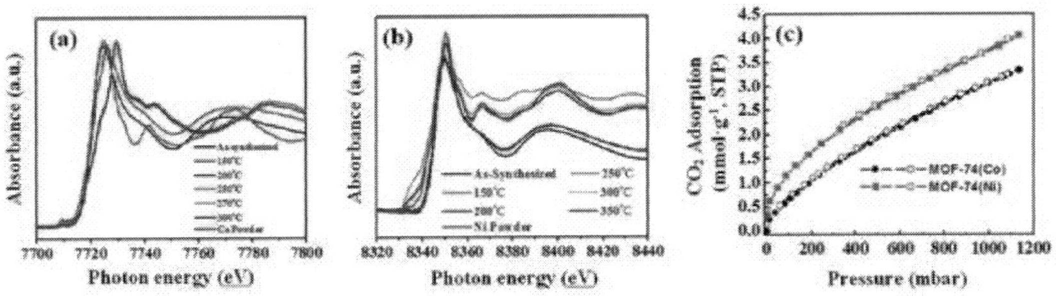

Figure 5. (a) Co K-edge XANES spectra of MOF-74(Co) samples with Co(0) standard, (b) Ni K-edge XANES spectra of MOF-74(Ni) with Ni(0) standard, and (c) CO_2 adsorption isotherms of MOF-74 samples measured at 273 K. Filled and open circles represent the adsorption and desorption, respectively.

Table 2. Fine structural parameters of MOF-74 samples analyzed by Co and Ni K-edge EXAFS.

Sample	Shell	$CN^a \pm 0.5$	$R^b \pm 0.002$ (Å)	$\Delta\sigma^{2c}$ (Å^2)
MOF-74(Co)	Co—O	4.27	1.953	0.0037
MOF-74(Ni)	Ni—O	4.23	1.974	0.0038

[a]Coordination number; [b]Bond distance; [c]Debye-Waller factor.

The CO_2 adsorption capacities of the synthesized samples are shown in Fig. 5c. The steady ascending slope of the CO_2 adsorption isotherms represents strong adsorption of CO_2 molecules having quadrupoles onto the open metal sites in MOF-74 samples. The CO_2 adsorption capacity of MOF-74(Ni) measured at 273 K and 1.1 bar was 4.07 mmol·g^{-1}; significantly higher than MOF-74(Co) which was 3.35 mmol·g^{-1}. However, both results are much higher than other prospective adsorbents of CO_2 such as Zeolite 13X [3] and ZIF-96 [4]. The elevation in CO_2 adsorption in MOF-74(Ni) may attribute to the higher surface area than MOF-74(Cu) and more importantly the strong affinity to CO_2 molecules due to different electronic configurations. It can be seen that the adsorption and desorption cycles are almost reversible for both samples. This integrity was also corroborated by the N_2 adsorption. The steady uphill slope of the adsorption curves indicates that higher adsorption capacity may obtain at higher pressure.

4. Conclusions

The MOF-74 samples were capable to store significant amount of CO_2 at relatively lower pressure. The difference in CO_2 adsorbed amount between two samples may attribute to the different affinity of cobalt and nickel ions to CO_2 molecules as well as the specific surface area. The moderate thermal stability confirms that these MOFs may apply to control the CO_2 gas from the emission of refinery or other industries.

References

1. A. Torvanger, M. T. Lund, N. Rive, Mitig. Adapt. Strat. Gl. 18 (2013) 187.

2. T. G. Glover, G. W. Peterson, B. J. Schindler, D. Britt, O. M. Yaghi, Chem. Eng. Sci. 66 (2011) 163.

3. J.-S. Lee, J.-H. Kim, J.-T. Kim, J.-K. Suh, J.-M. Lee, C.-H. Lee, J. Chem. Eng. Data 47 (2002) 1237.

4. W. Morris, B. Leung, H. Furukawa, O. K. Yaghi, N. He, H. Hayashi, Y. Houndonougbo, M. Asta, B. B. Laird, O. M. Yaghi, J. Am. Chem. Soc. 132 (2010) 11006.

Treatment of Flowback Water from Hydraulic Fracturing with Biochar

Maoqi (Mark) Feng

Steven Cooks

Zhigang Feng

April 2, 2014

AIChE 2014 Spring Conference, New Orleans

Chemical Engineering Dept., Div. of Chemistry and Chemical Engineering, Southwest Research Institute, San Antonio, TX 78238

Outline

- Introduction
- Technical Approach
- Experimental Work
- Summary & Conclusions

2

Water Demand in Hydraulic Fracturing

- Water demand is huge: 100 billion gallons of water used for 39,000 wells in 2011
- Fracking accounts for 50% of total water consumption in some Texas areas

3

Water Requirement in Shale Gas Production

- Water requirements for hydraulic fracturing, by shale plays in Texas

Water Requirements (gallons per well)	Low Value	Median Value	High Value
Barnett Shale	<1 MM	2.6 MM	>8 MM
Haynesville and Bossier Shale	< 1 MM	5.5-6 MM	>10 MM
Eagle Ford Shale	1 MM	6-6.5 MM	13 MM
Woodford, Pearsall, and Barnett-PB Shale	<1 MM	0.75-1 MM	<5 MM

4

Flowback Water Recycling

- Half of fracturing wells are in water-stressed areas
- Flowback water reuse is needed

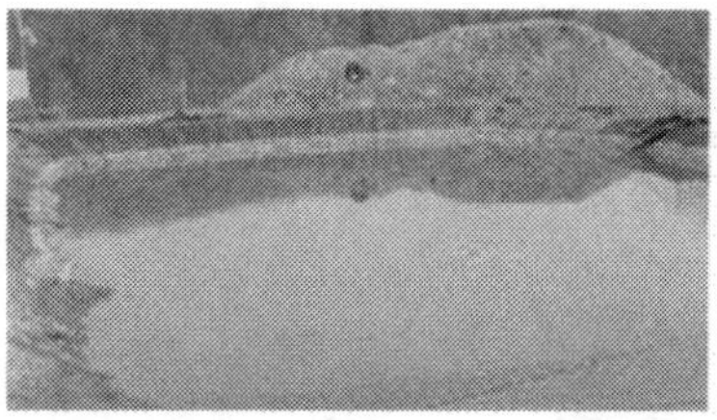

5

Reusable Water Quality

- Parameter Range

TDS at 180°C, mg/l	9,000–16,000
Turbidity, NTU	0-5
pH	6.5-8
Iron, mg/l	1-10
Chloride, mg/l	5,000-10,000
Potassium, mg/l	100-500
Calcium, mg/l	50-250
Magnesium, mg/l	10-100
Sodium, mg/l	2,000-5,000
Boron, mg/l	0-20

6

Flowback Water Treatment

- Membrane method is expensive and cannot handle solids and hydrocarbons
- Zeolites generate more solid waste
- Bioremediation is not efficient
- Nanofiltration is easily plugged
- Activated carbon adsorption is more expensive

7

Why Biochar

- Biochar is a good sorbent
- Green
- Biodegradable
- Soil compatible and amendment
- Easy to prodcue
- cost effective
- Mass production available

8

Biochar vs. Activated Carbon

- Biochar is produced at lower temperature
- Biochar is a precursor for activated carbon
- Activated carbon needs an activation step
- Biochar is more feedstock flexible

9

Biochar Production

- Biochar production from wood chips by pyrolysis
- Temperature 350°C – 550°C

10

Biochar Property

- Pine wood biochar carbon content vs. pyrolysis temperature

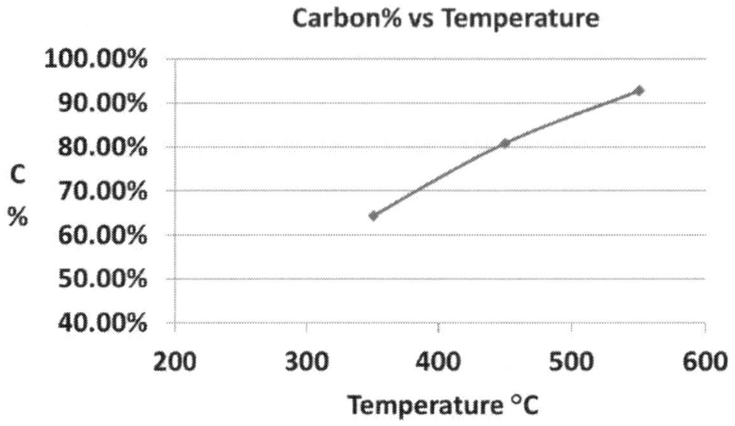

11

Biochar Property

- Oak wood biochar carbon content vs. pyrolysis temperature

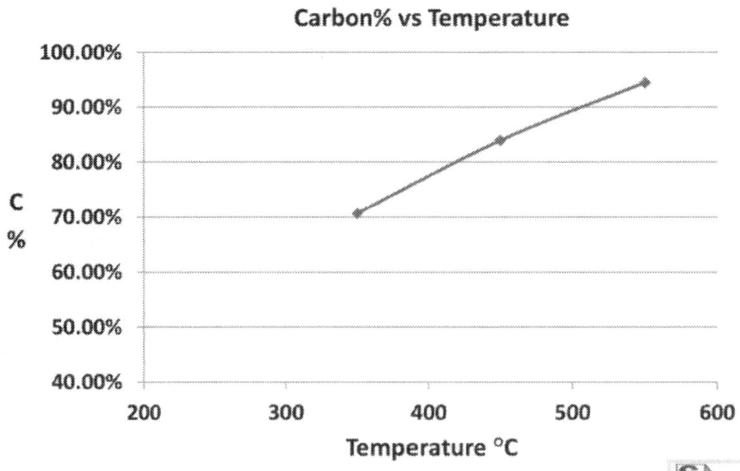

12

Biochar Preparation

- Biochar prepared from different sources at 350°C

Feed	Carbon%
Cellulose	90.09%
Pine Wood	66.43%
Oat Hulls	58.59%

13

Batch Adsorption

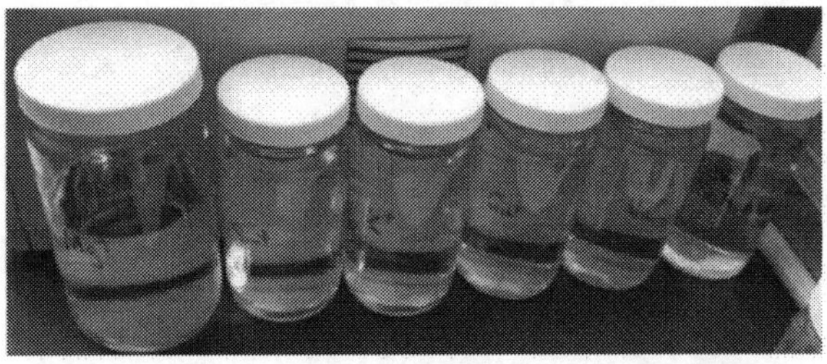

14

Batch Adsorption

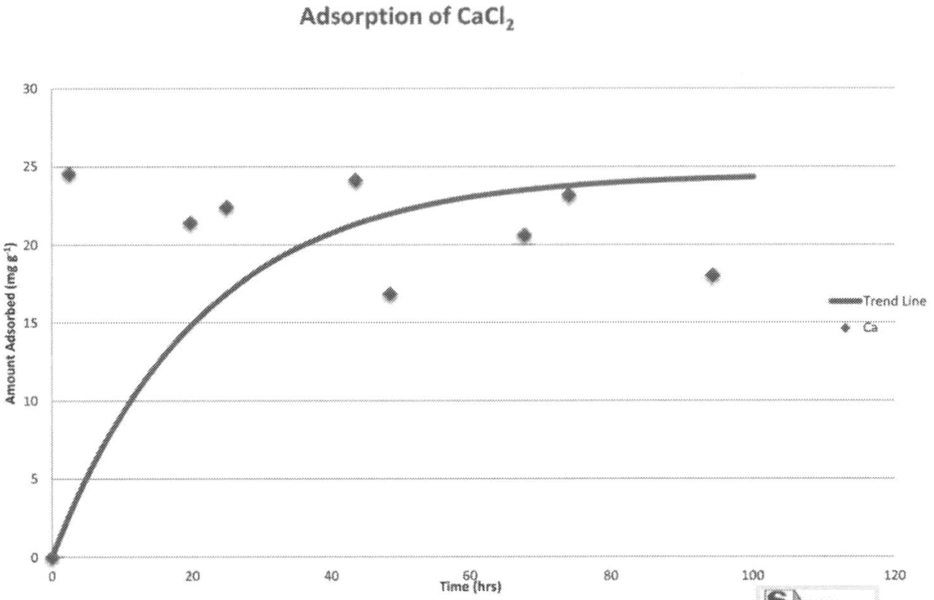

15

Sorbent Bed Evaluation in Continuous Flow Mode

- Quick evaluation of sorbent capacity in continuous flow mode – more efficient than batch mode

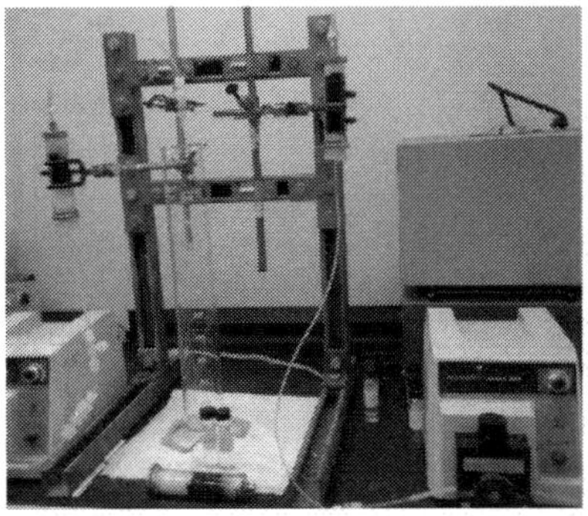

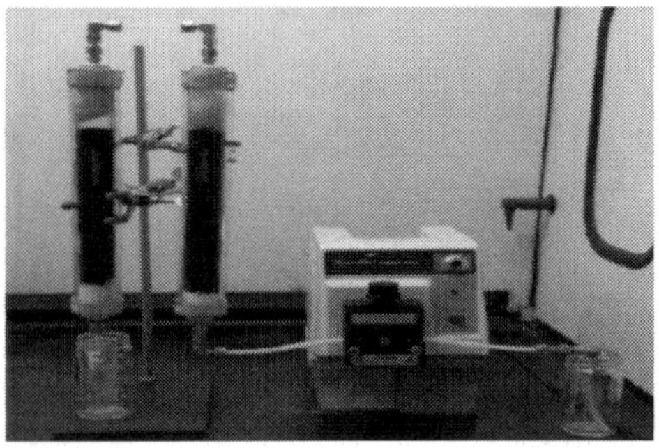

16

NaCl Removal

- NaCl removal

Time (hrs)	NaCl Mass (g)	Water Mass (g)	NaCl Concentration, wt%	NaCl wt% Adsorbed
0.00	12.75	247.5	4.90%	0.00%
2.62	0.8921	18.5788	4.58%	0.32%
19.53	0.9598	19.3488	4.73%	0.17%
24.57	0.9886	19.7544	4.77%	0.13%
43.37	0.9908	20.1408	4.69%	0.21%
48.32	0.9065	18.6041	4.65%	0.25%

17

Mg Removal

- $MgCl_2$ removal

Time (hrs)	$MgCl_2$ Mass (g)	Water Mass (g)	$MgCl_2$ Concentration, wt%	$MgCl_2$ wt% Adsorbed
0.00	4.8754	38.5514	11.23%	0.00%
0.5	4.1715	38.4528	9.79%	1.44%
1.0	4.3688	39.0205	10.07%	1.16%
1.5	4.6866	39.086	10.71%	0.52%
2.0	4.3229	38.8608	10.01%	1.22%

18

Ca Removal

- $CaCl_2$ removal

Time (hrs)	$CaCl_2$ Mass (g)	Water Mass (g)	$CaCl_2$ Concentration, wt%	$CaCl_2$ wt% Adsorbed
0.00	3.975	38.7583	9.30%	0.00%
0.5	3.9614	39.2739	9.16%	0.14%
1.0	3.936	38.9826	9.17%	0.13%
1.5	3.9684	39.5214	9.12%	0.18%
2.0	3.9067	39.0601	9.09%	0.21%
2.5	3.9021	39.369	9.02%	0.28%

19

Sr Removal

- $Sr(NO_3)_2$ removal

Time (hrs)	$Sr(NO_3)_2$ Mass (g)	Water Mass (g)	$Sr(NO_3)_2$ Concentration , wt%	$Sr(NO_3)_2$ wt% Adsorbed
0.0	0.4834	20.7043	2.28%	0.00%
0.5	0.4557	20.665	2.16%	0.12%
1.0	0.4491	21.1634	2.08%	0.20%
1.5	0.4191	20.3263	2.02%	0.26%
2.0	0.4114	20.5186	1.97%	0.32%
2.5	0.4031	20.1342	1.96%	0.32%

20

Inorganics Removal

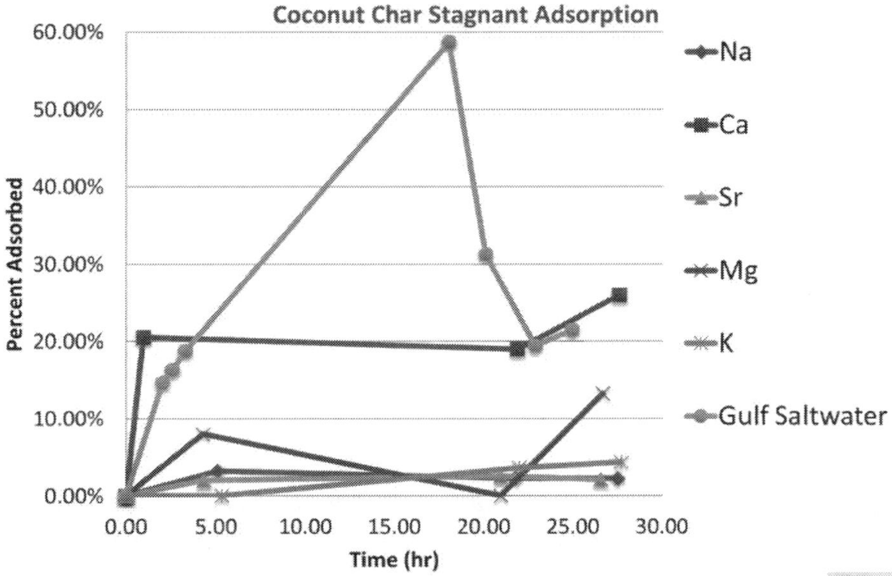

21

Hydrocarbons Removal

- Removal of diesel hydrocarbons in water by biochar adsorption
- Per pass removal 87%
- Higher adsorption capacity for biochar produced at higher temperature

22

Solid Particulates Removal

- Removal of solid particulates in flowback is necessary
- Pump damage
- Long settling time
- Fine solids removal is still in progress

23

Summary

- Biochar production from wood chips
- Quick evaluation of sorbent capacity in continuous flow mode
- Successful removal of inorganic salts: NaCl, K^+, $MgCl_2$, $CaCl_2$, Sr^{2+}
- Successful inorganics removal from sea water
- Successful removal of hydrocarbons from simulated flowback water

24

Some Thoughts

- Biochar adsorption bed can be used with other methods, such as ultrafiltration, reverse osmosis, and ion exchange units
- Biochar sorbent bed can be used as a pretreatment or post-polishing step for other methods

25

Acknowledgement

Thank you!

Questions?

26

352927 Hazards Management for Building Locations through Air Dispersion Modeling

Abstract:

Oil, gas, and petrochemical industries have reached a critical era involving its timeworn infrastructure. Previously, facilities degraded with minimal effort allocated for updates that ensure personnel safety (i.e., process inventory isolation, automative control systems, etc.). More recently, industry as a whole is progressively adopting a more safety conscientious attitude about personnel and the environment. These personnel and environmental concerns are driving new codes, standards, recommended practices, and regulations to be developed and implemented. Safety studies are being performed which assess facilities to determine if these new regulations are met, while providing key insight to eliminate any deficiencies that are discovered. Available advanced air dispersion models estimate or predict the concentration of air pollutants or toxicants emitted from various sources such as industrial plants, road traffic vehicles or accidental chemical releases. Facility siting assessments are amongst the required studies used for managing the risk to on-site personnel located in new and existing buildings intended for occupancy, where air dispersion models can be used to determine the consequences of accidental releases of hazardous or toxic materials. This paper discusses case studies carried out for hazards management for building locations with different air dispersion modeling software tools including: SCREEN3, AERMOD, ALOHA, and PHAST.

353139 H2S Removal from Biogas Using an Ionic Liquid as Physical Absorbent

Abstract:

Biogas is nowadays a valuable renewable energy carrier which can be used directly as a fuel or as a raw material for the production of synthesis gas and/or hydrogen. Methane (CH_4) and carbon dioxide (CO_2) are the main constituents, but biogases also contain significant quantities of undesirable compounds (contaminants), such as hydrogen sulfide (H_2S) and ammonia (NH_3). The existence and quantities of these contaminants depend on the biogas source (i.e., landfills, anaerobic fermentation of manure). Their presence constitutes a major problem because (1) they can adversaly affect any biogas thermal or thermocatalytic conversion device (e.g., corrosion, erosion, fouling); and (2) they produce harmful environmental emissions. It is therefore important to develop a reliable and efficient biogas purification process. This work is aimed at presenting an alternative removal approach for H_2S from biogas based on physical absorption using an ionic liquid as solvent.

HYDROGEN GENERATION PLANT DESIGN

UNIVERSITY OF HOUSTON
SPRING 2014
Cathy Cortez-Lambert, Diseye Komonibo, Guoda Liang,
Jose Tusell, Sukhpal Singh, Miguel Fleischer

Outline

- Introduction
- Plant Specifications
- Health, Safety & Environmental
- Equipment Specifications
- Heat Integration
- Cost Analysis
- Recommendations
- Questions

Introduction

- **Chemical Engineering Design I at UH (fall)**
  - ➢ The objective is to design an entire plant given a feedstock and product specifications
  - ➢ Calculations are done without the use of a computer-aided design software package
  - ➢ Teams are selected by the professor (usually 4 or 5)
  - ➢ Industry representatives review the design twice in the semester to provide feedback
    - ➢ Fluor is the current sponsor
    - ➢ Previous sponsors: Dow, ExxonMobil, Fluor, Oxy, Sabic, Shell (2)

- **Chemical Engineering Design II at UH (spring)**
  - ➢ The exact same project is assigned to basically the same teams
  - ➢ Calculations are done with the use of Aspen to optimize the fall design
  - ➢ Each team is assigned a different location in the world
  - ➢ Profitability calculations are also done to compare the various locations
  - ➢ Industry representatives review the design twice in the semester to provide feedback

Introduction

Project

Design a Hydrogen Generation Plant based on Steam-Methane Reforming

Plant Specifications: Block Diagram

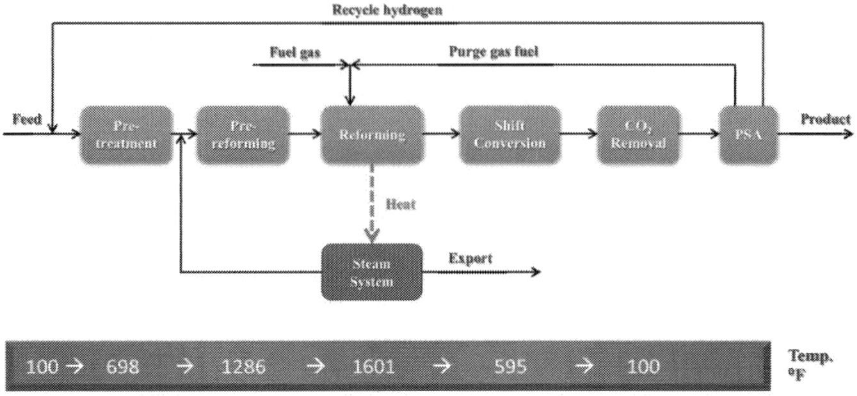

Plant Specifications: PFD Drawing

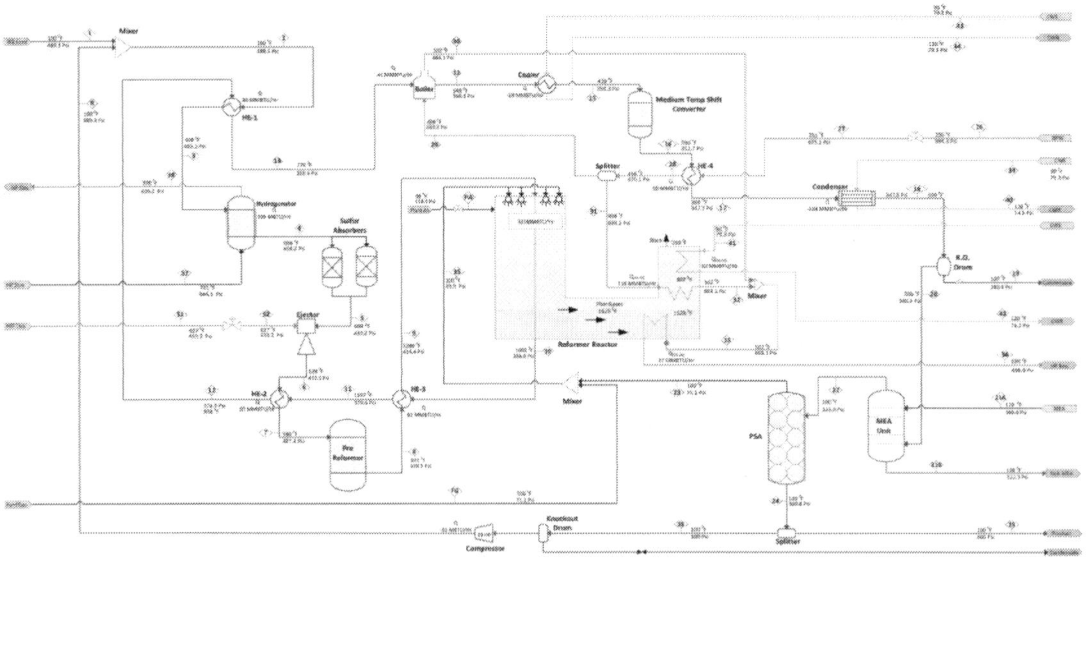

Health, Safety & Environmental

- **Environmental**
  - Emissions
    - Stack Gas
      - CO_2 Tax Permit
  - ZnS Containment
  - Cooling Towers
    - Monitor Emissions

- **Safety**
  - OSHA Standard Requirements
  - Process Hazard Analysis (PHAs)
  - Inspections & HAZOPs
  - SOPs and Training Enforcement
  - Strict PPE

Hydrogenator

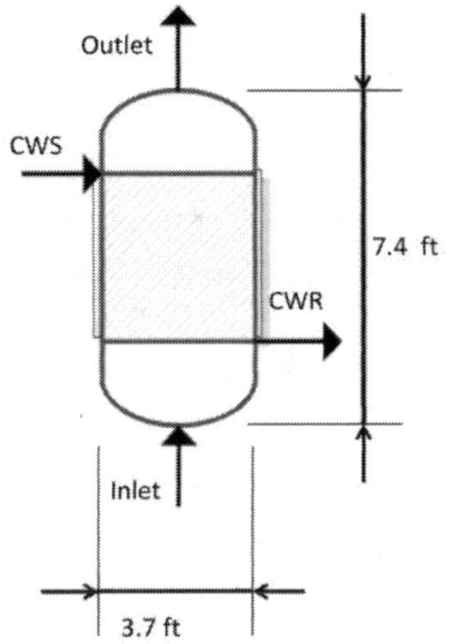

Operating condition: 708°F, 639psi
Residence time ---- 2.6 s
Over 99.5% conversion
Co-Mo/Al_2O_3 as catalyst
Reactor size---79.8 cubic feet
Constructure material: SS316l

Element	316L
Chromium	16.0 min. – 18.0 max.
Nickel	10.0 min. – 14.0 max.
Molybdenum	2.00 min. – 3.00 max.
Carbon	0.030
Manganese	2.00
Phosphorous	0.045
Sulfur	0.03
Silicon	0.75
Nitrogen	0.1
Iron	Balance

Sulfur Absorber

Residence time ---- 1.8 s
Reactor size -- 57.2 cubic ft
Recycle period ---- 20 days
Two ZnO beds

HS&E:
SS 316l
Sealed storage
Containment tank
Personal protective equipment

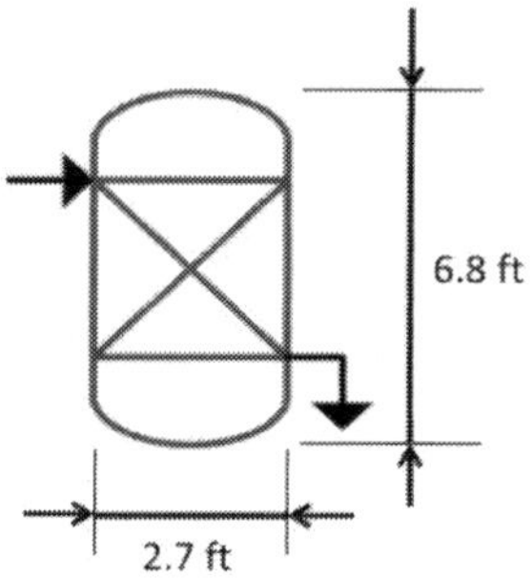

Pre-Reformer

- Capacity Increase
- Energy Savings
- Crack $>C_1$ Hydrocarbons
- Larger Operational Flexibility

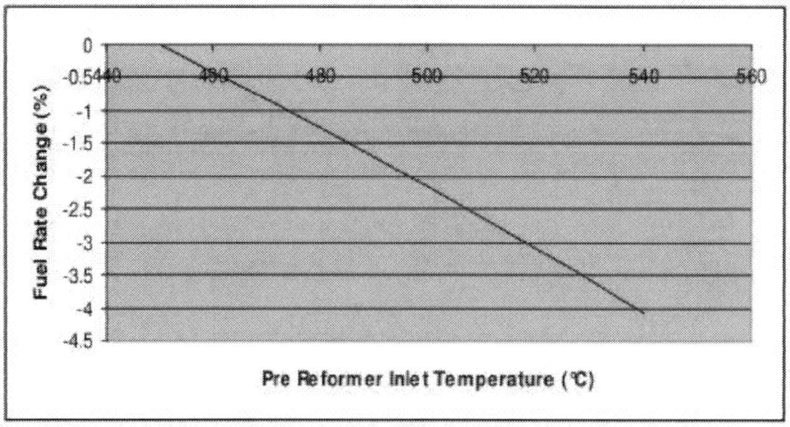

Pre-Reformer

$286679\frac{lbm}{hr}$ @980°F, 427.4PSIA

Packed Bed ADIABATIC RX with ΔT=139°F

- ❖ $C_nH_m + nH_2O \rightarrow nCO + \left(n + \frac{m}{2}\right)H_2$
  - $C_2H_6 + 2H_2O \rightarrow 2CO + 5H_2$
  - $C_3H_8 + 3H_2O \rightarrow 3CO + 7H_2$
  - $C_4H_{10} + 4H_2O \rightarrow 4CO + 9H_2$

  } X 100%

- ❖ $CH_4 + H_2O \rightarrow CO + 3H_2$
- ❖ $CO + H_2O \rightarrow CO_2 + H_2$

  } X 7.5%

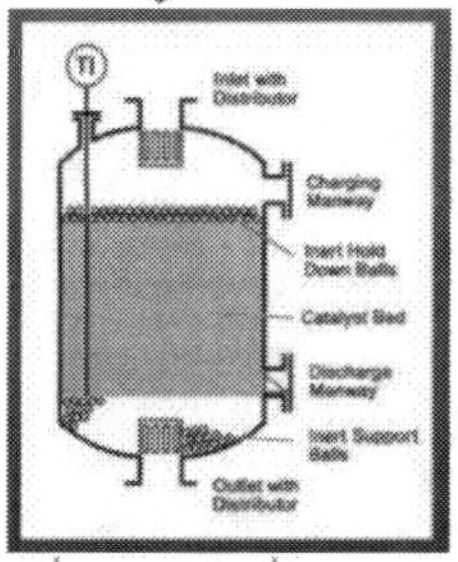

$\tau = 2\ sec$

15′2″

5′1″

MOC: SS 405

$286679\frac{lbm}{hr}$ @ 841°F, 419.5PSIA

Reformer

Reactions

$CH_4 + H_2O \rightarrow CO + 3H_2$

$CO + H_2O \rightarrow CO_2 + H_2$

} **X = 78.7%**

Requires
ΔT=317°F
Q = 325MM$\frac{BTU}{HR}$

SIZE	
Heat Flux	≈ 17,000$\frac{BTU}{Hr\ Ft2}$
Volume	= 11,860 GAL
Tubes Diameter	= 4.5″
Tubes Length	= 43′0″
# of Tubes	= 334
MOC	= HP-MA

$286679\frac{lbm}{hr}$ **@1284°F**, 414 PSIA

- Excess 1.2% Air
- Plant Fuel
- PSA Purge Fuel

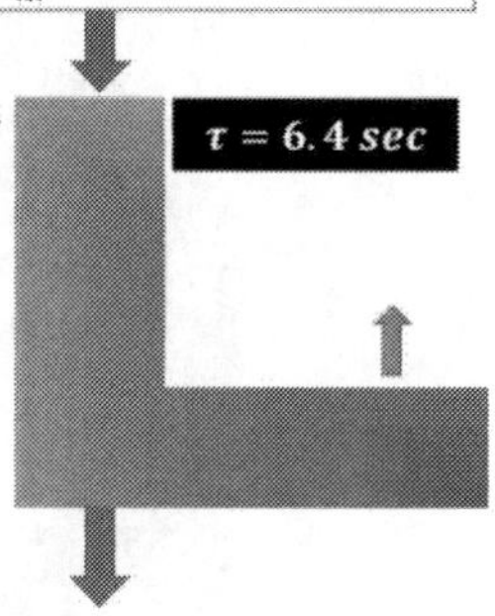

$\tau = 6.4\ sec$

1601°F, 384 PSIA

Medium Temperature Shift Converter

Adiabatic Operation

Conversion: 95%

Residence Time: 4.5 s

Reactor Size: 646 cubic feet

H/D Ratio: 2

Cost: $ 188,000

Material of Construction: Inconel® 600

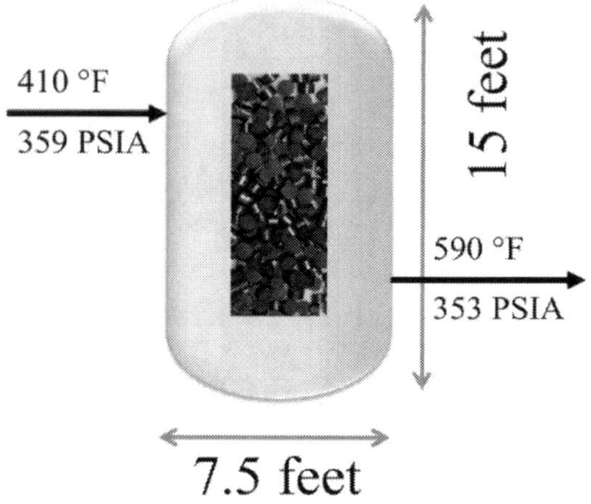

MEA Unit

Acid Gas Absorber			
Service:	CO2 Absorber	Tag:	MEA Absorber
Ave. Process T:	110 F	Average Process P:	323PSI
Process Stream:	Hydrocarbon	Absorbent:	MEA + Water
Column Diameter:	8.38 FT	Number of stages:	20
Column height:	54 FT	Tray spacing:	24"
MOC:	316 stainless steel	Loading Factor:	0.1 CO2/MEA
ΔH:	Adiabatic	Tray type:	Sieve trays
Cost		$1,018,019	

Pressure Swing Adsorber

- **Operating at** 100°F, 300 psia
- 5A zeolite, activated carbon
- Efficiency
  - 10% H_2 adsorbed
- 10 beds
  - 1 adsorbent bed per column
  - Cycle Time: 12 minutes
  - Column Volume: 954 ft³
- Purge Tail Gas
  - ~ 99.9% CH_4 removal
- Cost
  - Carbon Steel
  - $332,615 per bed
  - 10 beds: $3,326,149

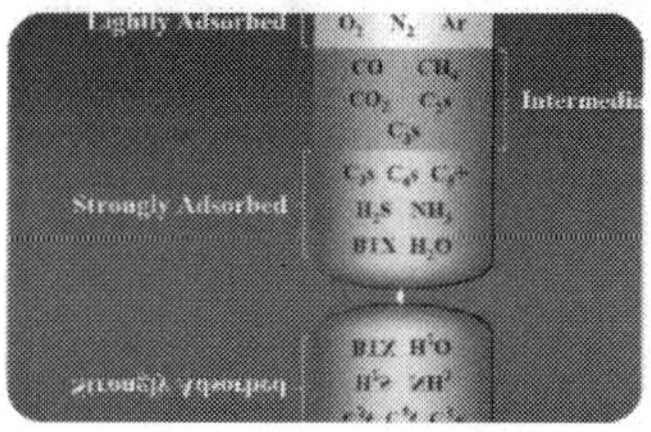

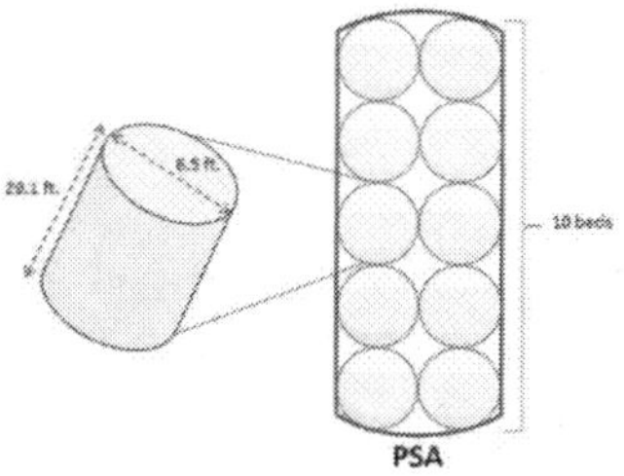

Heat Integration

Reformer Efficiency			AIR			Temperature of Flue Gases
Heat			20 % Excess Air & RH 68%			°F
Provided	560	MMBtu/Hr	N_2	281982	lbm/Hr	3434 Adiabatic Flame Temp.
CH_4 Cracking	325	MMBtu/Hr	H_2O	5817	lbm/Hr	1629 Convection Section
Saturated Steam	115	MMBtu/Hr	O_2	85617	lbm/Hr	1528 Convection Section
Superheated Steam	17	MMBtu/Hr	Total Air	373416	lbm/Hr	807 Convection Section
Stack Cooler	82	MMBtu/Hr	Total Fuel	37273.284	lbm/Hr	250 Atmosphere Stack
Total Heat Used	457	MMBtu/Hr				

Reformer Overall η	Furnace Tubes η
81.62%	58.00%

Total Steam Export
217,773 lbm/hr @ 600 °F, 658 PSI

Plant Heat Overall Efficiency
60%

Cost Analysis

Equipment name	Temp. range (°F)	Max P (psig)	MOC	Diameter (ft)	Volume (ft^3)	Heat Transfer Area (ft^2)	Power (BTU)	Quantity	Cost
HE1	[100-938]	474	SS416			1539		1	$ 235,686
HE2	[698-725]	418	SS405			4530		1	$ 388,390
HE3	[679-1197]	405	SS302			5179		1	$ 417,198
HF4	[841-1601]	660	CS			26641		1	$ 512,475
Reformer Tubes	[1286-3434]	400	HP-MA			16911		1	$ 1,276,463
Steam SH	[498-776]	648	SS302			530		1	$ 174,321
Boiler	[90-548]	654	CS			4486		1	$ 152,502
Steam Boiler	[251-590]	654	SS302			4972		1	$ 436,182
Cooler	[90-269]	350	CS			785		1	$ 66,254
Stack Cooler	[502-1629]	65	CS			4676		1	$ 130,916
Condenser	[498-1528]	333	CS			36319		1	$ 604,150
Hydrogenator	[90-807]	630	SS316l	3.70	79.8			1	$ 63,207
Pre-reformer	[410-590]	417	SS416	5.01	295.3			1	$ 173,958
MTSC	[841-980]	345	Inconel 600	7.44	646.0			1	$ 187,602
KO POT 1	[100-100]	319	CS	5.96	500.0			1	$ 199,688
KO POT 2	[100-100]	285	CS	3.02	54.3			1	$ 27,678
ZnO Absorber	[100-100]	443	SS316l	2.72	57.2			2	$ 228,916
PSA Unit	[698-698]	308	CS	6.46	954.2			10	$ 3,326,149
MEA Unit	[100-138]	326	SS416	8.60	998.5			1	$ 1,018,019
Compressor	[106-192]	475	CS				40	2	$ 68,539
Reformer Furnace	[1286-3434]	100	Various				220201	1	$ 15,618,714
Total								32	$ 25,307,008

Cost Analysis

Direct Costs	% of CBM	Amount
Equipment Purchases	1	$ 25,307,008
Equipment Installation	0.47	$ 11,894,294
Instrumentation and Controls	0.36	$ 9,110,523
Piping	0.68	$ 17,208,765.51
Electrical Systems	0.11	$ 2,783,771
Buildings	0.18	$ 4,555,261
Yard Improvements	0.1	$ 2,530,701
Service Facilities	0.7	$ 17,714,906
Indirect Costs		
Engineering and Supervision	0.33	$ 8,351,313
Construction Expenses	0.41	$ 10,375,873
Legal Expenses	0.04	$ 1,012,280
Contractor's Fee	0.22	$ 5,567,542
Fixed Capital Investment	4.6	$ 116,412,237
Working Capital	0.89	$ 22,523,237
Sub-Total Capital Investment	**5.49**	**$ 138,935,474**
OSBL	0.3	$ 41,680,642.35
Contingency	0.2	$ 27,787,094.90
Total Capital Investment		**$ 208,403,212**

Recommendations

- Improve heat integration
- Replace MTSC with low and high temp. shift converters
- Continuous improvement on process
- Strive to achieve excellence in quality

Questions?

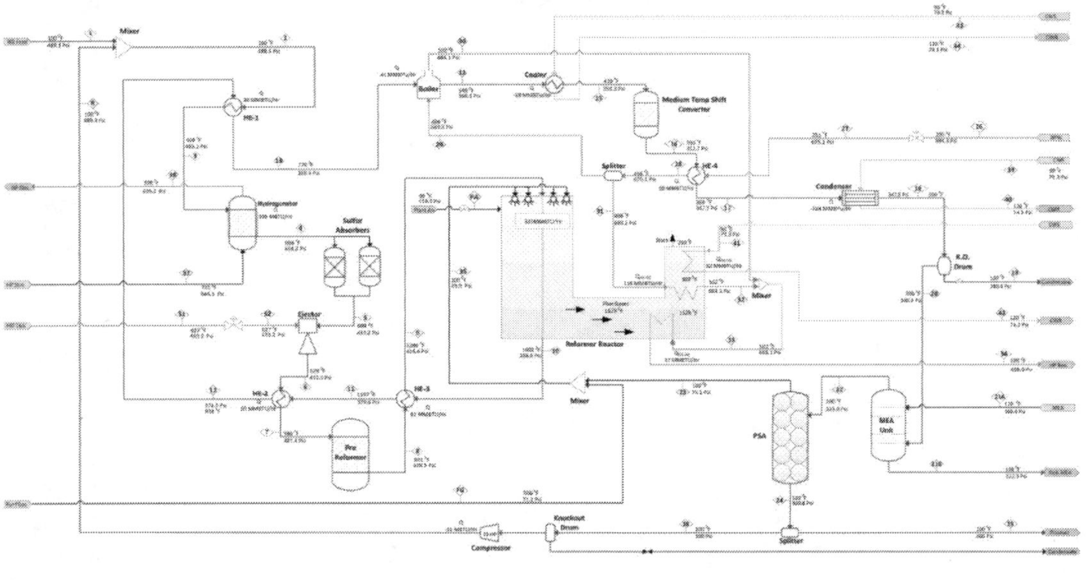

Ethylene Oxide Production

CENG: 4310 Final Design Project

Andrew Jones · Erica Diamond · Alex Tir · Morgan Weber

Tulane University

Goal

- Production Rate of 120 kTa of Ethylene Oxide
- Evaluate Process Design Options
  - Isothermal Reactor vs. Adiabatic Reactor
  - Oxygen vs. Air

$$C_2H_4 + 0.5\,O_2 \rightarrow C_2H_4O$$

$$C_2H_4 + 3\,O_2 \rightarrow 2CO_2 + 2H_2O$$

$$C_2H_4O + 2.5\,O_2 \rightarrow 2CO_2 + 2H_2O$$

1

Ethylene Oxide Production

- Isothermal Reactor
- Oxygen

- Annual Net Sales: $210 million
- Annual Net Cost: $137 million
- Annual Net Profit: $73 million

12 Year Cumulative Cash Flow:
$352 Million

2

Process Overview

- Process Description
  - Reactor Feed Prep
  - Reactor
  - Separation
  - Purification
- Safety
- Economics
- Recommendations
- Conclusion

3

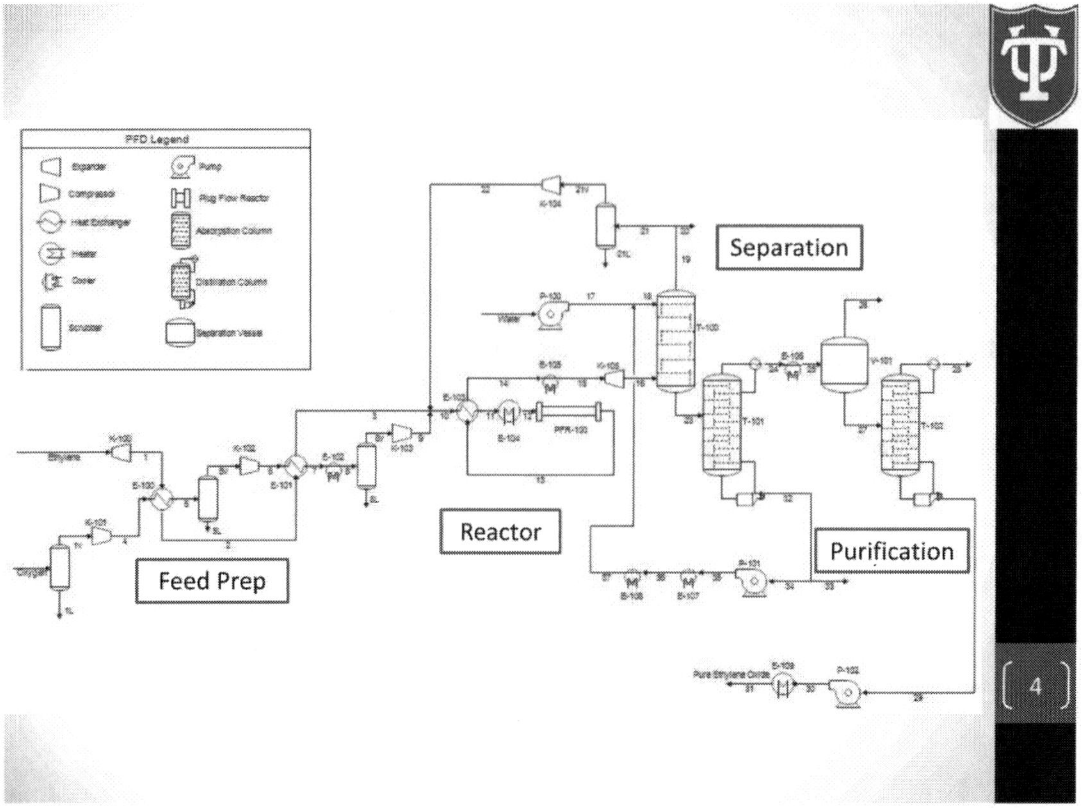
PFD Legend
Expander
Compressor
Heat Exchanger
Heater
Cooler
Scrubber
Pump
Plug Flow Reactor
Absorption Column
Distillation Column
Separation Vessel
Separation
Reactor
Purification
Feed Prep
Ethylene
Oxygen
Water
Pure Ethylene Oxide
4

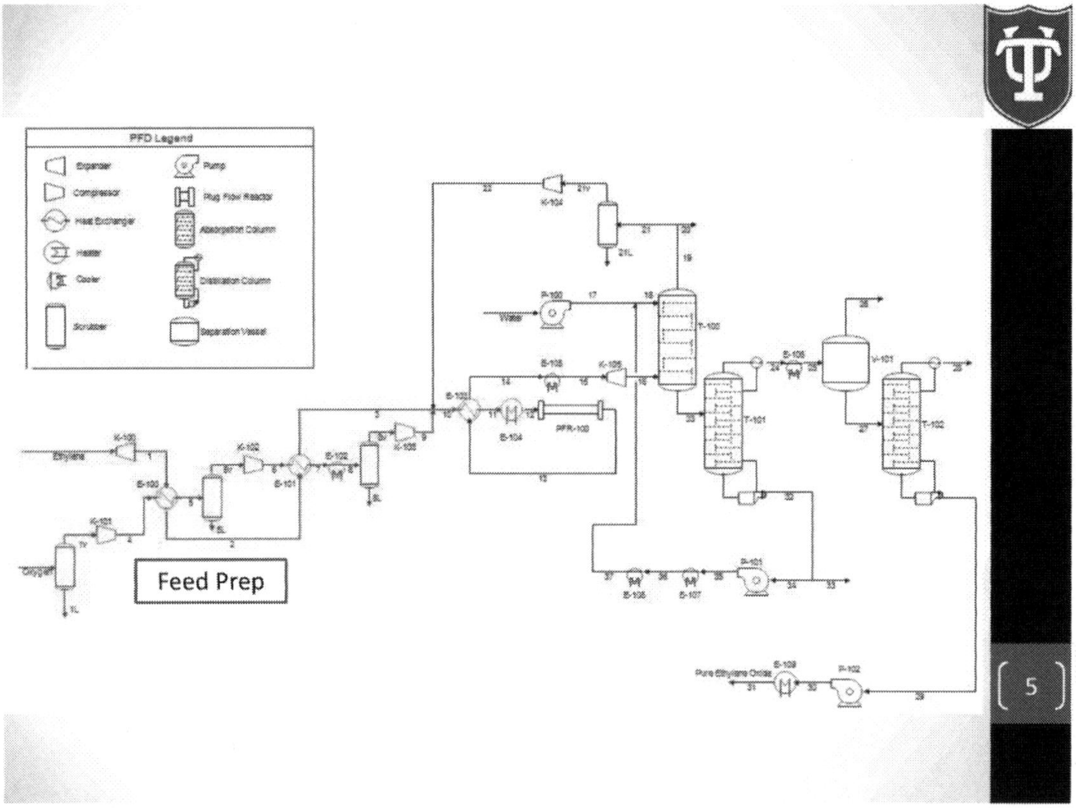
PFD Legend
Expander
Compressor
Heat Exchanger
Heater
Cooler
Scrubber
Pump
Plug Flow Reactor
Absorption Column
Distillation Column
Separation Vessel
Feed Prep
Ethylene
Oxygen
Water
Pure Ethylene Oxide
5

Reactor Feed Prep

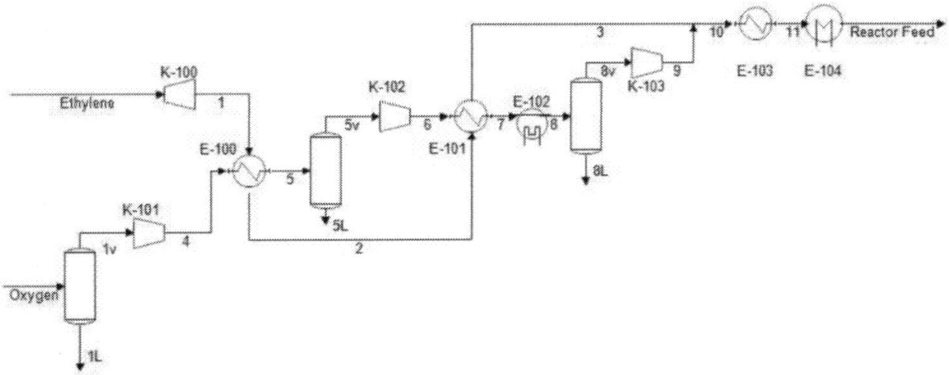

6

Reactor Feed Prep

Ethylene Depressurization

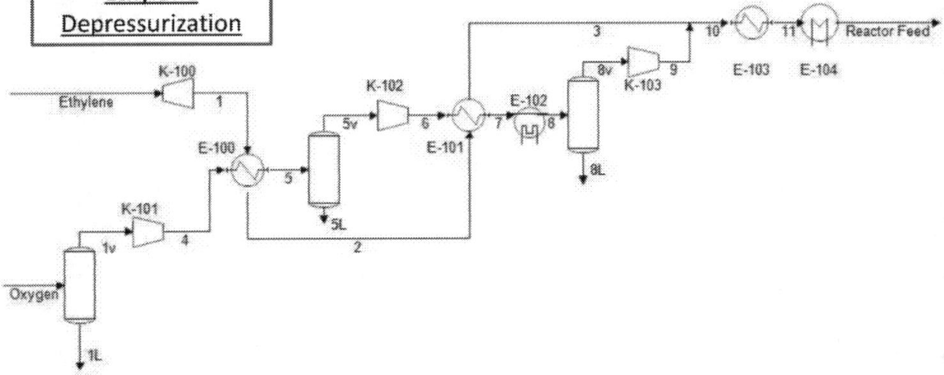

7

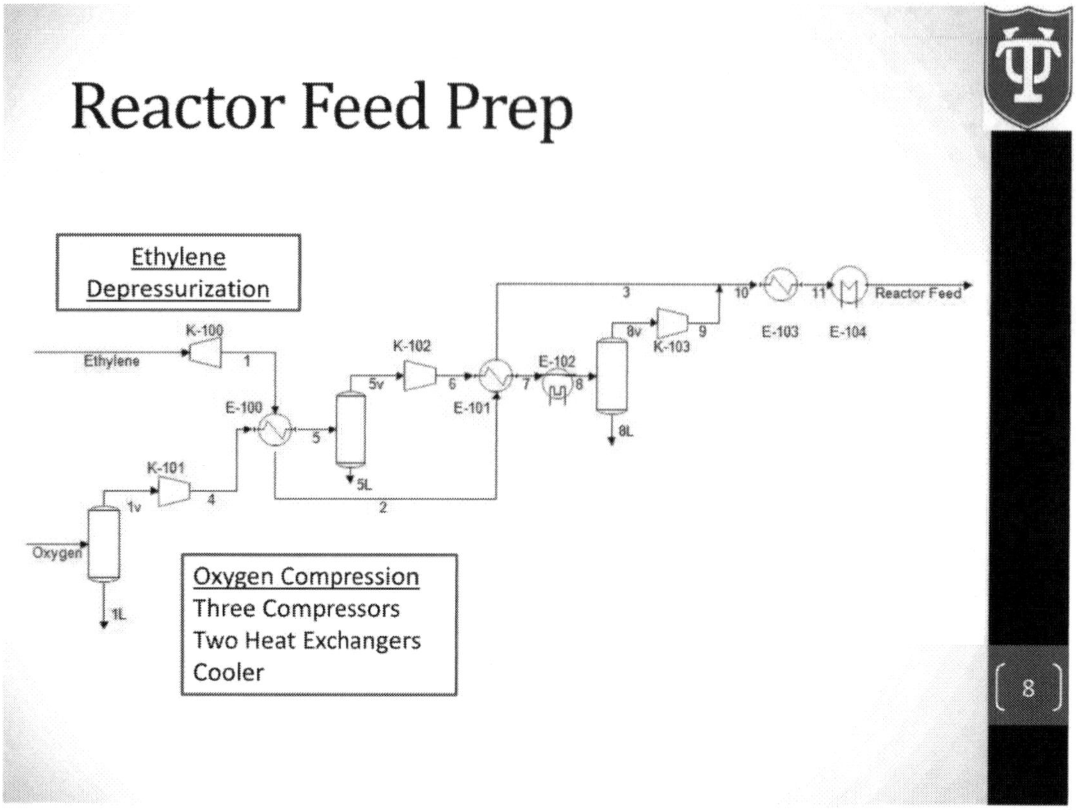
Reactor Feed Prep
Ethylene Depressurization
K-100
Ethylene
1
E-100
K-101
1v
4
Oxygen
1L
5
5L
2
K-102
5v
6
E-101
7
E-102
8
8L
8v
K-103
9
3
10
11
E-103
E-104
Reactor Feed
Oxygen Compression
Three Compressors
Two Heat Exchangers
Cooler
8

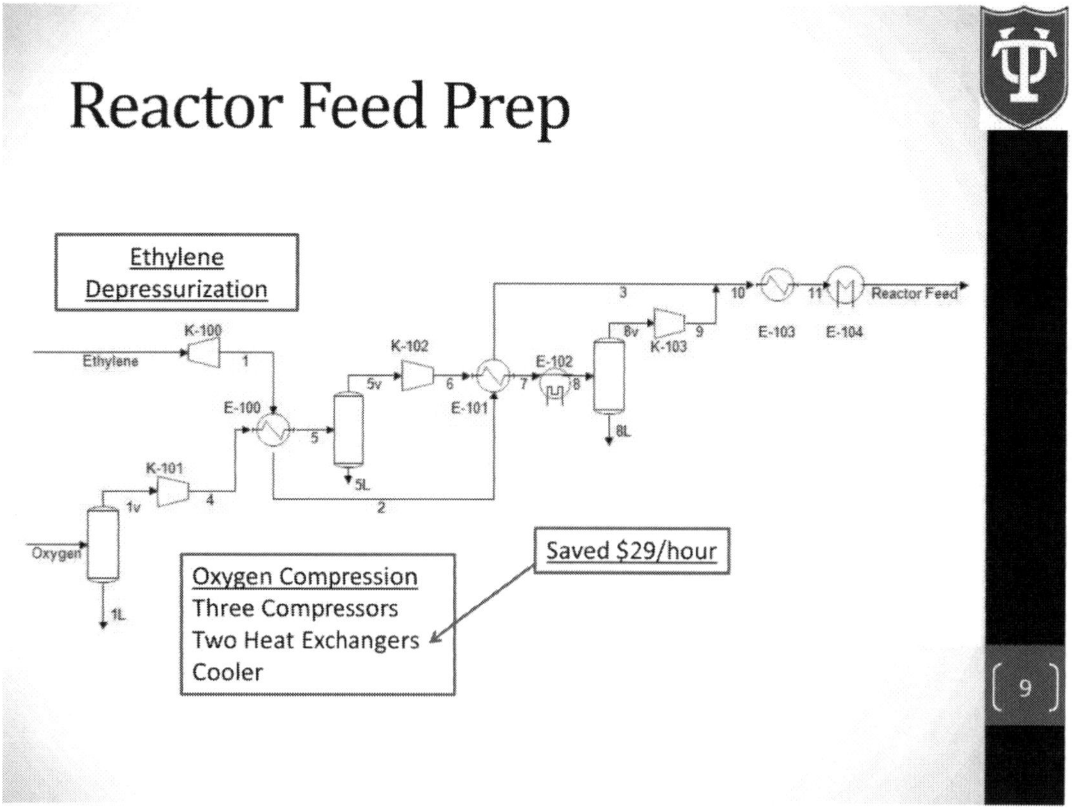
Reactor Feed Prep
Ethylene Depressurization
K-100
Ethylene
1
E-100
K-101
1v
4
Oxygen
1L
5
5L
2
K-102
5v
6
E-101
7
E-102
8
8L
8v
K-103
9
3
10
11
E-103
E-104
Reactor Feed
Saved $29/hour
Oxygen Compression
Three Compressors
Two Heat Exchangers
Cooler
9

Reactor Feed Prep

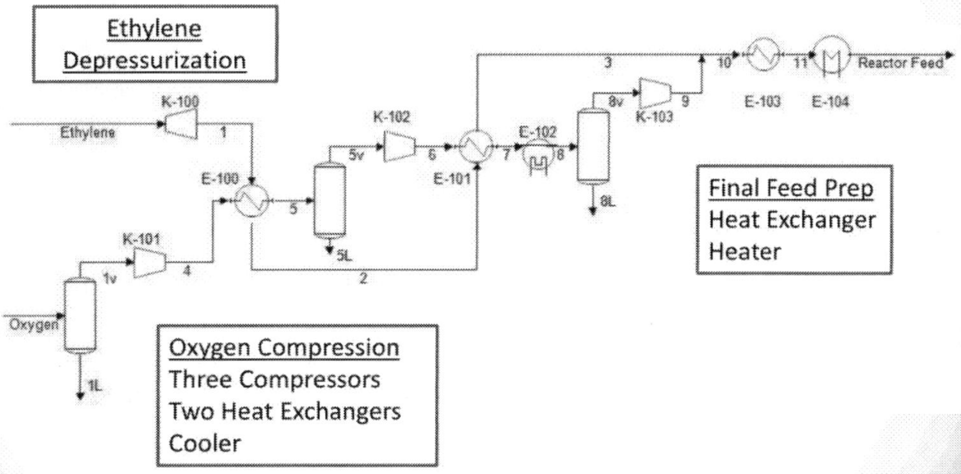

10

Oxygen vs. Air

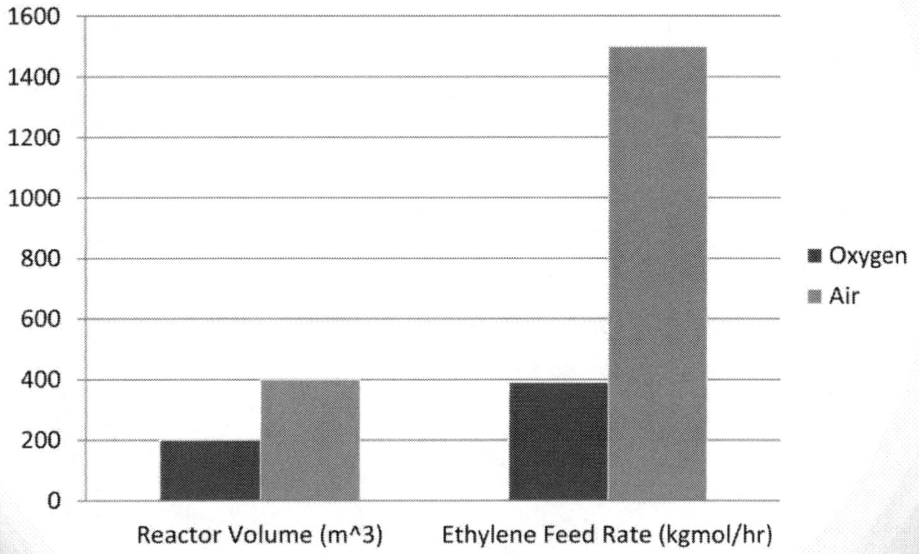

11

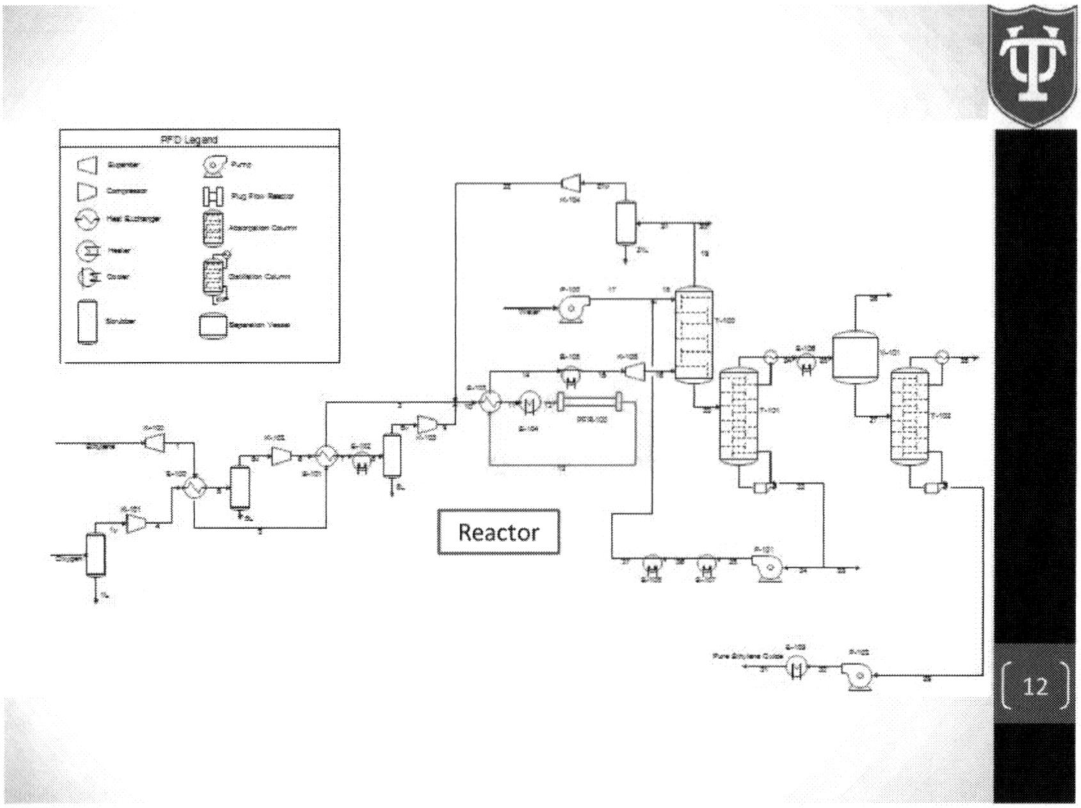
Reactor
12

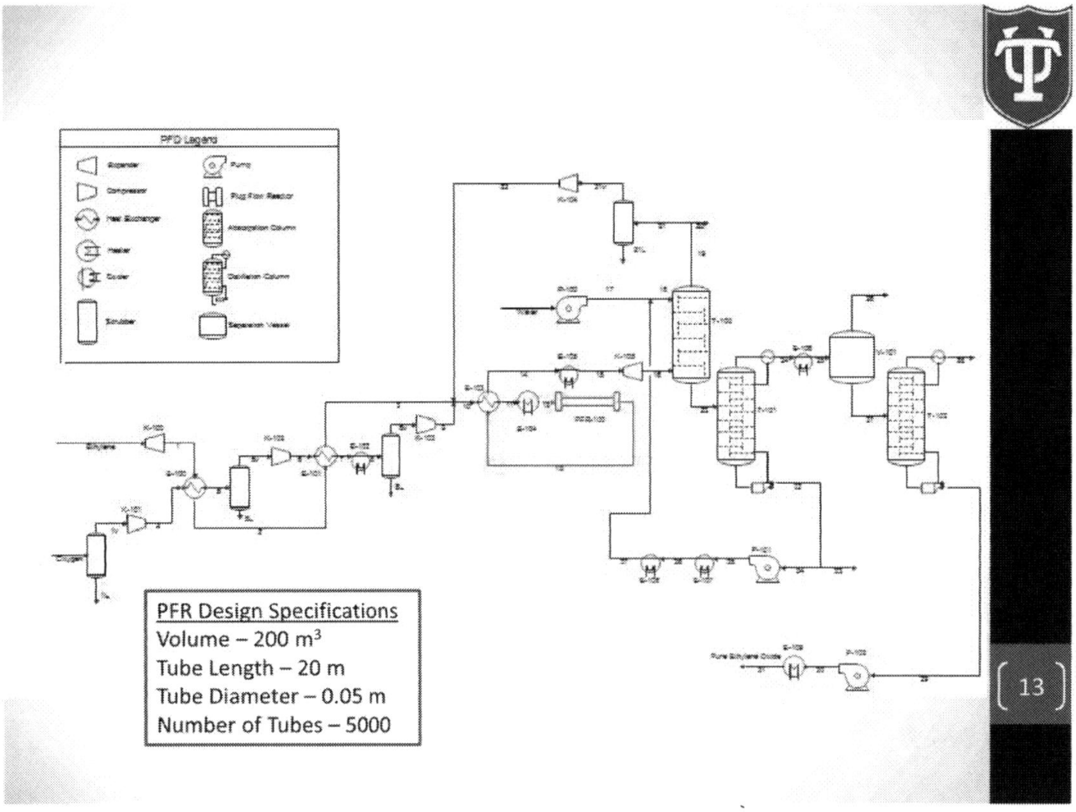
PFR Design Specifications
Volume – 200 m³
Tube Length – 20 m
Tube Diameter – 0.05 m
Number of Tubes – 5000
13

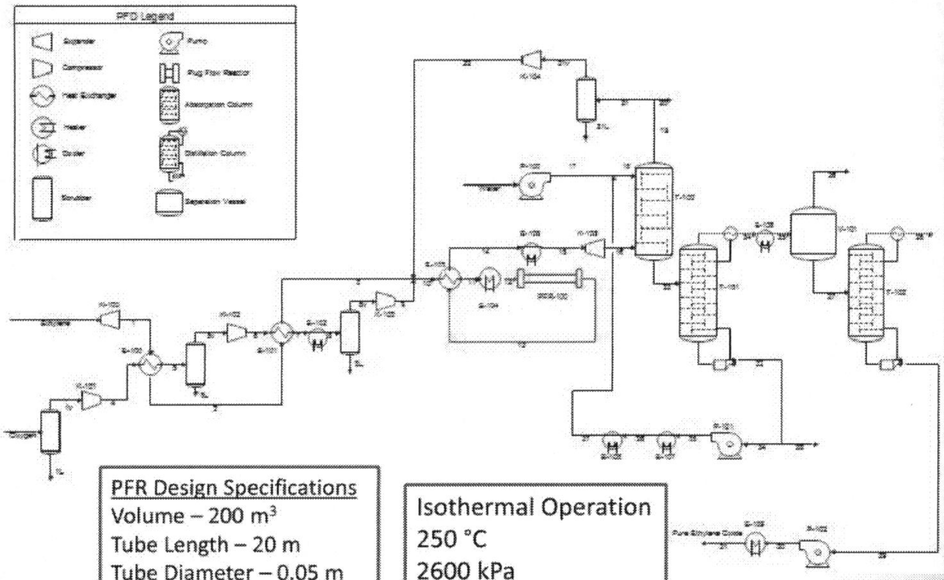

PFR Design Specifications
Volume – 200 m^3
Tube Length – 20 m
Tube Diameter – 0.05 m
Number of Tubes – 5000

Isothermal Operation
250 °C
2600 kPa

14

Case Study: Reactor T & P

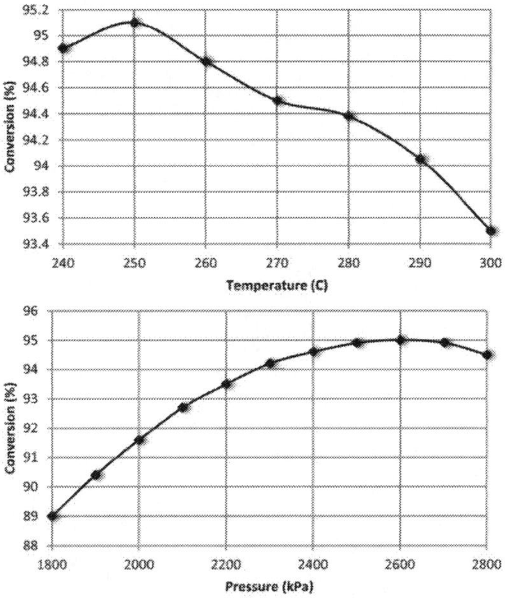

15

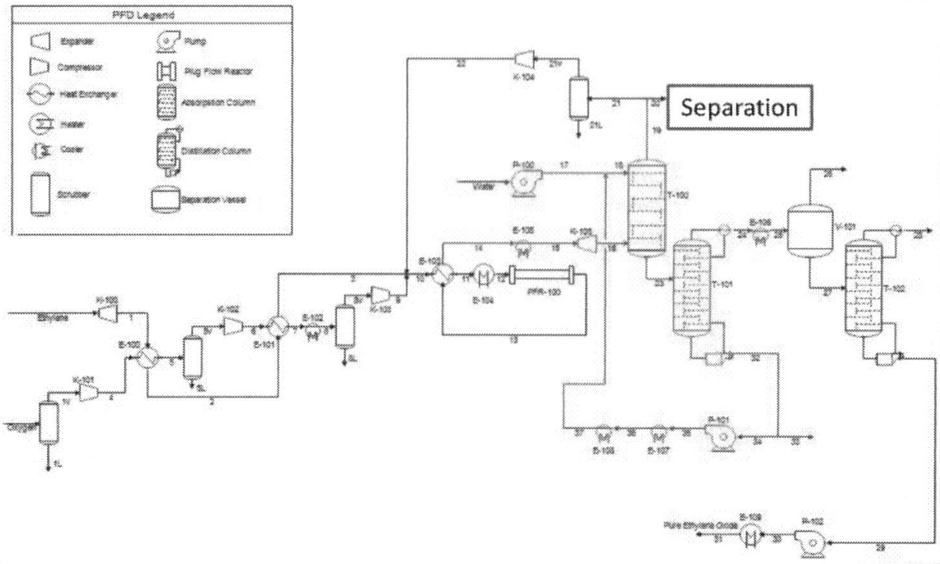

16

Separation

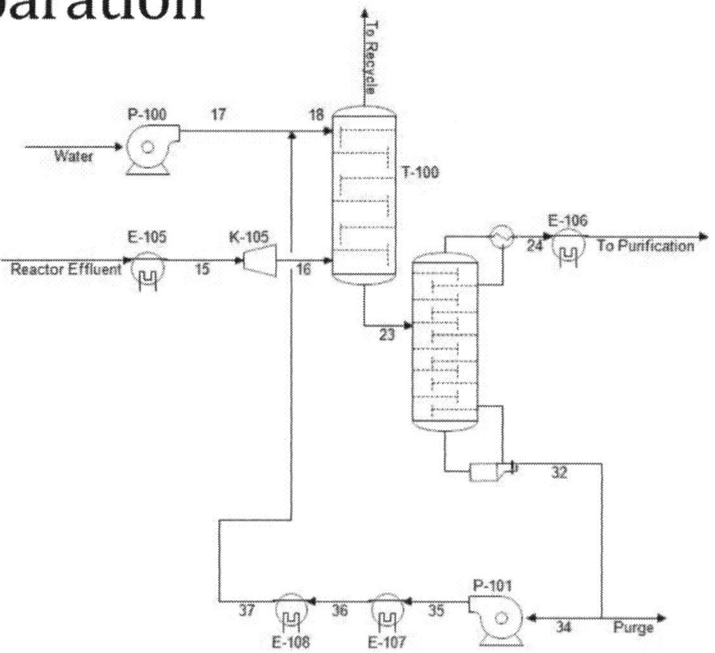

17

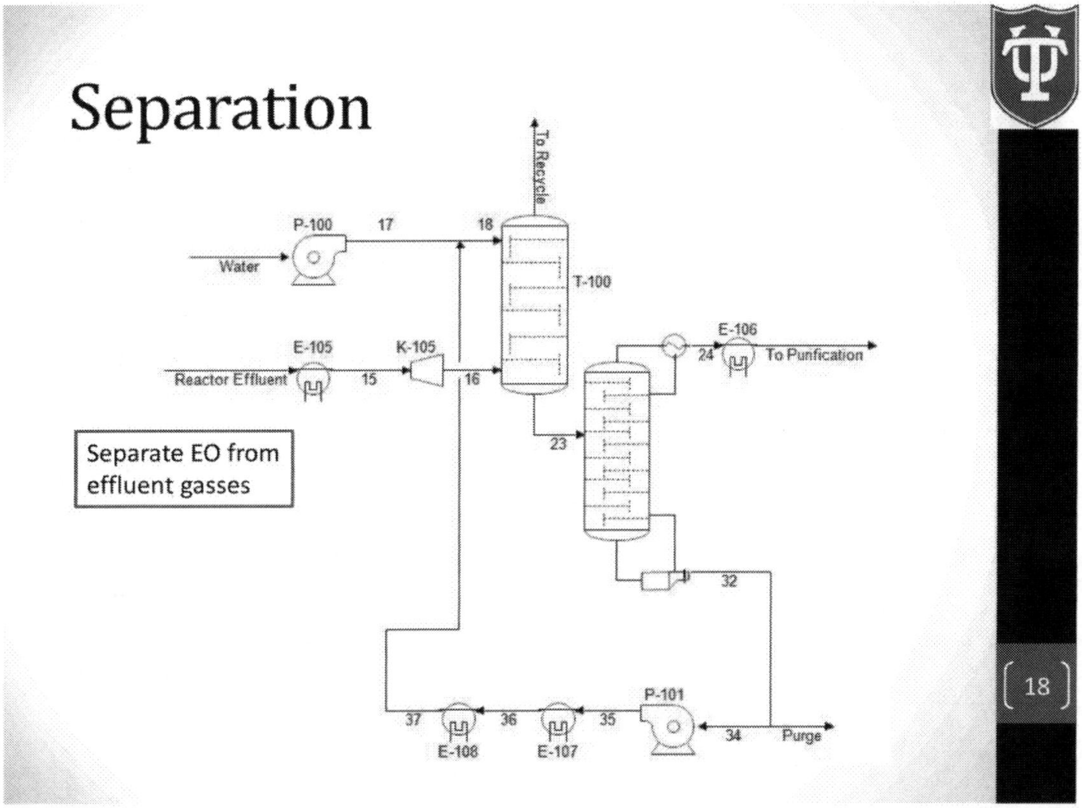
Separation
To Recycle
P-100
17
18
Water
T-100
E-105
K-105
E-106
24
To Purification
Reactor Effluent
15
16
Separate EO from effluent gasses
23
32
P-101
37
36
35
34
Purge
E-108
E-107
18

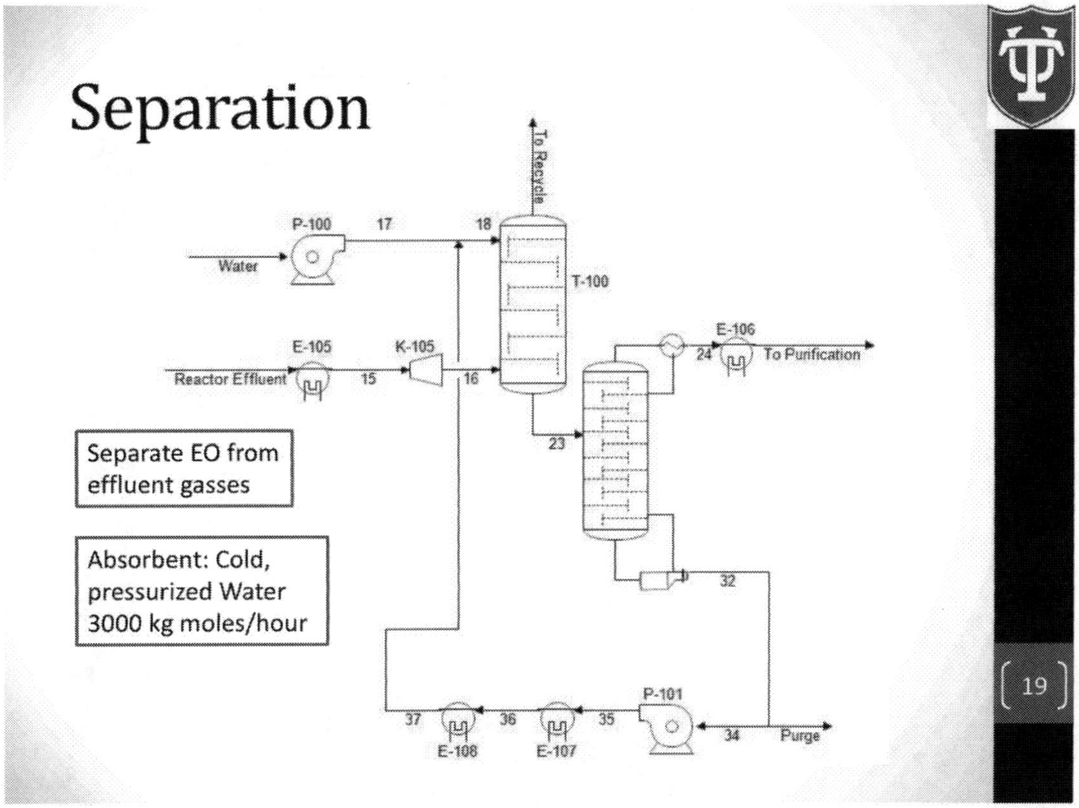
Separation
To Recycle
P-100
17
18
Water
T-100
E-105
K-105
E-106
24
To Purification
Reactor Effluent
15
16
Separate EO from effluent gasses
Absorbent: Cold, pressurized Water 3000 kg moles/hour
23
32
P-101
37
36
35
34
Purge
E-108
E-107
19

Separation

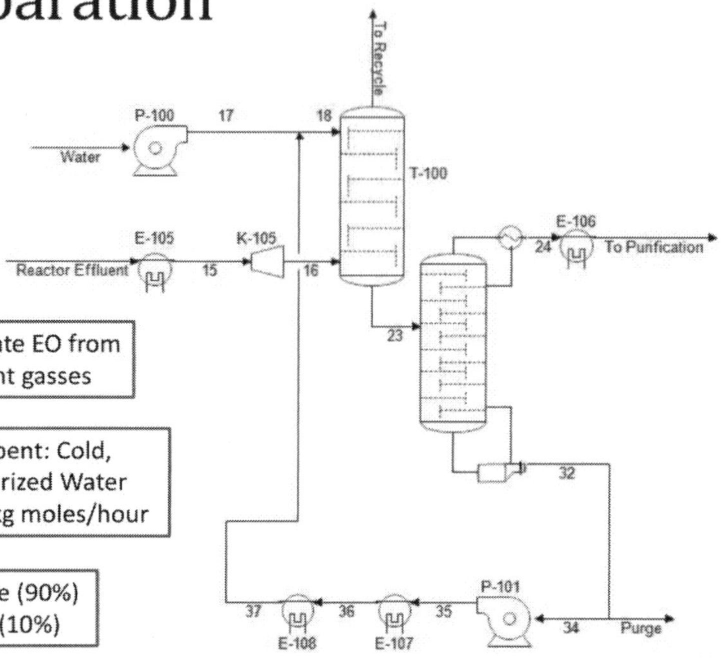

Separate EO from effluent gasses

Absorbent: Cold, pressurized Water 3000 kg moles/hour

Recycle (90%)
Purge (10%)

20

Separation

Separate EO from water

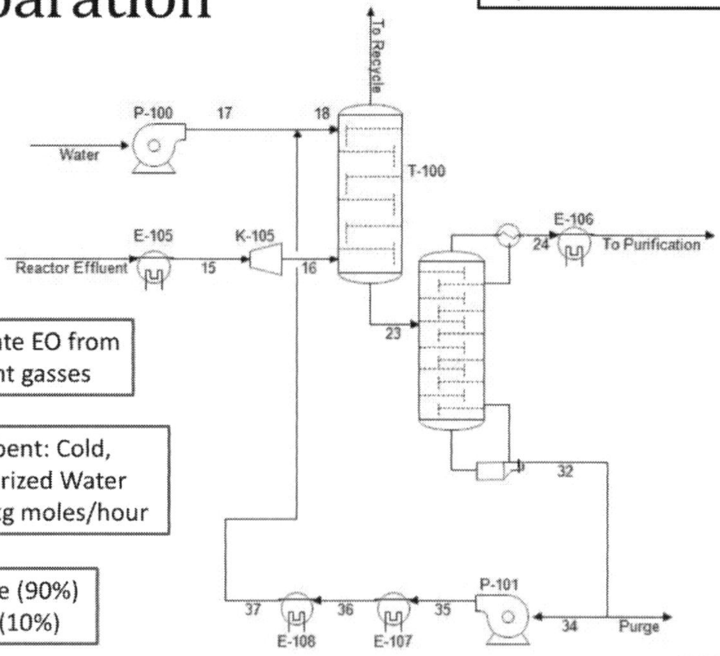

Separate EO from effluent gasses

Absorbent: Cold, pressurized Water 3000 kg moles/hour

Recycle (90%)
Purge (10%)

21

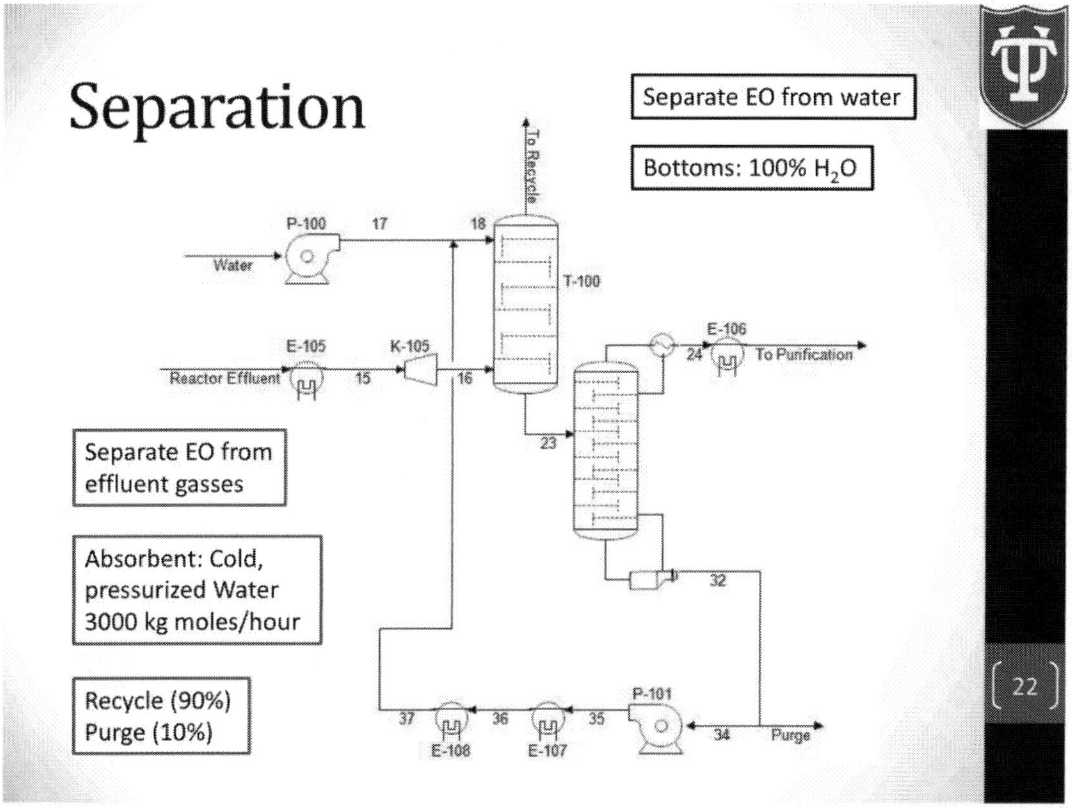
Separation
Separate EO from water
Bottoms: 100% H2O
To Recycle
P-100
17
18
Water
T-100
E-105
K-105
E-106
24
To Purification
Reactor Effluent
15
16
23
Separate EO from effluent gasses
Absorbent: Cold, pressurized Water 3000 kg moles/hour
32
Recycle (90%) Purge (10%)
P-101
37
36
35
34
Purge
E-108
E-107
22

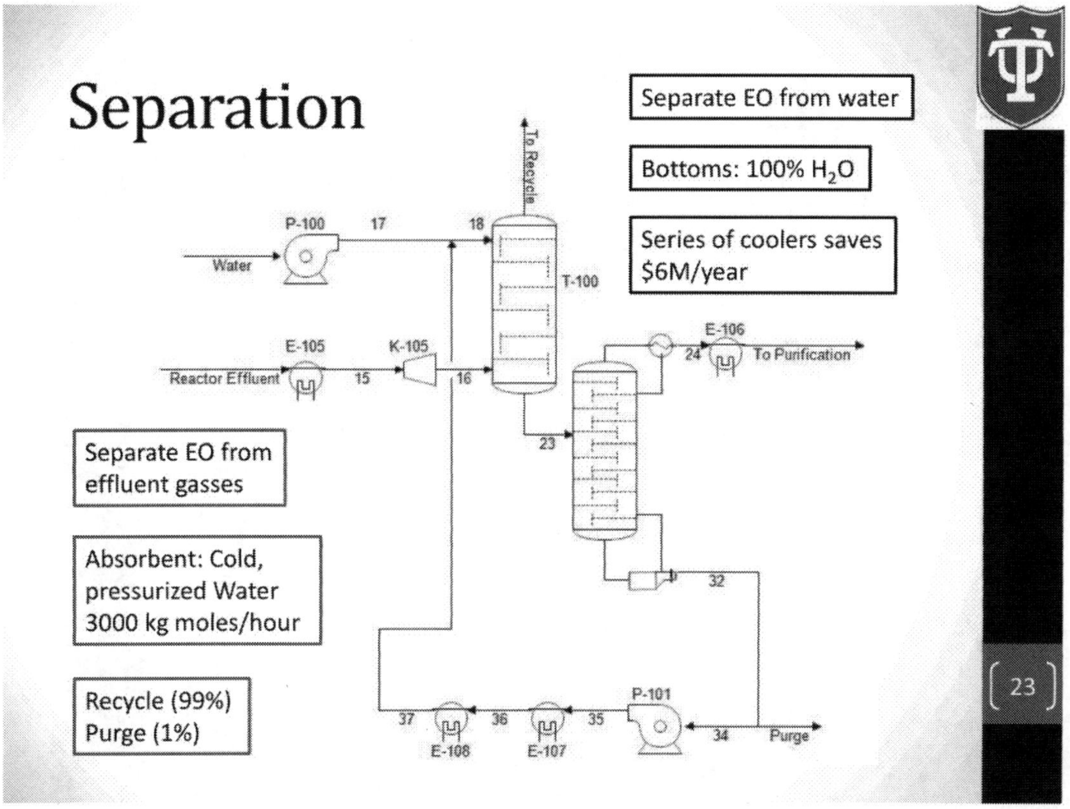
Separation
Separate EO from water
Bottoms: 100% H2O
Series of coolers saves $6M/year
To Recycle
P-100
17
18
Water
T-100
E-105
K-105
E-106
24
To Purification
Reactor Effluent
15
16
23
Separate EO from effluent gasses
Absorbent: Cold, pressurized Water 3000 kg moles/hour
32
Recycle (99%) Purge (1%)
P-101
37
36
35
34
Purge
E-108
E-107
23

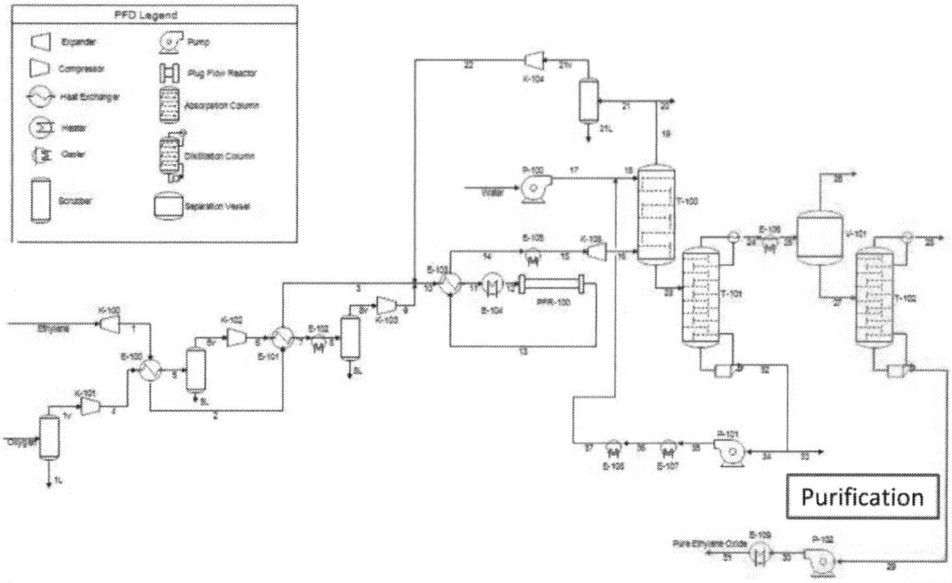

24

Purification

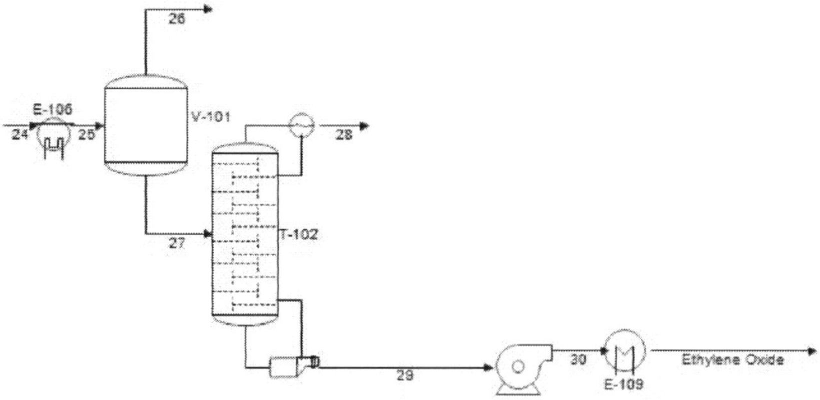

25

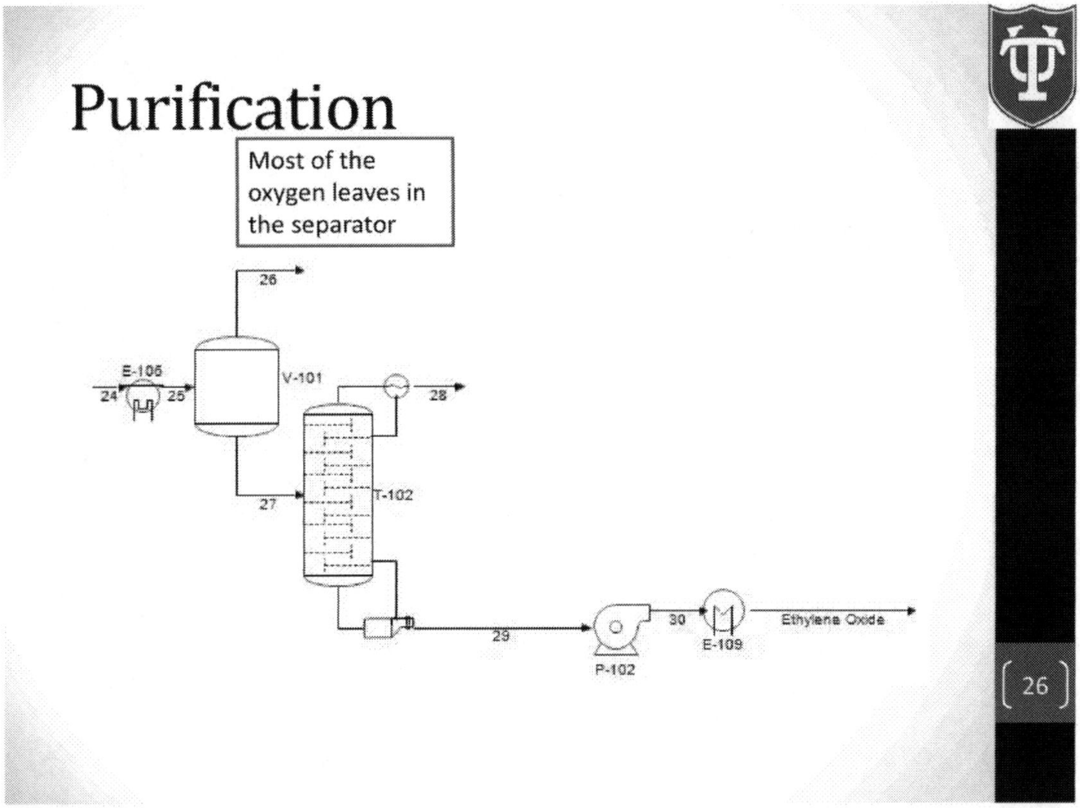
Purification
Most of the oxygen leaves in the separator
26
E-106
24
25
V-101
28
27
T-102
29
P-102
30
E-109
Ethylene Oxide
26

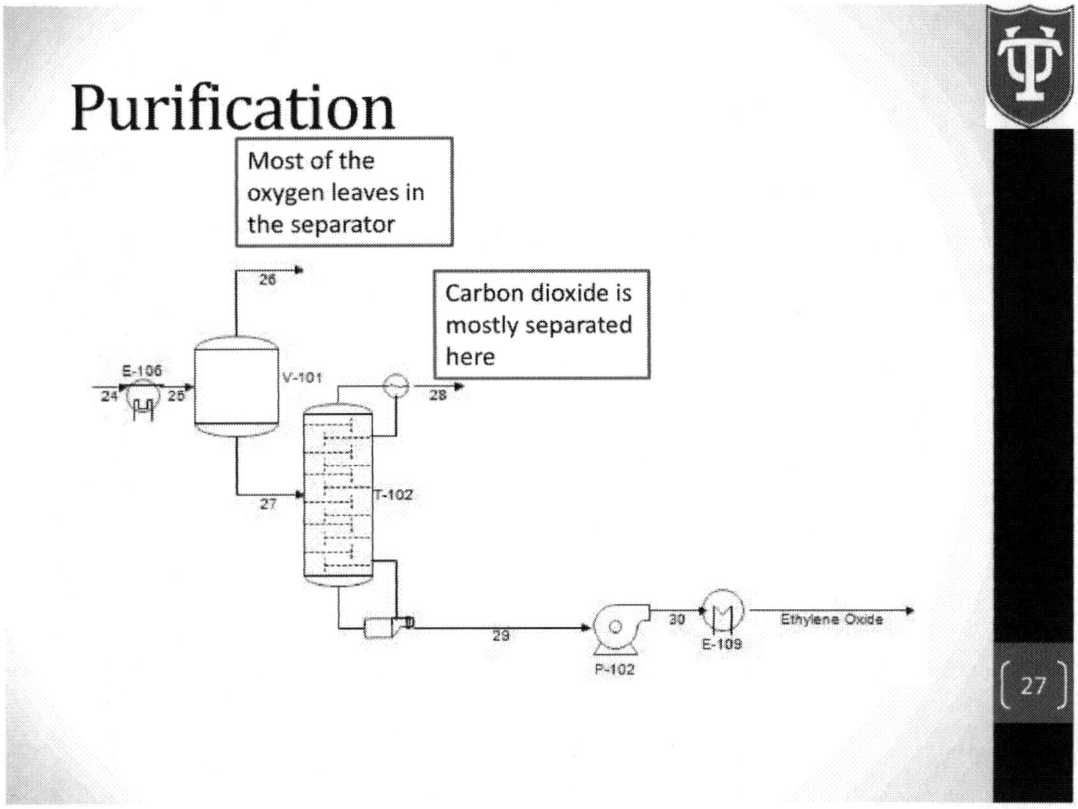
Purification
Most of the oxygen leaves in the separator
Carbon dioxide is mostly separated here
26
E-106
24
25
V-101
28
27
T-102
29
P-102
30
E-109
Ethylene Oxide
27

Purification

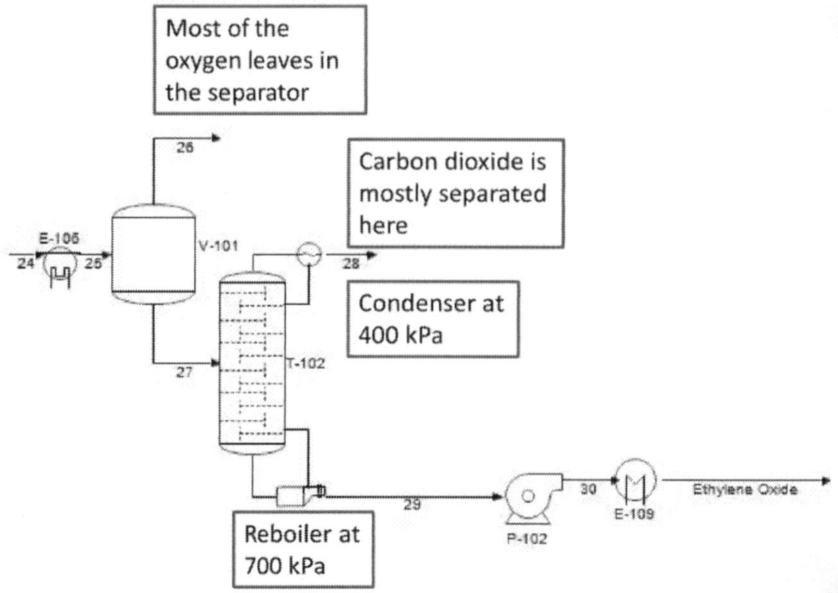

28

Purification

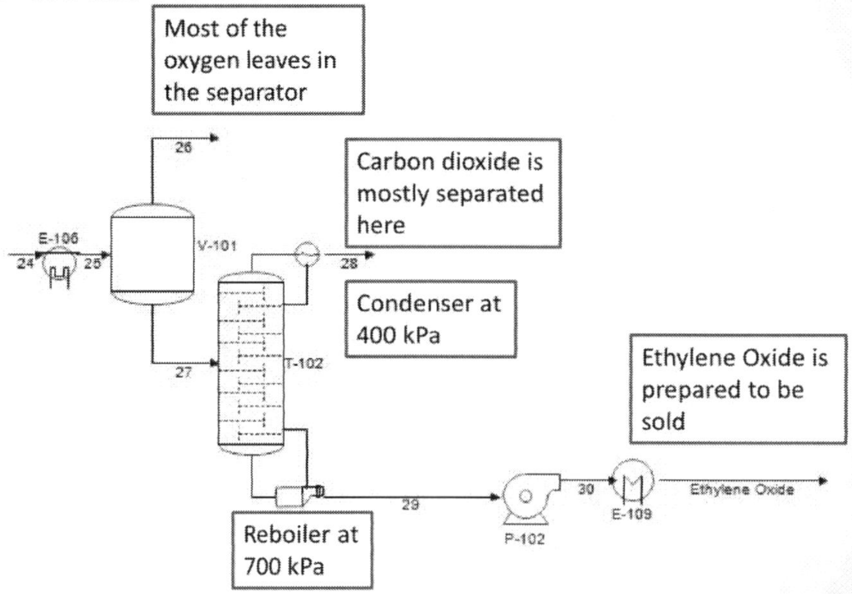

29

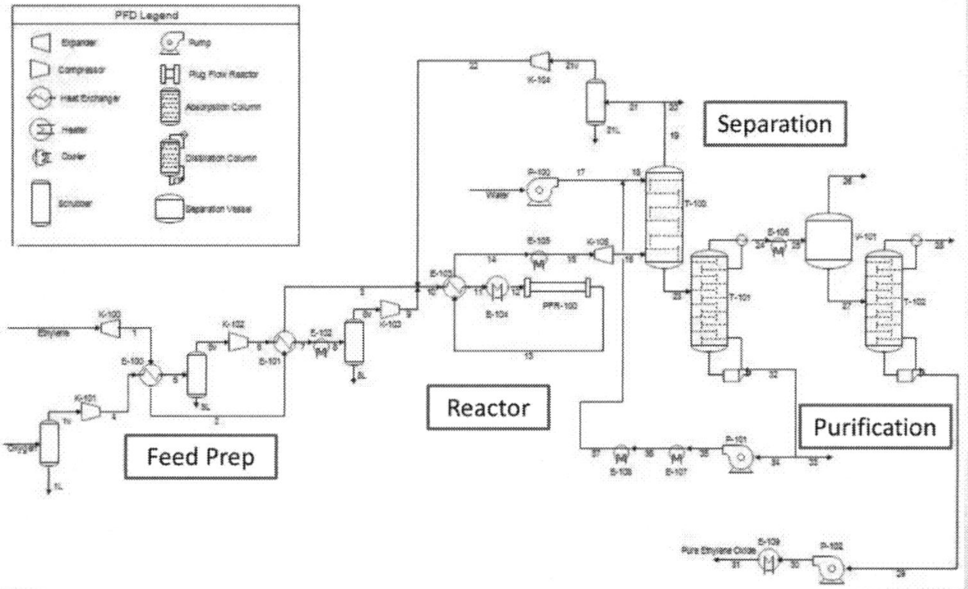

30

Safety

- Ethylene
  - Flammability limits
  - Safeguards
    - Oxygen shutoff system
    - Eliminate ignition sources
- Ethylene Oxide
  - Reactivity Hazards
  - Safeguards
    - Avoid reactive hotspots
    - Size relief systems for decomposition of ethylene oxide
- Pressure

Economic Summary

Total Capital Investment	$24,000,000

Economic Summary

Total Capital Investment Equipment and Construction Costs (10% overdesign) Working Capital Land	$24,000,000

Economic Summary

Total Capital Investment Equipment and Construction Costs (10% overdesign) Working Capital Land	$24,000,000
Yearly Product Revenue 10-Year Production	$210,000,000

Economic Summary

Total Capital Investment Equipment and Construction Costs (10% overdesign) Working Capital Land	$24,000,000
Yearly Product Revenue 10-Year Production	$210,000,000
Yearly Cost of Production	$137,000,000

Economic Summary

Total Capital Investment Equipment and Construction Costs (10% overdesign) Working Capital Land	$24,000,000
Yearly Product Revenue 10-Year Production	$210,000,000
Yearly Cost of Production Raw Materials Utilities Operation Labor Annual Capital Charge	$137,000,000

Economic Summary

Total Capital Investment Equipment and Construction Costs (10% overdesign) Working Capital Land	$24,000,000
Yearly Product Revenue 10-Year Production	$210,000,000
Yearly Cost of Production Raw Materials Utilities Operation Labor Annual Capital Charge	$137,000,000
Yearly Profit Margin	$73,000,000

Economic Summary

Total Capital Investment Equipment and Construction Costs (10% overdesign) Working Capital Land	$24,000,000
Yearly Product Revenue 10-Year Production	$210,000,000
Yearly Cost of Production Raw Materials Utilities Operation Labor Annual Capital Charge	$137,000,000
Yearly Profit Margin	$73,000,000
Cumulative Cash Flow (Present Value) Investment Payoff Inflation and Interest Rates	**$352,000,000**

Economic Summary

Total Capital Investment Equipment and Construction Costs (10% overdesign) Working Capital Land	$24,000,000
Yearly Product Revenue 10-Year Production	$210,000,000
Yearly Cost of Production Raw Materials Utilities Operation Labor Annual Capital Charge	$137,000,000
Yearly Profit Margin	$73,000,000
Cumulative Cash Flow (Present Value) Investment Payoff Inflation and Interest Rates	**$352,000,000**

Utilities and Optimization

Utility Type

Utilities and Optimization

Utility Type
Ambient Water
Refrigerated Water
Natural Gas
Electricity

Utilities and Optimization

Utility Type	Unit
Ambient Water	
Refrigerated Water	
Natural Gas	
Electricity	

Utilities and Optimization

Utility Type	Unit
Ambient Water	Coolers Reactor Condensers
Refrigerated Water	
Natural Gas	
Electricity	

Utilities and Optimization

Utility Type	Unit
Ambient Water	Coolers Reactor Condensers
Refrigerated Water	Coolers (below 25°C)
Natural Gas	
Electricity	

Utilities and Optimization

Utility Type	Unit
Ambient Water	Coolers Reactor Condensers
Refrigerated Water	Coolers (below 25°C)
Natural Gas	Heaters Reboilers
Electricity	

Utilities and Optimization

Utility Type	Unit
Ambient Water	Coolers Reactor Condensers
Refrigerated Water	Coolers (below 25°C)
Natural Gas	Heaters Reboilers
Electricity	Compressors Expanders

Utilities and Optimization

Utility Type	Unit	Optimization Method
Ambient Water	Coolers Reactor Condensers	
Refrigerated Water	Coolers (below 25°C)	
Natural Gas	Heaters Reboilers	
Electricity	Compressors Expanders	

Utilities and Optimization

Utility Type	Unit	Optimization Method
Ambient Water	Coolers Reactor Condensers	• Series of coolers • Controlled tops of separators • Heat exchangers
Refrigerated Water	Coolers (below 25°C)	
Natural Gas	Heaters Reboilers	
Electricity	Compressors Expanders	

Utilities and Optimization

Utility Type	Unit	Optimization Method
Ambient Water	Coolers Reactor Condensers	• Series of coolers • Controlled tops of separators • Heat exchangers
Refrigerated Water	Coolers (below 25°C)	• Ambient water first, then use refrigerated water
Natural Gas	Heaters Reboilers	
Electricity	Compressors Expanders	

Utilities and Optimization

Utility Type	Unit	Optimization Method
Ambient Water	Coolers Reactor Condensers	• Series of coolers • Controlled tops of separators • Heat exchangers
Refrigerated Water	Coolers (below 25°C)	• Ambient water first, then use refrigerated water
Natural Gas	Heaters Reboilers	• Controlled bottoms of separators
Electricity	Compressors Expanders	

Utilities and Optimization

Utility Type	Unit	Optimization Method
Ambient Water	Coolers Reactor Condensers	• Series of coolers • Controlled tops of separators • Heat exchangers
Refrigerated Water	Coolers (below 25°C)	• Ambient water first, then use refrigerated water
Natural Gas	Heaters Reboilers	• Controlled bottoms of separators
Electricity	Compressors Expanders	• Energy from expanders was recovered for use elsewhere

Sensitivity Analysis

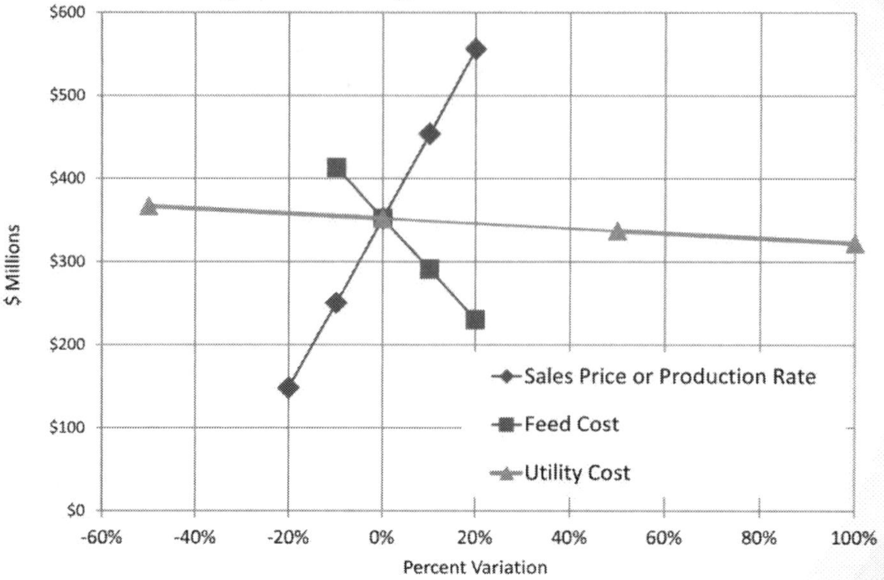

52

Optimization Comparisons

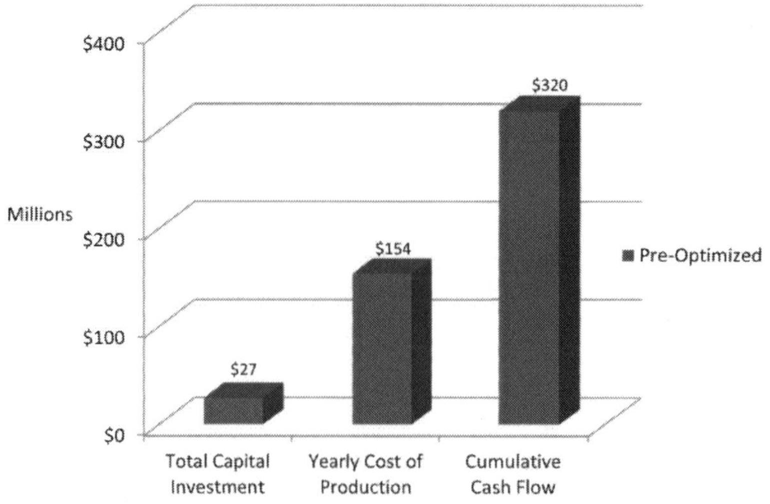

Optimization Comparisons

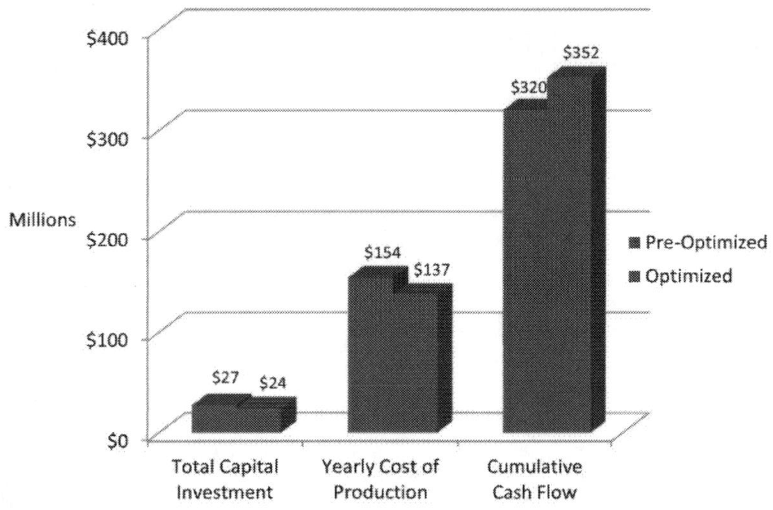

Recommendations

- Quality Check on lab data
  - Flammability envelope of ethylene in oxygen
  - Reanalyzing kinetics
- Ethylene Oxide market analysis→Higher production demands
- Valves vs. Expanders
- Higher Purity of Ethylene

Conclusion

- ✓Production Rate of 120 kTa Ethylene Oxide
- ✓Optimized Isothermal Reactor Operation
- ✓Pure Oxygen Feed

Conclusion

- ✓Production Rate of 120 kTa Ethylene Oxide
- ✓Optimized Isothermal Reactor Operation
- ✓Pure Oxygen Feed

- ✓Annual Revenue: $210 million
- ✓Annual Cost of Production: $137 million
- ✓Annual Profit: $73 million
- ✓"Worst-Case" Economic Studies

Conclusion

✓Production Rate of 120 kTa Ethylene Oxide
✓Optimized Isothermal Reactor Operation
✓Pure Oxygen Feed

✓Annual Revenue: $210 million
✓Annual Cost of Production: $137 million
✓Annual Profit: $73 million
✓"Worst-Case" Economic Studies

12-Year Cumulative Cash Flow: $352 Million

Acknowledgements

- Dr. David Wallace and Dr. Kory Blocker
- Department of Chemical and Biomolecular Engineering
- INEOS

59

Ethylene Oxide Production

CENG: 4310 Final Design Project

Erica Diamond · Andrew Jones · Alex Tir · Morgan Weber

352960 Economic Optimization and Hazard Analysis of a Hydrofluoric Acid Catalyzed Alkylation Process

Abstract:

A hydrofluoric acid catalyzed alkylation process was modeled using four series/parallel reactions starting with propylene and isobutane. A target of 5,000 barrels per day of 95% pure 2,2-dimethylpentane was specified. The product was also defined to contain less than 1% of heavies (1-nonene and 2,2-dimethyl-octane). The system was modeled using the process simulator ChemCAD and priced using cost equations from Turton, Bailie, Whiting and Shaeiwitz (2012). Both topological and parametric optimizations were considered. The recycle streams increased the complexity of the simulator significantly, but were required to accurately model the concentration build-up in the system. The impact of reactor residence time, separator recycle purity, and feed ratio on the economics will be presented. Process hazards will be identified and suggestions to mitigate hazards will be presented.

The Growth of Idea Ownership: Patent Infringement and Risk Management within the Chemical Process Industries

Paul LaVanway, Esq.
Fredrikson & Byron, P.A.

14AIChE Spring Meeting, New Orleans, LA

Fredrikson & Byron, P.A.

Overview of Presentation

I. Patent Activity Update

II. Practical Risk Management Techniques

III. New Tools

14AIChE Spring Meeting, New Orleans, LA

Fredrikson & Byron, P.A.

PATENT ACTIVITY UPDATE

14 AIChE
Spring Meeting, New Orleans, LA

Fredrikson & BYRON, P.A.

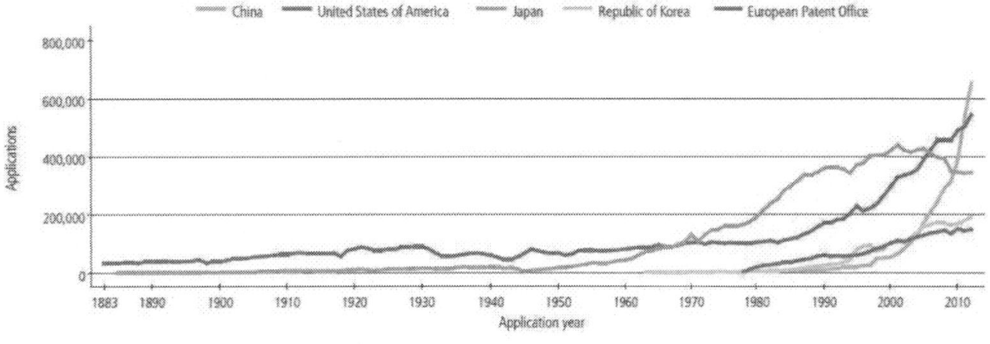

14 AIChE
Spring Meeting, New Orleans, LA

Fredrikson & BYRON, P.A.

Worldwide Filing Trends

Origin	Field of technology															
	Australia	Canada	China	Finland	France	Germany	Italy	Japan	Netherlands	Republic of Korea	Russian Federation	Sweden	Switzerland	United Kingdom	United States of America	Others
Chemistry																
Organic fine chemistry	1,164	2,798	18,900	539	20,317	38,897	4,436	38,136	6,148	10,424	2,185	5,375	17,365	12,314	75,204	30,237
Biotechnology	2,592	3,432	17,402	804	8,107	16,915	2,306	20,814	5,500	9,165	1,822	1,765	7,316	7,058	67,521	21,840
Pharmaceuticals	3,647	6,297	38,194	958	18,328	31,204	7,370	28,798	6,362	10,514	5,692	7,748	23,615	15,563	123,173	50,208
Macromolecular chemistry, polymers	382	816	11,135	2,173	4,439	20,009	2,446	45,323	4,576	7,906	935	325	3,439	1,714	31,810	7,538
Food chemistry	875	1,122	19,191	428	2,595	5,330	1,070	13,272	6,409	12,111	17,015	337	4,744	2,464	19,963	11,149
Basic materials chemistry	1,002	1,992	24,266	901	6,012	35,527	1,536	42,724	7,038	11,113	3,343	690	6,970	6,340	51,100	13,535
Materials, metallurgy	1,697	1,569	30,285	1,729	6,697	17,150	1,480	43,685	1,846	12,235	8,359	1,514	2,186	2,422	20,753	12,182
Surface technology, coating	670	1,386	11,959	1,185	4,907	15,954	1,642	52,591	1,933	10,246	1,779	1,200	2,604	2,498	36,858	7,336
Micro-structural and nano-technology	116	96	1,710	124	606	1,371	94	2,405	254	2,579	587	134	153	158	2,486	683
Chemical engineering	1,479	2,235	16,952	1,840	7,352	25,809	2,846	32,504	4,788	13,354	4,426	2,422	4,287	5,490	40,121	13,512
Environmental technology	762	1,514	13,620	855	5,042	13,594	1,481	27,259	2,533	12,888	2,303	1,116	1,765	2,719	20,263	8,002

14 AIChE®
Spring Meeting, New Orleans, LA

Fredrikson & BYRON, P.A.

Worldwide Filing Trends

Field of technology	Publication year					Growth rate
	2007	2008	2009	2010	2011	2007-11 (%)
Chemistry						
Organic fine chemistry	52,609	53,873	53,009	52,559	51,271	-0.6
Biotechnology	33,929	35,654	37,652	38,423	40,849	4.7
Pharmaceuticals	73,345	73,769	72,126	69,258	69,311	-1.4
Macromolecular chemistry, polymers	27,627	28,345	28,800	28,683	28,748	1.0
Food chemistry	21,690	23,582	27,245	27,889	30,858	9.2
Basic materials chemistry	38,920	41,085	42,322	43,795	45,115	3.8
Materials, metallurgy	30,200	33,775	34,968	36,988	38,542	6.3
Surface technology, coating	29,348	30,732	32,792	33,215	33,954	3.7
Micro-structural and nano-technology	2,302	2,478	2,829	3,094	3,203	8.6
Chemical engineering	33,263	35,087	35,766	36,618	37,991	3.4
Environmental technology	21,688	22,605	24,333	25,572	26,324	5.0

14 AIChE®
Spring Meeting, New Orleans, LA

Fredrikson & BYRON, P.A.

Comparative Trends

Field of technology	Publication year					Growth rate
	2007	2008	2009	2010	2011	2007-11 (%)
Electrical engineering						
Electrical machinery, apparatus, energy	101,013	105,591	111,876	116,009	122,697	5.0
Audio-visual technology	92,978	91,258	85,411	80,607	75,881	-5.0
Telecommunications	68,915	69,804	60,808	56,417	49,533	-7.9
Digital communication	57,696	64,237	68,191	74,389	79,726	8.4
Basic communication processes	17,208	17,734	17,164	16,515	15,554	-2.5
Computer technology	123,504	134,493	133,072	130,052	134,272	2.1
IT methods for management	19,370	21,684	25,022	23,261	23,532	5.0
Semiconductors	76,617	81,238	78,685	77,547	80,049	1.1
Mechanical engineering						
Handling	42,764	42,783	42,764	42,414	44,344	0.9
Machine tools	36,607	38,315	40,585	43,242	46,363	6.1
Engines, pumps, turbines	41,891	43,573	48,156	48,372	48,423	3.7
Textile and paper machines	35,815	33,681	32,251	30,633	30,364	-4.0
Other special machines	44,008	45,912	47,541	49,011	50,981	3.7
Thermal processes and apparatus	25,280	25,699	27,259	29,382	29,855	4.2
Mechanical elements	43,957	47,513	47,286	46,405	46,913	1.6
Transport	65,237	67,717	70,542	67,015	65,439	0.1

14 AIChE
Spring Meeting, New Orleans, LA

Fredrikson & BYRON, P.A.

Change in 5 year average over last 20 Years

Fastest Growing

Category	Growth Rate
Computer Software	+11%
Data Processing	+11%
Semiconductor Devices	+10%
Video Distribution	+10%
Computer Hardware	+8%
Chemical – Crystals	+8%
Nanotechnology	+8%
Information Storage	+6%
Communications	+6%
Design	+5%

Slowest Growing

Category	Growth Rate
Chemical – Purification/Evaporation /Distillation	-2%
Chemical – General Compounds	-3%
Time Measurement	-3%
Machine Mechanism	-3%
Chemical Manufacture Specific	-3%
Organic Compounds	-3%
Pipes & Joints	-4%
Education	-4%
Hazardous Waste	-4%
HVAC	-4%

14 AIChE
Spring Meeting, New Orleans, LA

& BYRON, P.A.

US Patent Filing Trends

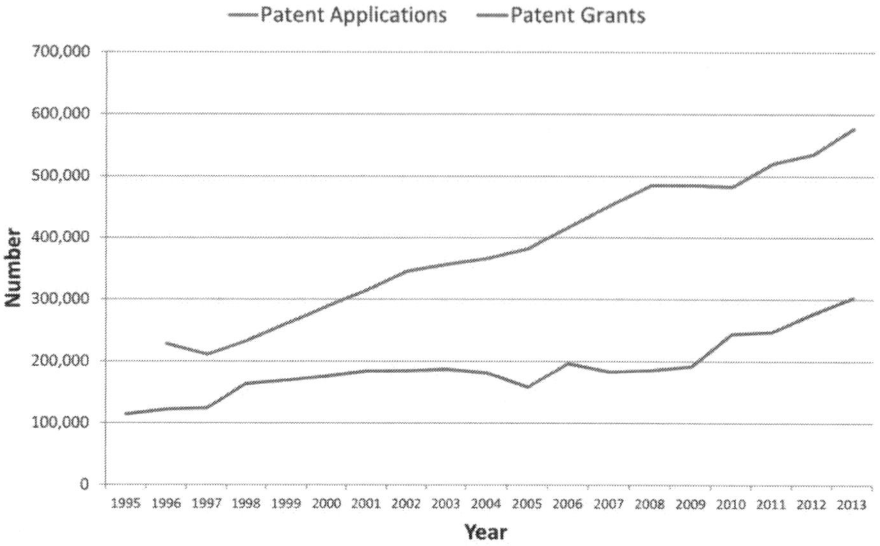

14 AIChE Spring Meeting, New Orleans, LA

Fredrikson & BYRON, P.A.

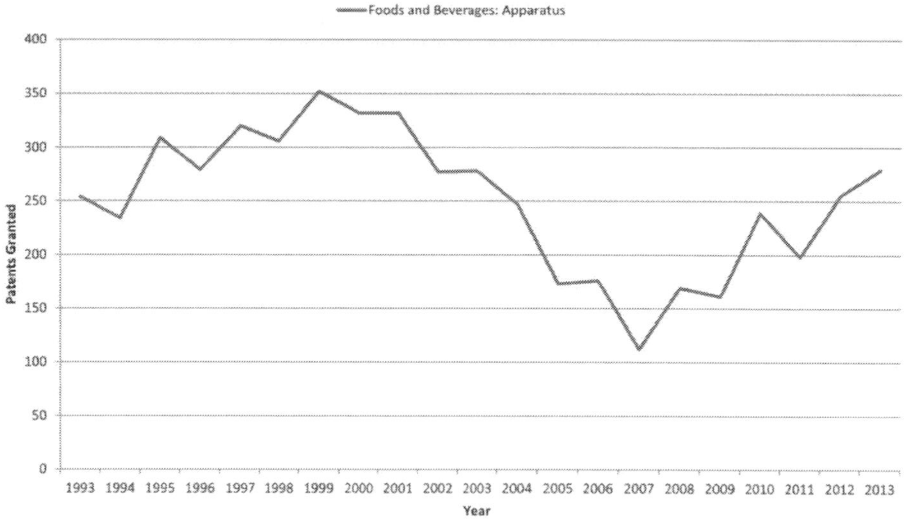

14 AIChE Spring Meeting, New Orleans, LA

Fredrikson & BYRON, P.A.

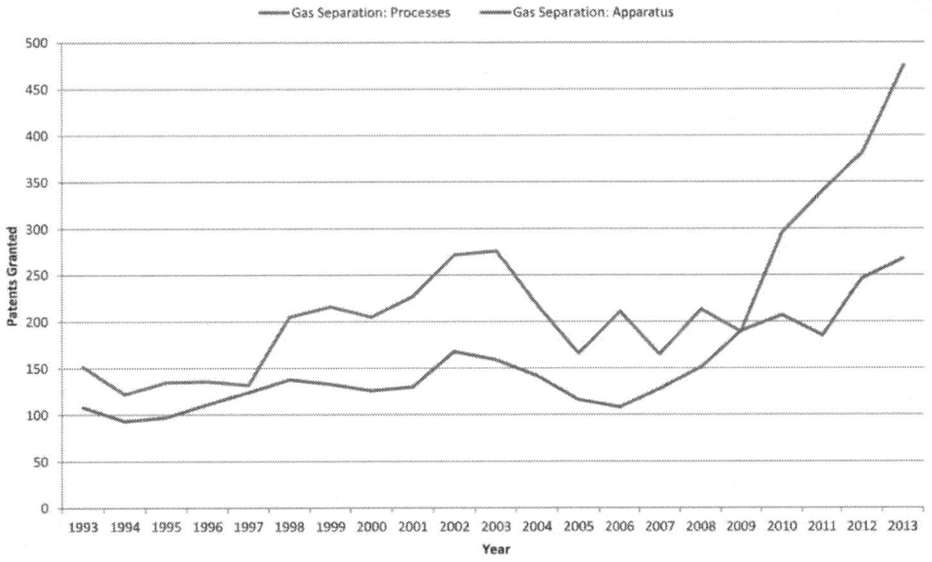

14 AIChE®
Spring Meeting, New Orleans, LA

Fredrikson & BYRON, P.A.

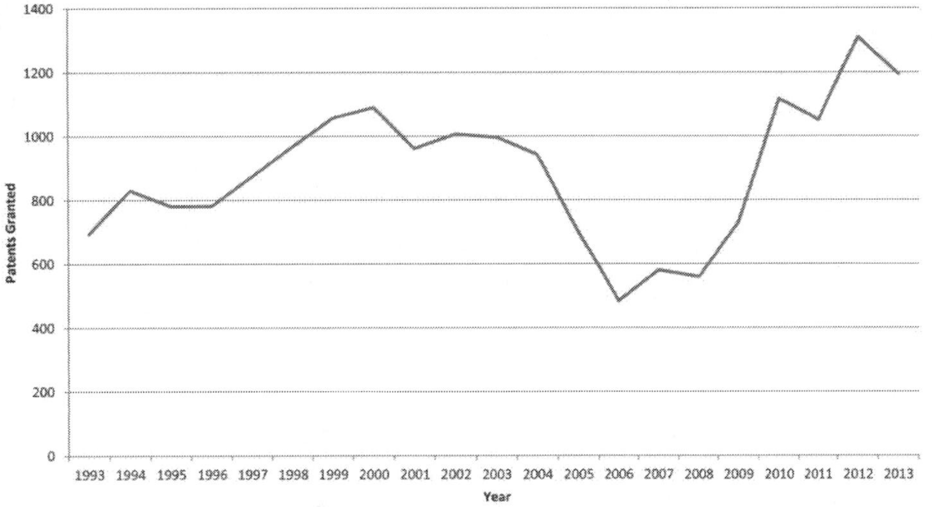

14 AIChE®
Spring Meeting, New Orleans, LA

Fredrikson & BYRON, P.A.

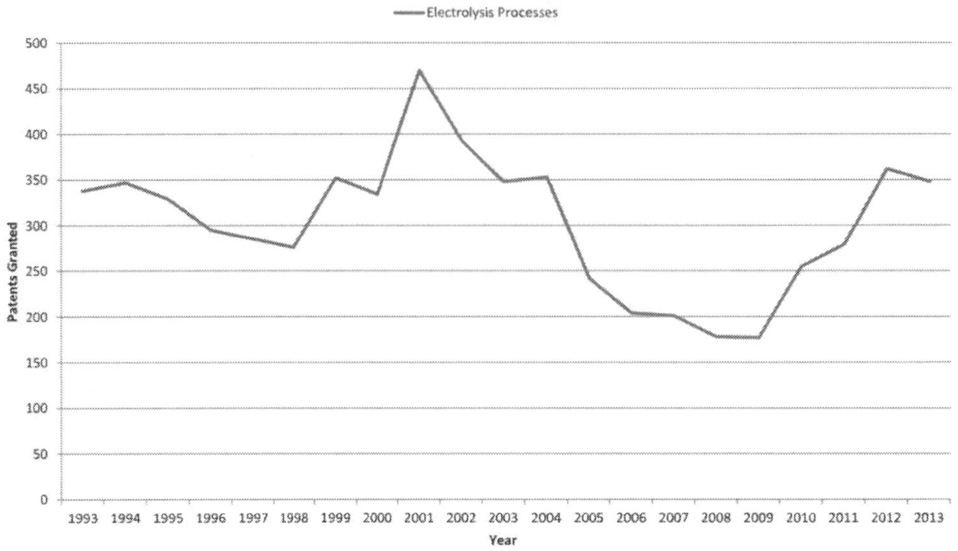

14 AIChE Spring Meeting, New Orleans, LA

Fredrikson & BYRON, P.A.

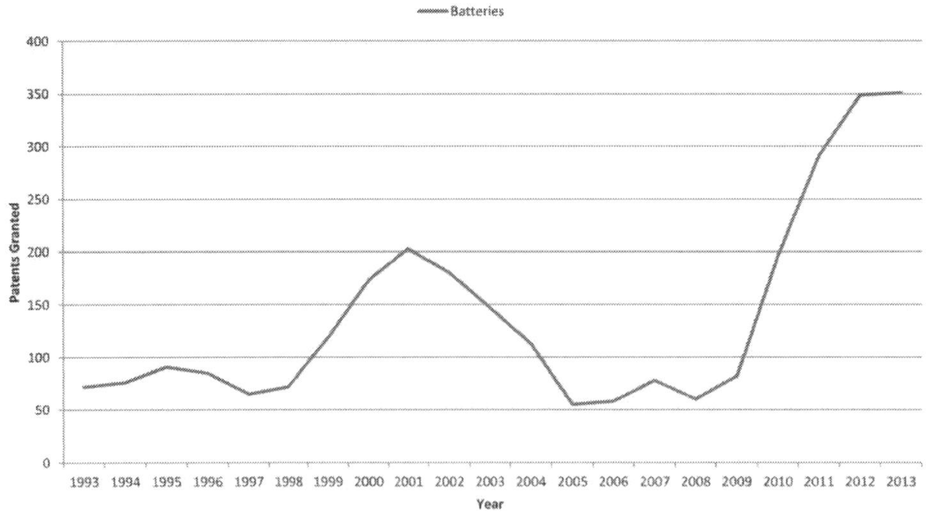

14 AIChE Spring Meeting, New Orleans, LA

Fredrikson & BYRON, P.A.

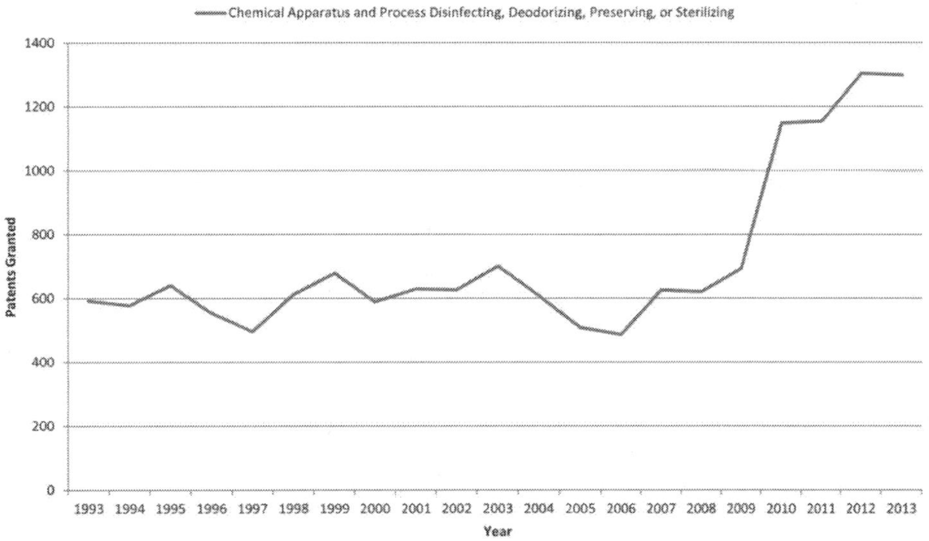

14 AIChE Spring Meeting, New Orleans, LA

Fredrikson & BYRON, P.A.

Patent Litigation Statistics

- Median damage award last 15 years all cases
  - \$1.9 M → 16.5 M
- Median over last 5 years
  - \$4.9M
- Chemical Industry (excluding bio/pharma)
  - Generally smaller median damages (less than \$3M) but characterized by outliers
- Disruptive
  - Median time to trial 2.3 years

14 AIChE Spring Meeting, New Orleans, LA

Fredrikson & BYRON, P.A.

Cost of Patent Infringement Litigation

Amount at Risk	Median Cost
Less than $1 Million	$700,000
$1-$10 Million	$2,000,000
$10-$25 Million	$3,325,000
$25 Million+	$5,500,000

14AIChE
Spring Meeting, New Orleans, LA

Fredrikson & BYRON, P.A.

Distribution of Patent Cases

Top Ten Industry	Percent of Cases
Consumer Products	18%
Biotechnology / Pharma	13%
Industrial / Construction	11%
Computer Hardware / Electronics	9%
Medical Devices	9%
Business / Consumer Services	6%
Software	6%
Automotive / Transportation	5%
Telecommunications	5%
Chemical / Synthetic Materials	5%

14AIChE
Spring Meeting, New Orleans, LA

Fredrikson & BYRON, P.A.

But . . . Looking at 11-15

Rank	Industry
11	Food/Beverages/Tobacco
12	Metals/Mining
13	Clothing/Textiles
14	Energy
15	Agriculture

14 AIChE Spring Meeting, New Orleans, LA

Fredrikson & BYRON, P.A.

RISK MANAGEMENT STRATEGIES

14 AIChE Spring Meeting, New Orleans, LA

Fredrikson & BYRON, P.A.

Understanding the Nature of the Risk

- “Fuzzy” boundaries Problems

<u>Features</u>	<u>Revised</u>
• ~~Back~~	• Seat
• Seat	• At least one base contacting the ground
• ~~Four~~ Legs resting on the ground	

14 AIChE Spring Meeting, New Orleans, LA

Fredrikson & BYRON, P.A.

Understanding the Nature of the Risk

<u>Revised</u>

- Seat
- At least one base contacting the ground

Legal Scope Not Necessarily Co-extensive with Thrust of Description

14 AIChE Spring Meeting, New Orleans, LA

Fredrikson & BYRON, P.A.

Managing New Product Risk

- Correlate Effort to Risk
  - Repeat Use of Old Technology in Existing Field
  - Old Technology in Different Field
  - New Technology in Existing Field
  - New Technology in New Field

Increasing Risk

14 AIChE
Spring Meeting, New Orleans, LA

Fredrikson & BYRON, P.A.

Other Factors to Consider

- Intensity of Competition within Field
- Past History Between Competitors
- Value of Marketplace and Opportunity
- Relative Size of Parties
- Monetization Proposition

14 AIChE
Spring Meeting, New Orleans, LA

Fredrikson & BYRON, P.A.

Managing New Product Risk

- Use a Phased Approach
  - Phase I
    - Basic Understanding of Landscape and Any "Show Stopping" Patents
    - Focus on Likely Competitors – Review Patents, Literature, Products
  - Phase II
    - Revisit Base Design and Evaluate Ancillary Features
  - Phase III
    - Check Any Design Modifications or Newly Issuing Patents

14 AIChE Spring Meeting, New Orleans, LA

Fredrikson & BYRON, P.A.

Managing Ongoing Risk

- Monitor Key Competitors / Inventors
- Monitor Technology Space
- Work with Sales and Marketing Personnel to Track New Product Introduction
- Watch Partners and Consultants

14 AIChE Spring Meeting, New Orleans, LA

Fredrikson & BYRON, P.A.

NEW TECHNIQUES FOR ACTION

14 AIChE Spring Meeting, New Orleans, LA

Fredrikson & BYRON, P.A.

Knowing What Tools Are Available

- Pre-Issuance Submission
  - Attack the Patent Application Before it Becomes a Problem
- Simple But Limited Patent Office Attack
  - Ex Parte Reexamination
- More Involved Patent Office Attacks
  - Inter Partes Review
  - Post Grant Review

14 AIChE Spring Meeting, New Orleans, LA

Fredrikson & BYRON, P.A.

CONCLUSIONS

- CPI Patents Generally Trending Upward
- Great Sector Specific Variability
- Effective Risk Management Requires Being Proactive and Deploying Resources Efficiency
- New Tools to Attack Problem Patents

14 AIChE
Spring Meeting, New Orleans, LA

Fredrikson & BYRON, P.A.

QUESTIONS?

14 AIChE
Spring Meeting, New Orleans, LA

Fredrikson & BYRON, P.A.

Chemical Engineering & the Law --Case Examples in Industrial Litigation

Paper# 352190

Satish Almaula
Senior Principal Consultant
Baker Engineering and Risk Consultants, Inc.
salmaula@bakerrisk.com

Presented at American Institute of Chemical Engineers Spring Meeting, New Orleans LA April 2014

Abstract

Chemical, petroleum, and allied manufacturing sectors involved in research, technologies, engineering and project management at times encounter issues and problems in successfully meeting a project's intended financial, technical, and operational goals and expectations. Significant deviations from goals and expectations can result in lawsuits between involved parties. Alternatively, parties file insurance claims, which can also result in lawsuits.

Research and technology issues can result in patent and intellectual property legal cases. In addition, issues can arise from engineering design, licensing, product liability, safety, regulatory compliance, construction and operational deficiencies, or an inability to meet contractual and performance commitments. Accidents and incidents at manufacturing or other facilities can result in damage to property and the environment, injuries and death as well as interruption to normal operation causing business loss. Such damages and losses also can result in lawsuits and insurance claims.

The need for collaboration between lawyers, scientists, engineers and other technical specialists is obvious when conducting legal, technical and business analyses to get to the root cause of the matter to support or defend a legal complaint or an insurance claim.

This paper presents a brief introduction to lawsuits and how legal and technical teams can collaborate with their respective experience and expertise in handling a lawsuit either as a plaintiff's or a defendant's team.

The main focus of the presentation will be case examples that demonstrate how principles familiar to engineers and other technical specialists play a part in litigation involving diverse issues in the chemical, refining and allied business sectors.

1 Introduction

Technical professionals, whether they are engineers, scientists or technical specialists often get drawn into legal disputes. Modern day media reports of claims and settlements tend to focus heavily on events in the electronic, information technology, pharmaceutical and drug industries, especially when the issues have a direct effect on consumers and involve huge sums of money.

Petroleum, chemical processing, and allied manufacturing sectors have their share of claims and disputes involving large amounts of money but do not always get the high visibility as the above-mentioned industries. Catastrophic incidents such as fires, explosions, release of hazardous materials leading to deaths, injuries and property damage do get expanded coverage and high visibility.

It is not surprising, therefore that chemical engineers, scientists and technology specialists are often drawn into insurance claims and legal disputes. Lawyers, insurers, scientists, engineers, and technical specialists often collaborate in such matters.

This presentation provides a brief background on typical issues leading to legal disputes followed by selected case examples. Names of organizations, involved parties, and individuals are not disclosed or discussed in this presentation. The selected examples are from the author's experience in lawsuits related to chemical processing and allied industries.

2 Insurance Claims and Lawsuits

Insurance claims and lawsuits can arise from diverse losses and disputes such as property damage, fatalities and injuries, negligent performance or, in extreme but uncommon instances, they can arise from intentional or criminal acts.

Losses alleged in processing industry claims and disputes tend to be related to: patent and intellectual property; engineering design; licensing; product liability; safety; regulatory compliance; construction and operational deficiency; and the inability to meet contractual and performance commitments.

Technical professionals including those in chemical engineering and similar fields can be drawn into insurance claims and lawsuits as fact or expert witnesses.

Fact witnesses are generally limited to testifying about facts that they know firsthand about an issue without expressing an opinion about them. Expert witnesses are generally technical professionals with extensive experience and in-depth knowledge in a particular subject matter. They analyze facts, evidence and information and give testimony on their findings as well as their expert opinions.

Given the diversity of training and experience of chemical engineers, they can be called upon to serve as experts in correspondingly diverse fields, some of which may or may not fall under a classic view of chemical engineering.

Collaboration between insurers, lawyers and technical experts typically involves, but is not limited to, the following:

- Developing a clear understanding of the claims and allegations.
- Knowing and understanding the involved parties and the type of business they are in.
- Gathering the information, evidence and facts necessary for legal, technical and business analyses.
- Performing a screening analysis of the strengths and weaknesses of legal, technical and business factors to support or rebut the claims and allegations.
- Determining the strategies for the approach and steps to be implemented by the legal, technical and business professionals in handling the matter.

3 Selected Examples

Example 1 -- Patent Related Lawsuit

A producer of asphalt shingle roof products hired a retired former senior executive as a consultant to reduce disposal costs of reject materials from production. Roofing shingles typically contain asphalt, filler material and granules evenly placed on a core material such as fiber glass or cellulose. The produced material is cut into standard sizes called "a course." The cutting produces some unusable material ("rejects" or "rejected material").

The consultant researched, developed and pilot-tested a method to recycle rejects back into production. The reject material, containing asphalt and filler material, was finely ground with limestone and then recycled back into production, saving costs of filler material and asphalt and avoiding costs associated with the disposal of the reject material. The tests were successful and easily scaled up to commercial level.

The consultant asked the company to file a patent for the new recycling method, but the company procrastinated and, ultimately, did not file for a patent.

Later, the consultant, after ending his work with the company, filed and obtained a patent for the new recycling method. Upon learning of the patent, the company sued the consultant/inventor for stealing proprietary information, which initially prevented him from offering the technology to others. The viability, cost effectiveness, and worth of the new technology as well as information protection were all issues raised in the lawsuit.

Analysis showed that the proposed retrofit of an average size recycle plant at 70 plants in the industry across the U.S. would roughly total $156,000,000 in capital costs. Estimated annual savings to the industry from reduced raw material costs and avoidance of disposal costs amounted to $190,000,000, yielding an attractive return on investment. So, the viability and cost effectiveness of the new technology were demonstrated in principle.

The worth of the new technology/licensing value was determined by financial and economic experts.

The matter went to trial, and the consultant/inventor won the case.

Principles familiar to technical specialists applied in this example: Manufacturing processes, unit operations, industry survey, capital cost estimation, annual cost estimation, return on investment, and protection of important information.

Example 2 -- Technology Transfer, Design-Build, Project Management Lawsuit

The owner of an operating plant wanted to convert by-product waste to useful and saleable products. The owner acquired a license for new technology to be used for the conversion of waste to intermediates. The owner also acquired established technologies to convert intermediates to saleable products. The new technology licensor represented to the owner that the technology was demonstrated and ready for commercial use. The licensor also offered used equipment in good condition, which the owner agreed to purchase.

A design-build contract for the plant and process for the new technology was awarded on a fixed price ($150 million) and fixed schedule basis to an architect and engineering firm ("A&E").

The technology licensor provided a Process Information Package (PIP) for design-build work as "frozen" --- that is, ready for detailed design and procurement. Numerous and untimely changes were imposed by both the technology licensor and the owner, causing delays in the detailed design, procurement and project completion. The A&E submitted change orders to the owner, who orally approved them.

The technology sold to the owner was actually its first application on a commercial scale in the U.S. The only other reported use of the technology on a commercial scale was in Japan, where several modifications were needed to make the process work correctly. The condition of the used equipment sold to the owner was not "good," as claimed. The equipment required significant cleaning and modification to meet design and code requirements.

Project completion was delayed by three months but, when completed, the system operated satisfactorily. The A&E submitted a Request for Equitable Adjustment ("REA") to the fixed price for the additional costs. In response, the owner sued the A&E for "Bad Project Management" claiming damages amounting to seven times the REA amount. Of note is that the technology licensor was not drawn into the lawsuit and was not held accountable for changes made to the "frozen" PIP.

The matter went to mediation, but did not settle. The owner later filed for bankruptcy, and the plant was bought by another party.

Principles familiar to technical specialists applied in this example: Review of process design, PFDs, P&IDs, impact of design changes and delays on construction, project management, and contractual performance.

Example 3 -- Process Technology Licensing Lawsuit

Two separate companies (A and B) had plants located at one site. Company A manufactured product X1 and sent its wastewater to Company B for treatment and discharge. Company A then started to manufacture another product, X2, at the same plant. Company A recognized that Company B's plant could not handle the additional wastewater created by the process to manufacture X2. Therefore, Company A decided to build its own new treatment plant on the site to handle the wastewater generated by products X1 and X2 combined.

Company A chose a wastewater treatment contractor (TC) to provide "turn-key" services, including technology, engineering, design, construction and startup of the new wastewater treatment plant. The selected technology was a "thermophyllic" biological treatment process, patented and licensed by a third party technology developer ("Licensor"). The license to operate using the technology was granted to Company A, and the technology and engineering details were given by the Licensor to TC.

The Licensor pilot-tested Company A's waste stream to confirm treatability and provided TC with process and engineering information tailored to treat the tested waste stream samples.

At startup, the new wastewater treatment plant failed to meet the treatment targets, and Company A filed a lawsuit against TC. The Licensor was drawn into the lawsuit as a third party defendant.

During the discovery phase of the lawsuit, the allegations against the Licensor were determined to be unfounded both technically and contractually. It was established that there were numerous deficiencies in Company A's design basis as well as defects in both the engineering design and installation by TC. There also were issues involving the performance warranties between Company A and TC.

The technology provided by the Licensor had been commercially proven at other facilities, and it was established that the Licensor had met its contractual obligations.

The matter was settled before trial.

Principles familiar to technical specialists applied in this example: Process evaluation, pilot testing, process design and engineering analysis, contractual commitments, review and understanding of Company A's and TC's obligations and failures.

Example 4 -- Plant Incident and Alleged Medical Injury

A manufacturing plant was undergoing a maintenance turnaround with contractor employees working on scaffolding near the top of a large carbon steel tank containing concentrated sulfuric acid. An undetected water leak into the tank diluted the acid and a relief valve on top of tank momentarily lifted and released vapors with a sulfurous odor. The contractor employees immediately evacuated the area and reported the incident. The turnaround work was shut down for the day, and all workers were sent home.

Reportedly, the contractor employee who was closest to the top of the tank went to a restaurant with his co-workers, where he collapsed. He was attended to and recovered. Later, the contractor employee sued the plant for negligence, alleging that the gas he was exposed to was hydrogen sulfide (H_2S). During the litigation, medical statements were produced stating that the contractor employee's nervous system was permanently damaged from exposure to H_2S.

The defense argued that the released gas was more likely to be sulfur dioxide (SO_2) and not H_2S. Reported public information showed that exposure to SO_2 could have bronchial effects but not permanent nerve damage.

A literature search as well as an examination of the conditions of the carbon steel tank (pressure, temperature, and acid concentration) on the date and time of the incident suggested that the gas released was as a result of corrosion, and it likely contained primarily SO_2 with trace amounts of H_2S.

Defense lawyers engaged an independent third party laboratory to test evolution of gases from reaction of concentrated and diluted sulfuric acid on carbon steel, thus simulating conditions in the tank on the day and time of the release. The tests showed that the evolved gas content was mainly SO_2 with likely trace amounts of H_2S.

A chemistry professor for the plaintiff testified in his deposition that the gas released at the time of the incident was primarily H_2S, but his testimony was strongly challenged by defense lawyers. A second professor from another university testified in his deposition that the plant violated OSHA's Process Safety Management (PSM) regulation. The second professor's testimony was challenged on the basis of his lack of experience with PSM.

Later, it was learned that the contractor employee who filed the lawsuit had been previously employed as a sheet rock worker and had developed indications of dust-related issues involving his bronchial system.

The matter was settled between the parties before trial.

<u>Principles familiar to technical specialists applied in this example:</u> Chemistry of reaction of dilute acid with carbon steel; examination of the incident timeline and possible cause(s) of water backflow into the acid tank; basic metallurgy of carbon steel; review of third party test results; and review of the plant's PSM program.

4 Observations and Conclusion

1. Insurance claims and lawsuits can arise from a variety of reasons and situations, and outcomes are not easily predictable.
2. Lawsuits and insurance claims can arise from unwritten "perceived expectations" by the involved parties.
3. Besides the terms and conditions defined in a contract, other factors, such as standard of care and generally accepted good practice in a given industry, are often encountered in allegations.
4. In addition to directly involved parties, all other entities that are in any way associated with the matter may be drawn into the lawsuit - especially those perceived to have "deep pockets."
5. Claims of advantages and success using new technology on a commercial scale should be thoroughly evaluated before acquiring the license for its use.
6. Clear agreements between a company, its consultants or involved third parties regarding ownership and rights regarding intellectual property need to be executed in advance.
7. Applying fixed price and fixed schedule terms to design-build contracts in capital projects (or any) in which a firm and definitive scope cannot be described clearly in the contractual documents will most likely result in cost overruns and schedule delays, prompting lawsuits.
8. A technical professional can take steps to minimize liability, although it is doubtful one can completely avoid it. Some exemplar suggestions are:
    - Try to clearly understand the technical and (to the extent possible) commercial needs of an assignment.
    - Take on only what you are capable of performing on a sound technical basis. Do not make unsupportable assumptions or try to "wing it."
    - Speak up if you are not the best person to take on the assignment. Have it passed on to someone else who can best perform the work.
    - Keep good written factual records of your work and your communications about the work you are doing.
    - Most of all, be factual and truthful in all you do.

Determining The Liability Of A Fire Starting Product

C. J. Abraham
Scientific Advisory Services Ltd
Great Neck, New York

Abstract

Natural Fire Starter is an ethanol based fuel gel which was designed specifically for wood and wood pellet stoves. It is sold in a plastic bottle with a child proof closure. This product had never been cited in any reported accident until a 7-year old was badly burned when he was allowed, unsupervised, to toast marshmallows in a wood burning fireplace. The subsequent events reported by the child, his parents and brother are discussed as well as the experts and lawyers claiming the liability of the defendants. Said defendants include the manufacturer of the formulation, the bottle manufacturer, manufacturer of the closure and the packager.

In order to reconstruct the accident and determine the cause and liability involved, this investigator had to consider the testimony of the injured child, his parents, brother , the ENT personnel, medical personnel, plus testing and testimony of all of the experts as well as the statements made by the plaintiff's attorney.

Determining The Liability Of A Fire Starting Product

As simple as the product appears, over 50 characteristics and variables of the bottle of fire starter were considered in reconstructing the accident.

After extensive testing by this investigator and all of the experts on both sides, analysis of all of the documents and videotaped depositions, the conclusion of this investigator indicates that the case, as presented by the plaintiff's attorney, is highly suspect and the opinions of the plaintiff's experts are unreliable and without scientific substance.

Background to the accident

On or about February 2010, Plaintiff "A" received burn injuries at his home located in West Virginia. According to the Complaint, he was using a product called Natural Fire Starter Gel, which is an ethanol-based fuel gel.

According to testimony of "A" and his parents, a fire had been burning in the fireplace on the lower level of the house for much of the day. The Natural Fire Starter Gel was near the hearth. At some point in the evening, A was given permission by his mother to go to the lower level of the house unsupervised to roast a marshmallow prior to taking a bath He was naked,.

According to the complaint, A applied the Natural Fire Starter Gel to kindling wood in the fireplace and the Natural Fire Starter Gel container suddenly failed near the lower portion of the bottle, projecting flaming gel onto "A", resulting in injury.

Natural Fire Starter Gel is designed specifically for pellet and wood burning stoves. It is manufactured and assembled by Packaging Service Company (PSC) using a plastic bottle manufactured by C.K.S. Packaging, Inc. and a cap manufactured by Stull Technologies, Inc. The product is distributed by Hearthmark, LLC. The sole purpose of the gelled product was that the product was to be used for only wood pellet and wood burning stoves.

The subject bottle was purchased by Plaintiffs at a local Walmart store. According to the Complaint, Plaintiffs' allege the Natural Fire Starter Gel was not reasonably safe for its intended use, was defectively designed and manufactured, and carried inadequate warnings.

The undersigned reviewed over 10,000 pages of documents obtained through discovery in addition to performing, examining, or analyzing the following:

1. **the subject bottle and exemplars**
2. **Testimony of injured party**
3. **View of fireplace and living room after the accident**
4. **Analysis of the PVC container**
5. **Analysis of the stopper for the PVC container**
6. **Proving or disproving the testimony of the plaintiff, his parents and brother**
7. **Testing and testimony of plaintiff's experts**
8. **Testing and testimony of the defense experts**
9. **Deficiencies in testing by defendant's experts**
10. **Deficiencies in testing by plaintiff's experts**
11. **What are the scientific basis, from a mechanical engineering stand point why either the plaintiff or defense experts were incorrect in their analysis?**
12. **Testimony of all experts.**
13. **Testing by all experts**
14. **Medical documents**

In addition to the above, it was important to evaluate all of the chemical, physical, thermal, and flammable properties of the product in addition to evaluating the warnings and instructions.

1. **Accident Reconstruction**
2. **Annealing**
3. **Basics of blow molding**
4. **Basics o f ethanol based fires**

5. Basics of gelled fire starters
6. Bottle closure
7. Brittle versus ductile fracture
8. Compatibility of a solvent and chemicals with P VC containers
9. Compatibility of ethanol and PVC
10. Consumer Product Safety Commission Reports
11. CT scans (Computerized Tomography Scan)
12. Curvature of bottle at shoulder
13. Deflagration, detonation, explosion
14. Destructive testing
15. Design of PVC container
16. Expansion o f PVC plastic
17. Explosion limits
18. Flame arrestor technology
19. Flame speeds
20. Flashback
21. Flashover
22. Heat damage
23. Heat exposure and PVC
24. Laminar flame speeds of ethanol-air mixtures
25. Material Safety Data Sheets (MSDS)
26. Napalm
27. Non destructive testing
28. Origin of fracture
29. Pictorials
30. Plastic bottle rupturing
31. Plasticizers
32. Pressure release valves
33. Prior reported accidents
34. Recalls on subject product and similar products
35. Scanning electron microscope
36. Seams in plastic container
37. Similar products
38. Single holes versus multiple holes
39. Stains in plastic bottle
40. Stress in plastic containers
41. Stress cracking
42. Stress whitening
43. Specification of PVC container
44. Supersonic versus subsonic
45. T ensile strength of PVC
46. The effect of dye in ethanol in PVC bottle
47. The effects of gelling ethanol
48. Thermal gravimetric analysis
49. Viscosity
50. Volatility

51. Warnings and instructions

The Plaintiff, 7 year old boy "A"\ (Deposition summary with comments)

At the time of the incident, "A" was 7 years old and claims in his deposition, that he shoots his own gun. He started shooting at the age of 5. He stated in his deposition that he has never started a fire without an adult present. His mother told him not to use the gel on a burning fire because it was a waste . *He stated his mother told him that he could start the fire that day. Contrary to that statement* "A"'s mother testified that she started the fire prior to taking the family to church.

"A" stated that he "*used the gel to start a fire and also to restart a fire*." This *admission is important and* more than documents the fact that "A" squirted a lot more of the Diamond product into the fire than he testified to. *On the day of the incident his "mom's back was out and he was sick with pneumonia."*

He stated that he was getting ready to roast marshmallows. However, his mother told him to take a bath. He asked her if he could roast a marshmallow. She said yes. He went to the kitchen, opened up the cabinet *and got a bag of marshmallows*. The alleged bag of marshmallows was not located or identified in any of the documents reviewed or photographs taken at the scene of the accident.

According to "A", the fireplace was in the basement and his mother, father and brother were upstairs. He walked down the stairs, went to the fireplace and stated that there was no fire . "A" testified that the coals were black with white powder over them, and he thought the fire had gone out. *If there was white powder all over the coals, how could he see the color of the coals?*

He then stated that he set the marshmallows down and then put one on the stick and laid the stick right by him. He then went and obtained the fire gel. *He moved the screen and put it on the side*. Moving the screen left him without any protection whatsoever. He put some wood up in a teepee and started squirting gel on it. He testified that the gel landed on a coal and started lighting. The heat emanating from the coals is over 400 degrees Fahrenheit and impossible for anyone to expose any part of their body especially an individual's hands. He then stated that he had the bottle up about a foot and the fire went up 6 or 7 inches. This was not accurate based upon the testing performed by the undersigned.

"A" stated that he held the gel bottle about a foot from the teepee made of wood. He was on his knees andwas holding the bottle and squeezing it angled downward . He then stated thatas he tilted it up and and backward, the gel expelled out of the bottom.He stated it exploded and his brother heard the noise upstairs while he was watching television.

He then stated that he was forced backwards onto the couch and got up and ran into the bathroomon the same floor. He grabbed a cup of water and threw it on theback

of his neck in an attempt to put the fire out. He then ran upstairs to his mother. He stated, contrary to what "T" testified to, that his father grabbed him) and threw him in the bath tub and started running the water.

"A" testified that his mom and dad taught him how to start a fire using a teepee. Contrary to that statement, his father stated under oath that it was "A"'s idea to use the teepee. ***Regardless, starting that type of fire did not include using the gel.***

Although he had observed his mom's friend start a fire using lighter fluid, he has never observed any adult pour lighter fluid on a pre-existing fire.

On the day of the fire, his mother started the fire prior to going to church. The fire was burning most of the day and went out approximately 1 or 2 o'clock in the afternoon. He stated the accident occurred at approximately 8:30 p.m. If that is true and no one replenished the fire with additional logs, it would be questionable if any embers still existed. ***Note - How would*** **"A" would not be able to build a teepee if the embers were still hot** ***and not all ash?*** **He stated that there were no coals in the fireplace, there were just ashes. He also stated that matches were "right there." He also stated that if the coals were hot, he would have just put paper on the coals and not used the fire starter gel.** ***("A"'s testimony does not make sense. If he didn't use paper and he just observed ashes how could he have built a teepee if there was sufficient heat to ignite the gel? There was a reply by "A" that he did not feel any heat while he made the teepee.***

At the time he was squirting the gel into the fire, he had to stand on the ledge (hearth extension) and lean in to pour the gel onto the wood. He was able to get up on the hearth using his hands and knees. He set the gel down, climbed up on the ledge and started squirting all around the teepee. He ***stated he squirted all around before there was a fire.. He also stated that he was still on his hands and knees when the fire came at him.***

Comment: Reconstruction of the accident by the undersigned with an exemplar fireplace documents that the fire does not flash back ***but goes directly in the vertical direction.*** **Fire travels directly in the vertical direction.** ***At this time, while on his hands and knees he tilted the bottle back and the bottom came out with a noise.*** **Contrary to the analysis of the plaintiffs' experts, this more than documents that their analysis is without merit and without a scientific basis.**

The amount of squeezes the plaintiffs' expert had to complete prior to meeting within the explosive limits of the ethanol does not coincide with "A"'s testimony or the real-world.

"A" stated that when he first picked up the gel bottle, ***it was not warm. As he fell backwards, he stated that he threw the gel bottle onto the couch. (If there was any leak or opening in the bottom of the subject container, the gel would have***

squirted out the leaking end as well as the opening at the top. "A" never testified that as he was squeezing the gel into the fireplace that any other part of the container was leaking.)

He had no clothes on when his mother told him to take a bath. However, his older brother, "T", stated that his mother used the wet towel from "A"'s bath that he took prior to roasting the marshmallows in the fireplace.

Further in his testimony he stated that he used a lighter to start fires and would get gel and squirt it on the wood and light it. When he was finished lighting the fire, he would close the cap and set it down about 10 feet from the fireplace. He would then put more wood on the fire. , "A" also stated he was told to be careful of the gel. *He knew that the gel "could light on fire" and it was dangerous.*

"A" stated that he thought the fireplace was still lit. He had observed that fireplace with the wood burning hours before that time. *(If he thought the fireplace was still burning, why didn't "A" first try paper on the coals as he was taught?)*

He stated that he used the gel even though the ashes were still warm because he wanted to get it started.* He described the ashes as whitish, black and grey. He states that he did not see any red.. *However, he previously testified that the coals were not warm.

Further on in his testimony on "A" stated that the fire was out about 1 p.m. and he went down to roast marshmallows around 8:30 p.m. (the incident occurred at 7 p.m.). He did not poke a stick in the ashes to see if there were any embers. *(His parents stated that they taught him how to start a fire and use the gel which is not consistent with "A"'s behavior and actions.)* "A" was allegedly taught how to start a fire and use the gel, which is an unreasonable misuse when it involves a child.

In addition, based on "A"'s stated awareness and knowledge, and based on the fact his mother taught him not to put the gel on an existing fire or hot embers, "A"'s actions were not reasonable. Moreover, it is not reasonably foreseeable that a 7-year-old child will use the fire starter gel to start or accelerate a fire.

He reiterated the statement he made previously. "A" stated that he squirted the fire starter gel on the hot coal and it flamed up only about 6 or 7 inches. *(However, he previously stated that the coals were not warm.)* As he tilted the container it came closer to the fire and then "it exploded." This statement cannot be accurate, because the air space contains close to 100% ethanol vapor. The explosive limits within the container are not reached until "A" squirts the gel into the fireplace a number of times so that air can replace the ethanol vapor and bring the concentration of the ethanol to within the explosive limits. *Hypothetically, assuming, for example, that there was a leak in the bottom seam, deflagration could not*

occur because the opening in the bottom would have released all pressure created by the reaction taking place.

Even though the bottom came closer to the fire, he never saw the flame come in contact with the bottle. "A" reiterated again that once he was on the hearth, he was just on his knees with his hands on the bottle. ***(Note - It would be impossible for the bottom of the container to come near the fire because "A" was squirting into the fire with the nozzle facing the fire and not the bottom of the plastic container. In addition, if "A" was squirting the bottle as he testified to, the gel would have leaked out through the small opening at the bottom if it did exist. There was no testimony by "A" that, as he was squeezing the gel into the fire, that any of the gel came onto any part of his body from the opening in the base of the container.)***

Comment - If "A" was on his knees on the hearth, there should not be any burns below his knees. It would have been impossible for him to be thrown back onto the couch. The burn diagrams from Shriners Hospital show that he is not burned a few inches above both knees and down to his toes in the front of his body except for the inside of his right calf.

The history as recorded by Shriners Hospital for Children states that "patient is a 7 year old white male with recent past history sustained on 08-28-2010. ***The patient had been roasting marshmallows inside when an accelerant for the fire caught on fire and exploded onto the patient. <u>The patient then fell backwards into the fire before being recovered by his father."</u>*** **The areas that were not burned were his scalp, his back and left side. There is no mention of a couch. If he fell backwards into the fire, his back would have been severely burned. It was not. However, it was possible that he fell forward into the fire. That would account for the intense burns on the front part of the body.**

The fact that "A"'s scalp and face were not burned refutes his testimony that he raised the container above his head and it exploded onto his body. If that were accurate, he would have received second and third degree burns on his scalp and his eyes and eye site would have been considerably damaged. ***(This also refutes the plaintiffs' contention that there was a small opening in the bottom of the container prior to the incident.)***

"A"'s mother, in her deposition, stated that she had problems recalling what the warnings stated on the subject container. However, she did remember that the product was flammable based on her experience. If that is true, how could she expect "A" to recall or fully understand any part of what she read and understand the meaning and intent of the warning?

"Father"

"A"'s "Father" also stated that he taught "A" at the age of 5 or 6 to make a fire. He taught him to place a piece of towel or tissue under the small kindling and light it. "A", he stated, came up with the teepee approach. ***(This is contrary to "A"'s testimony.)***

"Father" stated that he used the gel less than 10 times to start a fire and "A" watched him do it. He never saw "A" start a fire with the gel. More importantly, he told "A" that he was never to use the product because it was dangerous. Also, he never put the gel on the mantel and never saw it there. "Father" started the log fires with starter logs.

He purchased only one bottle of the Diamond product at Wal-Mart. That was the only time he ever purchased the product. This comment is contrary to "K"'s comment stating that she purchased it. He stated that he read the label and it said "keep out of the reach of children." That is why he told "A" not to use it. He does recall that it said to "store away from flame or heat sources." That is why he kept it on the floor which, in his opinion, was the coldest point in the room.

It should be noted that, contrary to the plaintiff's allegations, if it was kept on the floor, that would mean that it was well below the hearth and not be exposed to any heat from the fireplace which, the plaintiff's claim caused a leak in the container that ended up causing the fire and the deflagration.

The statement by the father also contradicts "T" who stated that the gel was kept on the mantel. "Father" always let the fire die down in the late afternoon and early evening. He again contradicted "T" when he stated that he tried to smother the fire with a towel.

When he went down to the basement, he does not remember seeing the gel bottle that day or after the fire. He did hear that "A"'s grandmother found the bottle of gel outside. I was under the couch. After the accident, his wife obtained a restraining order preventing him from seeing the children.

"Father"' stated in his deposition where, he quoted an entry in "K"'s (his wife) diary. It is alleged that after she got home on the evening of March 10 and checked the fireplace where the incident occurred, she ***discovered that "A" did not squirt the lighter fluid*** **(The gelled product)** ***into the fire as she first thought. The bottle was sitting close to the fire and appeared to have a melted crack near the bottom of the bottle allowing it to spray into the fire causing "A"'s burns. By that time, she had already retained a lawyer to sue.***

Question: "K"'s entry into her diary contradicts everyone else's testimony under oath. Her entry infers that "A" was not on his knees on the hearth holding the container as he testified to under oath. It also contradicts "Father"'s testimony. However, it does fit her attorney's theory and that of

the plaintiff's experts. This author, having over 40 years of experience in evaluating similar cases, the statements made by the family members are incredulous.

"Father" was not aware that "K", his wife at the time, allowed "A" to use the gel under her supervision. He did not expect "A" to start a fire on the evening of the incident. He also stated that before the incident, he thought he heard a popping noise. The question is whether you can actually hear a popping noise in the basement while he is on another floor watching and listening to a TV program.

"K" (mother of "A")
"K" stated n her deposition that "A" should not start fires by himself. She also stated that she never *caught him starting a fire*.

Contrary to "Father"'s testimony, she stated that *she bought the fire starter* because it was thick and it was right beside or right near the fire logs in the Wal-Mart store. <u>Without reading the warnings</u>, *she felt that the product was safe and could be used to continue the fire.* She also knew to keep it out of the reach of children and that the product could explode if it got too close to the heat source.(This contradicts "A"' testimony where he stated that his mother read the warnings to him in the store.)

It should be noted that there is not one word about explosions on the Diamond product label. **When "A" and his mother returned from Shriners Hospital in May he told her that *he was squirting the gel on the fire*. She also told aa registered nurse in the emergency department in the hospital that "*The patient squeezed the accelerant into the flames and exploded on the patient.*" At her deposition, she stated that she didn't recall telling anyone that statement. However, the statement made and recorded in the hospital notes fit the reconstruction of the cause of "A"'s burns and contradict the statements made by "A" and "K" under oath and also contradicts the opinions set forth by all of the plaintiff's experts.**

She also claimed that she saw a hole in the bottom of the bottle and assumed that it must have sat near the fire because it is only that it could have gotten there, because it got hot and melted.

She does recall seeing the marshmallow on the end of the fork .A small residue of what might have been the remains of a marshmallow was noted in one of the photographs. However, that evidence was never retained for examination. The "bag" of marshmallows were never found.

"K"'s mother found the bottle of the gel outside of the front yard under the couch. There is no explanation how her mother was able to remove the plastic container from under the couch. *It should be noted that some of the deformation on the subject container was due to the weight of part of the couch on the container caused deformation after the incident.*

She does not recall reading part of the instructions where it stated that one should "Close bottle and remove from immediate area as vapors are flammable." However, she knew that it was flammable.

When asked if she ever read the instructions on the Fire Starter Gel to "A", she replied "I do not recall." However, a question was then asked again of the reading of the label and she stated that she read it to both "A" and "T" and told "A" that he cannot use the product .

"K" stated in an answer to interrogatories that "A" had previously used the Fire Gel) under her supervision). When "K" allowed him to use it under her supervision, there was no fire. His father objected to allowing "A" to use the product. "K" does not recall how many times she allowed "A" to use the product. This contradicts her prior statements under oath. In fact, "A"'s father told "A" not to use the product.

- **"D" (registered nurse)**
Donielle is a registered nurse working full-time for the emergency department at the Memorial Hospital and the County EMS.

Her intake states that the patient was roasting marshmallows in the fireplace. *Patient squeezed an accelerant onto flames exploded on patient*. Patient had severe burns over entire body except for his back. Estimated 70% of second and third degree burns observed. No other injuries from explosion.

This recorded statement contradicts "A"'s testimony, his mother's, his father's and "T"'s.

- **"J" (registered nurse)**
"J" is an EMT-B. He also testified that he was told that "A" went down to the fireplace and *sprayed some type of aerosol or something on the fire and it blew back on him.*

Shriners Hospital (dictated by "R" MD in April of 1020)
***The patient had been roasting marshmallows inside when an accelerant for the fire caught on fire and exploded onto the patient. The patient then fell backwards into the fire before being recovered by his father.* This *more than* documents the fact that "A" sprayed an excessive amount of the Diamond product on to the existing fire. When he fell back, he was exposed to close to 1800 °F. It should also be noted the screen was not in place when this occurred.**

Critique of Dr. "G" (Expert for plaintiff)
Dr. "G"'s academic background and experience is excellent. However, his background in storage, transport and processing vessels for chemicals, gasoline and other flammable products in addition to the implementation of various flame suppression and arresting technologies for the safe handling of combustible products

in general and gasoline tank design and leak detection systems for gasoline stations and gasoline storage systems does not relate to the issues and the cause of the accident in the subject case. The "simple screen flame arrester for the inside of a gasoline tank" is not the solution in this specific case. The product in this case is thixotropic and the design of the opening in the cap must be at a size that would allow an individual to squeeze out a desired amount with minimal pressure on the container.

Dr. "G" conveniently forgets about the fact that the plaintiffs testified that "A" was going to roast marshmallows. There was no evidence found of any marshmallows in the area of the fireplace or the living room except for the small amount of white residue on the fireplace fork. There was some white color or remnant on the end of a fork in one of the photographs. However, that was never examined by any of the investigators.

He goes on to state in his report that that "A" "holding the bottle with both hands with the nozzle pointed down and pored the kindling wood when the fire gel hit the kindling/embers and then the heat source ignited the bottle due to the flashback."

Dr. "G" never attempted to reproduce the subject accident with an exemplar fireplace and the subject Diamond formulation. He accepted the questionable testimony of "A" and his mother when designing and performing his testing.

How could **an explosion or flashback cannot occur with one squeeze when the head space contained almost 100% ethanol vapor and testing by the plaintiffs experts documented that a large number of squeezes were necessary to reduce that concentration of ethanol vapor and have sufficient enough air in the head space so that the vapors could be ignited. The experiments performed using an exemplar fireplace by the undersigned disproved all allegations made by the plaintiff and the plaintiff's experts.**

Dr. "G" neglected to state that "A" testified he was on his knees on the hearth and that he allegedly made a teepee out of kindling and squeezed the bottle on the teepee. He also neglected to state what the rest of the family testified as to what occurred as well as other recorded evidence. He also did not mention the fact that there was no evidence of any kindling in the area of the fireplace or the fire. ***If "A" constructed a teepee out of kindling, where was the kindling wood located? Not one of the photographs contained any evidence of kindling wood.***

There is no evidence that the container caught on fire and exploded. If it did, it is more likely than not, that all of the contents would have been burnt up and there would have been extensive deformation of the subject container. His explanation of the deformation of the container is incorrect. The plastic container can be deformed, as it was in this case, when a weight of sufficient size is placed on the plastic for a period of time. The deformation was mechanical and had nothing to do with the incident.

More likely than not, the fire test performed by Dr. "D", another plaintiff's expert , documented that the container had to be placed in a specific position for an extended period of time before it would catch on fire and deflagrate. The amount of time that Dr. "P" had to squeeze the container before anything was notable was well over two dozen times.

Analysis, methodology and opinions (Part of Dr. "G"'s report)

A. "G" Opinion: The rupture to the subject bottle is consistent with an internal explosion.

Comment: The rupture to the subject container is not consistent with an internal explosion. ***The rupture was more consistent with the deflagration. It also required a very large number of squeezes at a specific angle over a long period of time before the fire was spread inside the container before the rupture occurred, and this was not from a "flash back" occurrence. Tests performed by the undersigned with the Diamond container filled with the thixotrophic formulation and refute the experts theory and opinions.***

B. "G" Opinion: The burns on "A" are consistent with burning gel being expelled from the bottom of the ruptured bottle and spraying on his body.

Comment: The burns on "A" are also consistent with "A" spraying an excessive amount of gel in many areas ***(into the fireplace and onto his body).*** **It is also consistent with "A" falling** ***back*** **into the fireplace** ***(see medical documentation).*** **This can expose "A" up to temperatures reaching up to 1,800 degrees F. No one knows exactly how many times "A" squeezed the plastic container.** ***For example, it is obvious that he squeezed a large amount of the gel directly in front of the hearth. There is also evidence that there were many squirts all around the fireplace, on areas surrounding the fireplace, the walls and ceilings.*** **As proven by the undersigned, in all of the experiments performed, no flashback occurs at anytime no matter how much gel is squirted into an open flame.**

C. "G" Opinion: Explosion tests were conducted with 5 exemplar bottles of Natural Fire Starter Gel. Ten tests were conducted with single hole caps by squeezing the bottles of gel over a candle flame. In each of these tests, the bottle had an internal explosion. In one of these tests, the bottle ruptured. The ambient temperatures of these tests were between 55 and 65 °F.

Comment: The explosion tests that Dr. "G" conducted on five exemplar bottles did not involve methodology that was scientific. The tests conducted on 8/29 self extinguished. No explosion occurred. The container deflagrated.

The test conducted on 9/11 by "G" where the base of the container blew out did not, in anyway, reproduce the results produced in the exemplar container. The container was held in a horizontal position for an extended period of time which is unrealistic. The same is for the second test that was held on 9/11.. In addition, the initial ignition of the external vapor was snuffed out because the internal concentration of the ethanol was not within the explosive limits.

There is no documentation that the subject container was only partially filled containing a lot of head space. Flammability tests performed by the undersigned using an exemplar fireplace disproved the theory and opinions presented by Dr. "G". The thixotropic Diamond product was squirted directly in front of the fire exposing the spout, vapor and thixotropic liquid to over 200 degrees Fahrenheit and through a hole on the top of the fireplace exposing the spout, vapor and thixotropic liquid to over 300 degrees Fahrenheit. No flashback or internal deflagration occurred while the undersigned personally squeezed the plastic container. The spout was so hot that the undersigned was not able to close it until it cooled down

The additional test *performed by Dr. "G"* with a different cap are also non scientific and do not, in any way simulate the conditions that existed at the time "A" squeezed and squirted the gel all over the pre-existing fire, carpeting and walls.

Using the same formulation with a different cap documented what probably occurred on the original cap on the subject bottle. The subject bottle did not explode but the squirting of the gel all over the fireplace by "A", including his body on a pre-existing fire significantly contributed to "A"'s burns when he was ignited.

If "A" was really going to toast marshmallows, there would have been evidence of marshmallows that were roasted on the pre-existing fire and glowing embers. There was no evidence in the room where the fireplace was located that there was even a supply of marshmallows anywhere else in the house. "A" nor his mother never testified where the marshmallows were originally located. "A" never testified that he took the fireplace fork into any other part of the house and placed a marshmallow or marshmallows on the fork.

D. "G" Opinion: The concentration of the vapor space in the gel bottle was measured with an O2 sensor. This concentration was found to be in the explosive range for temperatures above 50 °F.

Comment: When the cap is first removed the headspace contains well over 19% ethanol vapor. When the container is constantly squeezed, external air constantly dilutes the concentration of ethanol vapor in the headspace to a point where it can fall within the explosive range of 3 to 19%. This can only occur if one squeezes the container a large number of times. That is the only way that the plaintiff's experts were able to create sufficient oxygen in the headspace when the container was constantly squeezed and held in a horizontal angle. When the concentration of ethanol vapor is above 19%, the flame front will be snuffed out.

E. "G" Opinion: The cap on the Natural Fire Starter Gel bottle had on 0.113 inches (2.87 mm) hole and with this cap all of the tests showed internal combustion.

Comment: The statement is not true. Most of the ignition tests were self extinguished *because the concentration of the ethanol vapor was above 19%.* It is only when the container was continually held in the horizontal position on a continuous basis where air was introduced into the air space and the concentration of the ethanol vapor falls within the explosive limits did Dr. "G" obtain internal combustion and not an explosion. If he recalls, "A" testified that he was holding the container in two hands and aiming downward. If that is the case, then there would be no head space and no internal combustion. In the tests that Dr. "G" conducted he should have directed the cap in the downward direction rather than horizontal. In addition, had he placed more gel in the container, it is more likely than not it would have no internal combustion taken place at all.

F. "G" Opinion: Four tests were performed with Kingsford charcoal lighter fluid caps (3 smaller holes) on a Natural Fire Starter Gel bottle. There were no internal combustion for these tests.

Comment: The tests conducted with three small holes were not conclusive and the methodology was not scientific. In addition, the opening in the subject cap was designed to expel a thixotropic formulation. Three holes could have been used for a liquid but not the thixotropic formulation. The use of a three-hole cap would not prevent the underlying incident and would not have acted as a flame arrestor. Moreover, a three-hole cap could expose users to more dangers because if they misuse the product near a flame or hot ember, the consumer has to position the bottle closer than the one with the one-hole cap. In addition, the end of the cap was held much further away from the candle than all of the prior tests with the standard closure.

G. "G" Opinion: The Consumer Product safety committee induced Anderson Chemical to recall 168,000 bottles of Eco-Lite Charcoal starter in 1991. The CPSC learned of five reports of the bottles exploding in consumers' hands when this charcoal lighter fluid was sprayed on hot charcoal. Four of these explosions involved serious injuries, including hospitalizations for burns.

Comment: The recall of the Eco-Lite Charcoal Starter in 1991 had nothing to do with the subject formulation. That product was a liquid formulation not a thixotropic and was designed for a different intended use than the product at issue. A red herring.

H. "G" Opinion: Eco-Lite (82% ethanol, 8% isopranol, 3.5% methanol) is made of a mixture of alcohols and has a flash point of 56 °F. The other more familiar charcoal lighter fluids are made of petroleum products and have a flash point above 100 °F. It should be noted that the Diamond Natural Fire Starter Gel (85-95% ethanol) is made of ethanol and Bitrix and it has a comparable flashpoint to the Eco-Lite mixture.

Comment: What Dr. "G" does not understand is whether the flashpoint was 56 °F or 150 °F, it wouldn't make any difference when the vapor is exposed to an ignition source of 1,500 to 1,800 °F. In addition, the flash point reported was incorrect. The closed cup flash point was significantly higher.

I. "G" Opinion: Caps with 3 small holes are commonly used in the industry and well known. A standard 3 hole cap, manufactured by Stull Technologies, would have prevented this incident as the smaller holes act as a flame arrestor. An igniting flame cannot pass through the small holes and into the container to cause internal combustion. Thus, an explosion cannot occur.

Comment: Dr. "G", in all his tests, did not reproduce the alleged blow (deflagration) out of the subject container in any of his tests. His statement regarding flame arresters is utter nonsense.

The undersigned has been involved with exploding batteries with micro porous flame arresters for over 30 years. The flame front goes through the flame arrester into ullage area much too fast for even the plastic to melt and snuff out the flame front before the battery explodes. The purpose of the flame arrester is to snuff out the flame before it reaches the internal head space of the flammable liquid. When it comes to polymeric materials, the flame front does not melt the opening fast enough to snuff out the flame. It is not foreseeable when one performs a risk analysis on the subject formulation that a foreseeable consumer would constantly squeeze the container when only a half a tablespoon would be used to start a fire. The product was not designed for a 7 year old to use the product as a toy. That is not a foreseeable misuse of the Diamond product.

J. "G" Opinion: Warnings do not alleviate the product manufacturer and retailer of the duty to design and sell a safe product. It is a basic, generally accepted mechanical engineering principle that performance of a risk assessment (e.g. failure modes and effects analysis) is necessary in product design. It is also a basic, generally accepted mechanical engineering principle to preferentially design out risks as opposed to warn against them. Manufacturers, distributors and resellers have a duty to consider anticipatable uses and misuses of a product.

Comment: The warnings and instructions on the subject container were well done for a pictorial of a wood pellet stove. *The product is not designed for 7 year olds. It is designed for a small segment of our population of adults who own wood pellet stoves.* It is not designed for fireplaces otherwise a pictorial of a fireplace would have been on the front panel of the product. The wording on the labeling on the container covers that specific area.

Dr. "G" fails to point out that there were starter logs on the side of the fireplace and that the parents knew that there was a safe method of starting a log fire in their fireplace. He also fails to point out that the parents knew that the subject fire starter was highly flammable and dangerous, must be kept away from children and was designed for wood pellet stoves as shown on the pictorial on the label of the product. The warnings were written for a person with a fifth grade education and would be obvious to any consumer that the small container, like gasoline, should be kept in the garage and out of reach of young children. A manufacturer is not responsible to follow a consumer to their residence and hold their hand while they use their product.

K. "G" Opinion: It is apparent that none of the defendant's *experts* in this case performed any tests of any kind to determine if the container and cap could be safely used by a consumer in the presence of a flame. Had the management or engineers for these defendants conducted appropriate testing of this product, they would have determined that the Diamond Natural Fire Starter Gel and its container pose a significant and unsafe explosion risk.

Comment: The defendants did not have to perform any tests because the properties of the gel are known as well as the use and application for pelletized wood stoves. The properties of the one hole cap and its integrity and use for thixotropic formulations is well known for many applications including harmless products such as soap. The same design is also used for squirting insect repelling products inside and outside residential and commercial buildings.

The public also knows that gasoline in any container poses a significant and unsafe explosion risk. The same goes for Drano which is composed of almost 100% of sulfuric acid. *These* products are not designed for 7 year olds.

What Dr. "G" fails to point out is that there have been no reported accidents with fireplaces and the subject product with young children. In addition, there have been no reported accidents with wood pellet heating stoves especially when only a tablespoon of the Diamond product was used to start the fire in a wood or wood pellet stove.

This author measured the force required to release the snap cap of the subject bottle. The tests indicate that it takes over 8-pounds of force to pull the cap off. The cap would be considered child proof. Therefore, it is questionable whether a 7-year opld would be able to easily open the cap with one finger.

L. "G" Opinion: It is also apparent that none of the defendants in this case performed a reasonable risk assessment of this product kind to determine if the container and cap could be safely used by a consumer in the presence of a flame. Had the management or engineers for these defendants conducted an appropriate risk assessment of this product, they would have determined that the Diamond Natural Fire Starter Gel, its cap, and its container pose a significant and unsafe explosion risk. Further they would have discovered the problem had been solved centuries ago and the solution implemented.

Comment: The problem that was allegedly solved centuries ago did not involve 7 year olds. The flammable tests using an exemplar fireplace performed by the undersigned refute all of Dr. "G"'s opinions. In addition, he is not in any position to come to that conclusion without performing his own tests with an exemplar fireplace and the subject Diamond formulation.

M. "G" Opinion: This product as packaged is unreasonably dangerous to be sold as a consumer product.

Comment: The product as sold to the consumer was not unreasonably dangerous because it was not designed or sold with the intent that a 7 year old would be squirting the contents of the bottle on a fire. *In fact, there is no evidence that "A" squirted the contents of the bottle on embers. There is evidence to the contrary that he actually squirted the contents of the bottle onto the pre-existing fire.* In addition, no one knows how many times or how much of the gel that "A" squirted onto the fire once it

got started. He also squirted additional amounts to restart the fire. ***We also know that he squirted the liquid in a large number of other areas. This is illustrative of the fact that "A" was playing with the Diamond product and did not use any of the warnings and instructions his mother stated she gave him previously.***

N. "G" Opinion: An internal explosion can occur in a Diamond Natural Fire Starter Gel bottle as it is being squeezed onto a flame or ember.

The statement that Dr. "G" made has a low probability ***and is totally false.*** **Experiments performed by the undersigned continuously documented that no explosion or deflagration can occur internally no matter how much gel was expelled into the fire from the front or directly on top of the fire. This was documented in the videotape test performed by the undersigned. It is more probable than not that "A" had to squeeze excessive amounts of the gel on the fire in front of him.** ***What no one can answer at this point is when the bottom seam separated and melted. It is more probable than not that the bottom separated well after "A" was on fire.***

In addition, not one of the experts for the plaintiff was able to reproduce the appearance of the subject container in anyway whatsoever. Similarities resulted by squirting the container excessively until sufficient air entered into the head space and reduced the concentration of the ethanol vapor to within the explosive limits. ***This illustration by the plaintiff's experts did not, in any way, simulate "A"'s testimony.***

Squirting the contents excessively into a burning fireplace caused a large flashback onto "A"'s exposed body, ***which had been previously squirted on by "A" himself,*** **to the elevated temperatures emanating from the fireplace while he was** ***allegedly*** **on his knees facing the unscreened fire.** ***The high temperatures emanating from the fireplace as measured by the undersigned in the tests performed, was more likely than not, much too high for anyone to withstand.***

Contrary to Dr. "G"'s statement, the subject product was not designed to be squeezed on glowing embers or pre-existing fire ***in a fireplace*** **on a continuous basis by an adult or by a 7 year old.** ***The product was created and sold for pellet and wood***

stoves only. This was open and obvious to any adult looking at the label on the container. **If several tablespoons were placed on the glowing embers or pre-existing fire, "A" would not have been injured.**

O. "G" Opinion: This internal explosion can occur when the ambient temperature is above 50 °F.

Comment: Dr. "G"'s tests were not supported by scientific testing and documents readily available in the subject case.

P. "G" Opinion: While the outside temperature was 24 °F, the accident happened in a finished basement of a house where a fire had already been burning to roast marshmallows so the temperature at the time of the incident was likely over 50 °F.

Comment: *There was no testimony or comment by "A" that, prior to the subject incident, he was able to safely roast and consume a marshmallow.* **Not one of the experts reported marshmallows being roasted or present in the area. If the experts continually believe the mother and "A", they will continue to make errors and produce results that are not scientifically acceptable. In addition, there was no evidence of a bag of marshmallows available anywhere.**

Q. "G" Opinion: The 5 previous explosions of the Eco-Lite Charcoal starter in 1991 are very similar to the explosion of the present case. The flash point for this previous Eco-Lite product and the present gel are both approximately 56 °F and the chemical compositions of the two products are substantially similar.

Comment: The formulation for the Eco-lite Charcoal Starter in 1991 was a pourable formulation and is not related to the gelled thixotropic formulation. These two formulations should not be intermixed.

R. "G" Opinion: The flashpoint for alcohol is particularly hazardous since at room temperatures, alcohol vapors are in the explosive range. Petroleum distillates like diesel fuel have a higher flashpoint and their vapors tend to be too lean at room temperatures to be explosive. Fuels like gasoline with a very low flashpoint have vapors that tend to be too rich at room temperatures to be explosive.

Comment: When alcohol is placed in a thixotropic formulation, it is much less

dangerous than the pourable liquid. The tests performed by the experts document this statement. Furthermore, the flammable tests performed by the undersigned using an exemplar fireplace and the thixotrophic formulation demonstrated the integrity and the formulation of the container.

S. "G" Opinion: A flame arrestor could prevent internal explosions of this alcohol gel. Flame arrestors have been proven to prevent the propagation of flame into a container. Even a cap with smaller holes, such as the Kingsford cap tested by BEAR, can act as a flame arrestor.

Comment: Dr. "G" failed to test the squirting capabilities of the smaller holes versus the one in the subject container. What he also is demonstrating is that the subject gel does not flow in comparison to a non-gelled liquid. If the product is so highly flammable, it would have emitted sufficient vapors and ignited when first exposed to the radiated temperature emanating from the fireplace.

What Dr. "G" fails to demonstrate is that the open holes have to be small enough to snuff out the flame front entering through the hole openings in the cap, or that the openings will melt and close fast enough to snuff out the flame front before it enters the head space area in the container. That is impossible.

In summary, I am sure that the billing for Dr. "G"'s work conformed to the inadequate testing over a long period of time.

There were many areas that proved that the subject product, when used in accordance with the pictorial that was on the front of the label was relatively safe and not designed to be used by 7 year olds who were suppose to be roasting marshmallows.

More probable than not, based upon over 50 years of experience in the area of flammable materials, fires and the reconstruction of accidents, "A" did a lot more than just squirt the thixotropic product on the fire. There is no doubt that he squirted the product into a number of areas ranging from the carpeting near the hearth, the hearth, the walls and other areas of the living room in addition to squirting a large amount onto a pre-existing fire. The pre-existing fire was a result of "A" starting the fire and then adding the thixotropic gel in large quantities in the area around the fire, the walls, the back of the fireplace and over parts of his body. Dr. "G" ignored the reality of what actually occurred.

Critique "D" (Another PhD plaintiff's expert)
"P" describes the accident stating that ""A" Nelson, a 7 year old boy, was attempting to ignite a fire in a home fireplace when the bottle of fire gel ruptured causing the napalm gel contained in the bottle to spray back onto him resulting in severe burns."

A. "P" states "that the bottle was an unsafe container for a napalm or napalm-like gel."

Comment - Napalm (naphthenic and palmitic acids) is a thickening/gelling agent generally mixed with gasoline or a similar fuel used in military operations. It is a tactical weapon to remove vegetative cover. ***The product at issue is not considered a napalm gel.***

"P"'s conclusion that "the bottle was unsafe" has no scientific or technical basis."

B. "P" stated "that the bottle was manufactured which is not ***fully*** **compatible with the hazardous contents (i.e., ethanol based napalm gel)."**

Comment - PVC containers have been safely used for ethanol and many other solvent and flammable products for generations. The use of the word "fully", does not, in any way indicate that the contents of the subject bottle are not safely contained for its foreseeable use and application. At room temperature, the container is completely compatible with ethanol if one performs a risk analysis of the foreseeable use and foreseeable misuse of the subject product, the container will not, foreseeably, be exposed to elevated temperatures. The undersigned has personally used PVC plastic containers for ethanol from 1955 to the present time. All of the research performed by the undersigned in graduate school that incorporated pure ethanol was always transferred from large containers from the supplier to small PVC containers. In undergraduate school, the undersigned worked in the chemical supply room for 4 years transferring 80 to 100% ethanol from large containers to a small PVC containers for students to be used for laboratory experiments. The undersigned had the same experience in industry. There was never a question as to the integrity of PVC and its use as a container for ethanol liquid.

Furthermore, this investigator had, in the 90's patented an ethanol based product in a PVC bottle that contained a non-toxic repellent. Not one accident in the years it was sold throughout the United stated, Europe and China.

C. "P" states "that the bottle wall did not have sufficient wall strength to accommodate the internal forces of an accidental internal vapor ignition."

Comment - What "P" does not admit was that there is no polymer container that can withstand an internal the deflagration or explosion. Why would he single out this specific polymer and not include every other type of polymer? The flammability experiments performed by the undersigned with an exemplar fireplace and exemplar Diamond containers of the thixotropic formulation refute "P"'s statement.

D. "P" states "that the bottle wall had too broad a wall thickness variation leading to stress concentration in the thin areas near the bottom."

Comment - As stated previously, there is no polymer container that can be designed with respect to the subject and similar ones that can withstand internal deflagration and/explosion. However, the flammability experiments performed by the undersigned using an exemplar fireplace and exemplar Diamond containers of the thixotropic formulation refute "P"'s statement.

E. "P" states "that the thinnest area (weakest and most likely point of rupture) in the wall was near the bottom portion of the bottle which was the area closest to the person holding the bottle."

Comment - What "P" fails to state or consider is that the head space above the ethanol gel contains close to 100% ethanol vapor or, in the alternative, well over the upper explosive limit of 19%. When the container is turned upside down and a small portion is poured out (a spoonful or two), no change takes place because the thixotropic material moves towards the spout and the head space moves to the rear or bottom of the container. It is only when an individual, such as "A", continually squeezes and squirts the contents many times causing air to then mix with the head space and reduce the concentration of alcohol vapor to within the explosive limits can the vapor ignite and eventually deflagrate internally. *This was proven and well documented by "P"'s tests. (See reply in I.)*

Furthermore, the differencews in thickness amounted the the thickness of an average human hair.

F. "P" states "that the appearance of the subject bottle is consistent with the testimony of "A" concerning how the explosion occurred."

Comment - "P"'s reliance on "A"'s deposition testimony versus everyone else's documented records is incredulous. *He never attempted to reproduce the subject incident using an exemplar fireplace with the actual product.* His testing does not in any way reproduce even "A"'s testimony assuming even part of what he testified to was true. His testimony contradicts the scientific analysis of the ignition and the injuries documented.

G. "P" states "there is no way to design and manufacture a squeezable blow-molded plastic container that can safely contain in the detonation of a napalm gel. However, it is possible to design a bottle to handle a deflagration/slow burn without blowing napalm back onto the consumer using the product."

Comment - The subject container and similar thixotropic products have been safely used by tens of millions of consumers for decades. He continually forgets that the

product was not in any way designed for 7 year olds to play with *or a fireplace*. The product was designed for a small segment of our population that used wood pellet or wood stoves.

Contrary to "P"'s opinion, if there was a practical and inexpensive design for the same container to handle a deflagration, slow burn without blowing "napalm" back onto the consumer it would have been produced and used in lieu of the subject one. If CKS was responsible, and they are not, in manufacturing a plastic container filled with thixotropic ethanol, that would be safe for a 7 year old, it would be supplied in a steel safe with no combination to open it.

If "P" performed a risk analysis relating to the foreseeable use and the foreseeable misuse of the subject product, he would be enlightened to know that the product was designed specifically for pelletized wood burning stoves and not fireplaces. He would also agree that the instructions state that only a tablespoon of the material must be placed within the pelletized wood stove burn pot, that an additional half tablespoon of gel is to be placed on top of the wood pellets, and that the product should not be squirted by a 7 year old all over a room including the fireplace where an excessive amount was squirted continually by "A". In fact, "P" reproduced how many times one has to squeeze the subject container before enough oxygen was drawn into the head space to bring the ethanol vapor concentration down to within the explosive limits. *That is not a foreseeable misuse of the product. It is not foreseeable that parents would allow a 7 year old child to use the product as a toy.*

H. "P" states "Based upon my experience I believe that bottle manufacturers should communicate with their customers to learn the intended use of their bottle. The bottle manufacturer and/or the customer should test the bottle to determine if the bottle is safe for the intended use. If the bottle is found to be unsafe for the intended use, the bottle manufacturer should refuse to market the bottle for that application."

Comment - This more than documents the fact that "P"'s experience is limited in reference to the real world of involving the manufacturing of plastic containers. The sealed area is usually the weakest part of both plastic and glass containers. "P" should also know that if he was capable of performing a risk analysis on the subject container, it would have been easy to *prove that it was not foreseeable that a 7 year old would be playing with the container as a toy*; holding it in a horizontal position adjacent to a flame for an extended period of time and then raising it above his head where he testified it exploded. It should be noted that there is not one other reported incident with the subject product where the incident was described similar to "A"'s.

I. "P" states "a PVC bottle is not appropriate as a container for napalm gel but manufacturing the bottle with the bottom being the weakest point (closest to the user) in the container was exceptionally reckless."

Comment - "P"'s comments again demonstrates his lack of expertise in the area of plastic containers. The seals of all plastic containers are the weakest part of the container. It is also the weakest part of a metal and glass containers. The subject container had seals at the base and on each side of the container. The seal that will be based on how the container was being held while the internal vapors inside the container was on fire. The seals could have failed at any of locations where the seals were located.

The subject plastic container in this case was designed to satisfy load drop-test and also satisfy the internal pressure produced by ethanol vapor. A drop test, in this connection, tests the ability of the bottle to withstand impact forces when the bottle is dropped on its mouth or neck portion because the mouth and/or neck portions are generally dimensionally the smallest parts of the bottle, any impact forces applied to those regions create maximum stresses in the walls of the container that tend to cause the container to burst or rupture at the weakest areas, typically the mold seams. Existing bottles, such as the one in the subject case, used by the chemical industry, have satisfied both requirements or specifications established by the various government authorities and transportation laws. It is important to note that a spherical surface where the seams are located in the subject container is the optimum design for the most resistance to failure.

PVC (Polyvinyl Chloride) is a softer plastic, which is relatively inexpensive. It is very durable and can be dropped without much of a problem. It also takes a lot of deliberate strength to tear PVC apart. The variation of the hardness of PVC is based upon how much of the plastic is put into a specific area. PVC can also be recycled. In addition, PVC has been used to contain ethanol and other alcohol for decades.

"P" also states in his opinion, that PVC is "marginally compatible" with ethanol and that is not true. The rest of his opinion on page 12 of his 28 page report is purely subjective without scientific documentation.

"P" also states that when he was with Dow Chemical, Dow was obligated to work with their customers to insure the safety of the consumer. In this case, CKS was not responsible to perform any risk analysis studies because they supplied the PVC containers as requested. However, he neglects to state that liquid ethanol and other alcohols such as isopropanol have been sold in PVC containers for decades. If that is true, there is no other obligation or responsibility on the part of CKS with reference to the end use application of the PVC container in this case.

If one examines "P"'s litigation experience on page 16 and his scientific presentations listed on pages 17-26 as well as the patents listed on pages 28-29, there isn't one reference that relates to PVC, ethanol, fires, explosions and deflagration of containers.

Lastly, **the undersigned assumes by choice, that "P" chose not to address the subject of marshmallows even though he had an opportunity to review the depositions of "A" and other members of the family as well as an opportunity to examine all of the photographs associated with the accident. He also failed to comment on the fact that if there was an opening at the base of the container that any squeezing of the container would have released the thixotropic ethanol through the opening in the top closure as well as the alleged opening in the base of the container.**

Furthermore, tests performed by the undersigned using an exemplar fireplace and exposing a half filled Diamond product container to well over 200 degrees Fahrenheit documented that the bottom seam would not rupture or melt and that the container actually expanded as shown in the videotapes and photographs taken by the undersigned during and after the tests.

Critique by "H" (Expert for the plaintiff)
Dr. "H" stated in her report that "The product is marketed as a fire-starting accessory, it is suggested for use with wood or pellet stoves."

From Section 3.0 (Background) to page 7 of Dr. "H"'s report, she re-iterates what is already known through prior publications and/or company literature. What is important on page 8 is that she measured head space of the volatiles in the PVC container and found that it was essentially ethanol. There was no report that air and/or oxygen was present in the head space. That is important because no explosion can occur if the concentration of the ethanol vapor is greater than the upper explosion limit.

Dr. "H" refers to the CPSC's 1991 voluntary recall of Eco-Lite charcoal starter. This is a product composed of ethanol and individuals were warned not to spray the material over a burning fire and/or charcoal because, like gasoline it may result in a flashback.

Comment**: The objective of the Natural Fire Starter Gel was exactly what the name refers. The name and pictorial on the label designate that the product is designed to be used for wood and pellet stoves and not fireplaces. The warnings Eco-Lite product was not thixotropic and designed to start coal fires for barbeques, as an example. The two products are not designed for the same application.**

It is open and obvious to a consumer that one does not squirt gasoline into any type of fire. Therefore, when the consumer is properly warned to keep the product away from children and a parent, as in this case, does not place the subject product out of the reach of their children, including "A", and claims they allow "A" to use the product. Manufacturers cannot be responsible for their actions when they violate the manufactures instructions. Manufacturers cannot be responsible for the parents'

actions when they decide, on their own, to use a product designed for wood and pellet stoves to be used in a fireplace. Dr. "H" ignores that issue.

When a product involves a fire starter of any type, there are no liquid or gelled fire starters that are inherently safe. However, in the subject case, even "A"'s father stated that he used the Safe Fire Starter logs that were on the side of the fireplace. Why any parent would use any other product when they have a safe way of starting a fire is questionable.

What Dr. "H" fails to also point out is that all of the other products do not designate that their products should be used for wood and pellet stoves with a pictorial on the front. *In the opinion of the undersigned, that significantly differentiates the subject product in this case versus all of the others.* The CPSC staff are not warning experts and never addressed that issue with reference to the fact that the Natural Fire Starter Gel with the pictorial and the fact that it was designed only for wood and pellet stoves distinguishes that product from all of the others.

Dr. "H" continually relates the Natural Fire Starter Gel to the Eco-Start Natural Lighter Fuel. They are two separate products designed for two separate objectives. She continually fails to recognize or even discuss the differences in the two and the pictorial of one product versus the other. In fact, on page 12 of her report, she does not feel that there are any differences in the two.

Dr. "H" relates the two by stating that they would act identically with a hot coal or a piece of burning wood "regardless of whether the products are used in wood stoves, barbeque grills, or fireplaces." Dr. "H" is not a warning specialist and, from that stand point alone, *there is a significant difference*.

The Natural Fire Starter Gel is specifically designed for wood stoves and pellets and not barbeque grills or fireplaces. In addition, from a risk analysis standpoint, it is not foreseeable that a parent would use a product designed for wood and pellet stoves or fireplaces when they already had purchased and are experienced in using the Safe Starter Logs. Furthermore, it is not foreseeable that a parent, under supervision or not, would allow a 7 year old to use the product with or without their permission.

"H"'s reports her comments relating to the determination of the subject fuel gel versus what was reported by the manufacturer and the CPSC may vary slightly but it is not an issue in this case. The temperature of the flame of a fire is in the range of 1800 °F and that of burning coals is over 900 °F. What difference would it make whether the flash point of the gel is 54, 73 or 130 degrees? Once the vapor is ignited in the fireplace, it would still ignite and flash back whether the flash point is 52 or 130 degrees.

Therefore, the comments Dr. "H"'s report do not make any sense, and in the opinion of the undersigned, ,the fire starter is safe when it is used in accordance with its instructions. for wood and pellet stoves.

Every consumer has the right to purchase any product they want in Wal-Mart or any other store. For example, many stores sell tanks filled with propane. Propane is highly explosive and has explosive limits between 2.1 and 10.1 percent. Gasoline is also explosive and has explosive limits between 1.4 and 7.6 percent.

When gasoline is used for gasoline lawn mowers, gasoline hedge cutters and emergency electrical power generator, the gasoline is usually dispersed from small container ranging from 1 to 10 gallons. There has never been a recall for any of those products even though they are more dangerous than the thixotropic ethyl alcohol. When the thixotropic ethanol, propane gas and gasoline are used for their intended purpose by adults, the risk of foreseeable injuries is reduced significantly. Dr. "H" fails to continuously point out that all of these products are not designed for parents allowing 7 year olds to use the product under any conditions, *especially when "A" used the product as a toy.*

Lastly, Dr. "H" spent a lot a time reproducing data that is readily available by the manufacturers of the products and documents readily available on line. There was no need to perform any laboratory tests proving what is already known and attempting to reinvent another wheel.

1 "W"J.D.(Expert for the plaintiff)

Mr. "W" defines his area of product safety management as a "system that a reasonable prudent manufacture puts in place before the first product is conceived and ensures that the final product along with its warnings, packaging and marketing materials is reasonably safe."

What Mr. "W" neglected to discuss is that every consumer has the right to purchase any product they want in Wal-Mart or any other store. For example, many stores sell tanks filled with propane. Propane is highly explosive and has explosive limits between 2.1 and 10.1 percent. Gasoline is also explosive and has explosive limits between 1.4 and 7.6 percent.

When gasoline is used for gasoline lawn mowers, gasoline hedge cutters and emergency electrical power generator, the gasoline is usually dispersed from small container ranging from 1 to 10 gallons. There has never been a recall for any of those products even though they are more dangerous than the thixotropic ethyl alcohol. When the thixotropic ethanol, propane gas and gasoline are used for their intended purpose by adults, the risk of foreseeable injuries is reduced significantly. Mr. "W" fails to point out that all of these products are not designed for parents allowing 7 year olds to use the product under any conditions.

The hazards were identified in each of those products; the risks were assessed and adequate safety measures were applied to minimize both hazards. The same approach was made by the manufacturers and distributors of the Natural Fire Starter Gel which is much less dangerous than propone and gasoline. *As stated previously,*

the undersigned has been involved with the use and application of 80 to 100% ethanol and PVC containers for close to 58 years.

It should be noted that Mr. "W" is not a warning specialist especially in the area of hazardous products and chemicals. He points out on page 7 of his report that he is developed on-products warnings and instructions for a number of manufacturers and distributors. Not one of them involved flammable chemicals with physical and chemical characteristics similar to the Natural Fire Starter Gel.

Vending machines, ATVs and pool products are not at the level of warnings and instructions required for products such as the Natural Fire Starter Gel. In contrast, the undersigned has created warnings and instructions for a number of corporations, presented OSHA safety meetings at a large number of construction sites, created Material Safety Data Sheets for flammable products, evaluated polymers from room temperature to hyperthermal temperatures, consulted to NIOSH (the research arm of OSHA) and performed many explosive tests on a variety of flammable products.

Mr. "W" also claims that he created warnings and instructions for fitness equipment in Dick's Sporting Goods. The undersigned has evaluated many accident cases involving fitness equipment. I have also invented sports products that were sold in Dick's. Contrary to Mr. Kitzes' claims, the warnings and instructions for fitness equipment are created by the manufacturers of the equipment and not the management of Dick's Sporting Goods.

Mr. "W"'s' report he points out the use of a pictogram but fails to point out the pictorial of a wood stove on Natural Fire Starter Gels label versus that of a fireplace. He also fails to point out that the words in the warning are at the level of a fifth-sixth grader. In addition, he failed to point out that the product must be kept away from children and is not designed to be used by a 7 year old.

The label of the Diamond product states in block letters "PERFECT FOR ALL WOOD AND PELLET STOVES." There is also a pictorial of a wood stove - NO FIREPLACE is shown or mentioned on the front or back labels of the Diamond product. Mr. "W" failed to address that issue which would be considered self-serving on his part by not commenting on that information on the label.

Mr. "W" describes "A"'s behavior prior to the accident without noting all of the different statements made to the EMTs and medical specialists at the hospital. He even failed to mention "A"'s brother's statement under oath where he stated that "A" had already taken a bath and his wet towel was upstairs in the bathroom. In fact, the description of the accident described by Mr. "W" is totally inaccurate and technically incorrect.

Mr. "W" makes a statement regarding "A" being an "avid outdoorsman, hunting, fishing and trapping." What he doesn't point out that "A" at the age of 7 was able to read and did well in school and was sophisticated for his age. ***is immature and***

irresponsible, based upon his behavior prior to and up to the time of the incident. In addition, contrary to Mr. "W"'s statement, it should be obvious to him that "K" did not read the Diamond Natural Fire Starter Gel label to "A" on prior occasions. That conclusion is obvious to the undersigned.

Although Mr. "W" had been associated in some way with the CPSC (not true), he is not a chemist, chemical engineer, safety specialist, safety engineer, professional engineer or certified in warnings and instructions. In addition, he is not qualified to perform a risk analysis, based on his background and experience on the product, Diamond Natural Fire Starter Gel.

The fact that "K", "A"'s mother, *could not read certain important parts of the label and, on top of that,* **disregarded the warnings concerning children, negates all of his opinions in regarding the subject warning. When you have a consumer such as "K" dismissing the dangers which she was aware of and the warnings concerning her own child, how could Mr. "W" criticize anything on the label of the subject bottle.**

Mr. "W" knew, or should have known, especially with his law degree, that when you have a recalcitrant worker on a construction job, who refuses to wear protective equipment and is injured, that worker is responsible for his own injuries. In this case, "K" openly refused, in spite of her husband's opinion, to follow the simple instructions and warning on the back label of the Diamond Natural Fire Starter Gel. Now that the horse has run out into the pasture, both Mr. "W" and "K" want to place a lock on the barn door. The undersigned handles similar situations like this each week when the son or daughter of an individual receives a brain injury in soccer and can't go back and play unless they wear his headgear (see www.forcefieldheadbands.com). By that time, it is much too late *to help the child reduce the risk of a brain injury.*

Although the warnings and instructions on a subject label were accurate based upon the manufacturer's intended use of the product in wood and wood pellet stoves, the instructions regarding children were completely disregarded by "K". If she just obeyed that one warning, "A" would never have been injured.

In spite of "K"'s behavior, Mr. "W" continues to quote various sections of the CPSC without addressing the real issues in the case. For example, on page 12 of his report he describes the vapor "may burn with near invisible flame." What does that mean to a 7 year old? The fact is that the warnings and instructions were never designed for a 7 year old to read and use the product. What Mr. "W" fails to recognize is that "K", the mother of "A", had complete control of the product and completely ignored the warnings.

There was nothing complex about the simple warning concerning children. There is nothing that a manufacturer can do if a consumer, such as "K" refuses to follow and ignores important instructions concerning the safety and welfare of her own child.

On top of all of the instructions that should have been followed if "K" used the product, no one used a tablespoon of the product at any time. If a tablespoon amount was placed on any fire, no injuries would have occurred.

The undersigned has been involved with warnings and instructions regarding all types of chemicals from the early fifties to the present time, from the warnings and instructions published by the Manufacturing Chemical Association to the current ANSI Z 129.1 which is the labeling standard for Hazardous Industrial Chemicals-Precautionary Labeling. It is important to note that this is a voluntary standard and not statutory. All of the other references Mr. "W" makes from do not relate to the real issues in the subject case.

Most important, the manufacturers of the product knew that the product will meet the needs of a small segment of the market which is wood and wood pellet stoves.

"W" states that "the highly flammable ethanol based gel with a nearly invisible flame, could flashback to the bottle, causing it to explode in the hands of the consumer and spread the flaming contents over the body of the user." This statement is totally inaccurate and false. It contradicts the tests performed by "P" and also documents the fact that Mr. "W" is limited with his lack of background in the areas of chemistry, chemical engineering, fires and explosions, deflagrations, flashbacks and why and how the container failed in the instant case. In addition, although Mr. "W" never addressed the subject, it is more likely than not, that no polymeric, glass or metal container could have withstood the deflagration that occurred.

What "W" is attempting to do is validate the defective scientific methodology used by the other plaintiff's experts that did not, in anyway, attempt to reproduce "A"'s description of his accident using an exemplar fireplace and the Diamond thixotropic product. The same group of "experts" chosen by the plaintiff are comparing the same inaccurate models and opinions with one another-but not versus the real world.

Furthermore, Mr. "W" never addressed the subject in which "A" irresponsible actions resulted in him squirting the contents of the Diamond Natural Fire Starter Gel in many areas on the carpet, the hearth, the walls and the firewood while it was burning. In addition he held the container for an extended period of time at a horizontal angle just like "P"'s test demonstrated before the vapor, emanating from the open container, caught on fire resulting in the deflagration of the container. ***This did not happen.***

Mr. "W" deficient background relating to the issues in this case is well beyond his capability of explaining what occurred in this case posing the subject incident and resulting injuries to "A". ***His reliance on the other experts inaccurate and inept attempt to reproduce the questionable testimony of "A" and his family rather than independently prove or disprove his case, defines his lack of***

experience and background in the technical issues associated in the subject case.

Fire and burning tests
Based upon over 50 years of experience in testing the flammability of a variety of solid and liquid chemical formulations in addition to testing the subject formulation in a simulated wood fireplace, it is the opinion of the undersigned, with a reasonable degree of safety engineering and scientific certainty, that "A" did not testify in his deposition as to his actual behavior prior to and up to the time he was injured. There is no question, based upon the testing by the undersigned and the plaintiff's expert's testing, that "A" did more than just squirt a little of the thixotropic alcohol into a burning fire. It is more likely than not that he squirted an excessive amount of the contents not only on the carpeting and the areas surrounding the fireplace such as the hearth, the back wall of the fireplace, the riser to the hearth, the walls, etc., but the straw that broke the camel's back was the excessive amount of formulation squirted on himself and the fire itself for an extended period of time until the vapors emanating from his body ignited due to the fire and heat radiating from the fireplace ending up igniting the vapors.

Experiments by the plaintiff's experts document the fact that an excessively large amount of squirts. The only way the vapors inside the container could be ignited was if "A" constantly squirted the contents out and air continued to enter the head space until there was sufficient air in the container to lower the concentration of the ethanol vapor to within the explosive limit so it could ignite. That was also verified in the video tape submitted by the plaintiff's expert.

In addition, the fact that "A" did not receive burns on the top of his head and back documents the fact that he did not raise the container above his head while the gel was coming out the bottom. That had to occur elsewhere such as on the or around the couch. (The resistance of the container from igniting and burning was well demonstrated in the fire test performed by the undersigned.)

Furthermore, this also verifies the fact that, in the opinion of the undersigned, "A"'s mother, "K", contrary to her sworn testimony under oath, never read the instructions to any of her children and never demonstrated that only a tablespoon amount of the contents should be placed on an unlit piece of wood in a wood stove. If she had, this would also document that "A" did not understand anything she attempted to warn "A" about and, as a 7 year old, he was irresponsible and disobedient.

. ***In addition to the above,*** **the undersigned squirted excessive amounts of the Diamond product into the burning fireplace from the front and from the top exposing the spout to well over 200 degrees Fahrenheit in the front and over 300 degrees Fahrenheit one foot from the flame from the top. All of these tests were videotaped and photographed. In both instances, the undersigned could not touch the closure on**

the top to close it. It was much too hot. In all cases, there was no indication of any flame front entering the container because it would be impossible for that to occur. Therefore, those tests disprove "A"'s account and testimony of what occurred prior to the subject incident.

The undersigned has consulted on several hundred burn cases to both children and adults over the last 44 years. Based on that experience and studies, I am able to convey that experience and observations to the following:

a. Explosions of natural gas, propane, gasoline and other flammable liquids, including ethanol, isopropanol, cause intense heat for a very brief time. Clothing, unless it ignites, is protective in flash burns. Hence, flash burns generally have a distribution involving all exposed skin, with the deepest areas facing the source of ignition. For the most part, flash burns reach progressive layers of the dermis in proportion to the amount and kind fuel that explodes.

b. In contrast to flash injuries, flame burns are invariably deep dermal if not full-thickness because of more prolonged exposure to intense heat.

c. Adding and squirting a voluminous amount of Diamond gel into the fireplace in all directions did not result in any flashback whatsoever. The heat emanating from the fireplace was too high in temperature for any human to withstand. If that is true, where was "A" actually positioned at the time his body caught on fire because of the large amount of gel all over his body? How did he really react once the gel all over his body caught on fire?

Based upon the extensive experiments performed by the undersigned on an exemplar fireplace squirting extensive amounts of gel on a burning fire, there was never a flashback experienced under any conditions. Any testimony regarding flashbacks squirting the Diamond product into a fire cannot be reproduced and would be incorrect.

d. With reference to the subject incident and "A"'s burns, it is highly probable based upon the amount of contents that he squirted into the fire, that he could have easily received a significant amount of his burn injuries from the gel that "A" squirted all over his body prior to getting on his knees in front of the fire.

. **Marshmallows**

There is some evidence of a white residue on a large metal fork in one of the photographs. *If "A" did roast one or more marshmallows and placed the fork down, this would then lend itself to "A" finishing the marshmallow(s) and then continuing to play with the Diamond product.* Squirting of the gel onto the carpet, hearth and other areas around the room, including his own body prior to kneeling before the unscreened fireplace and fire contributed to all of "A"'s burn injuries. All of the vapors emanating from his body was exposed to the heat from the fireplace. The temperature was more than enough to ignite the vapors. In fact, the temperature was more than enough to ignite most other solvents. No one ever found the bag of marshmallows "A" claimed he took downstairs to the fireplace.

Reconstruction of Subject Accident (Flammability Testing)

The undersigned conducted extensive fire tests in Boynton Beach, Florida by creating a exemplar fireplace using hickory wood chunks, paper and the Diamond Fire Starter Gel to start the fire. The following was observed:

a. **As the fire developed, the undersigned squirted large volumes of the gel onto the fire. No flash fire was observed. What was observed was a fire of higher intensity.**

b. **Large volumes of the Diamond product were squirted into the fire directly from the front. No flash fire was observed. What was observed was a fire of higher intensity.**

c. **A metal surface was covered with the gel and placed near the front of the fireplace. This was done to stimulate the gel covering "A"'s body that he previously squirted on himself. The surface burned with high intensity within a short period of time.**

d. **The surface of the 2" X 6" piece of wood, approximately 16" long was coated with the gel and placed along the floor surface adjacent to the front of the fireplace. Within a short period of time the vapors from the surface ignited and the gel on the surface burned.**

e. **Large volumes of the gel were squirted onto a blazing intense fire in the fireplace from a position directly above the fire and in front of the fire. No flashbacks occurred at any time. *The spout was exposed to temperatures ranging from 200 to 300 degrees Fahrenheit. The spout, containing the gel never caught on fire.* This contradicts all statements made by the plaintiff's experts and "A" in his sworn testimony.**

f. **The Diamond Fire Starter Gel container was placed directly in front of the fireplace. Nothing initially happened to the container. However, additional tests on the integrity of the container proved that it held up exceptionally well when** exposed to temperatures over 200 degrees Fahrenheit as documented in the videotapes and photographs taken by the undersigned during the flammability and integrity testing of the container. *The container was half filled and the only observation made while the container was exposed to over 200 degrees Fahrenheit was that the body of the container facing the fire expanded as shown in the videotapes and photographs.*

g. The Diamond Fire Starter Gel container with approximately 40 cc's of gel then thrown into the fireplace. There was no explosion and it took an extensive amount of time before the container started to burn documenting the fact that the integrity of the design of the container was more than satisfactory.

h. The undersigned squirted large quantities of the Diamond Natural Fire Starter Gel from the opening in the top of the fireplace shown in the photographs and videotapes. The temperature opening of the top of the fireplace was well over *300* degrees Fahrenheit. It was so hot that the undersigned could not touch the top of the closure and seal the top with my bear hands. The temperature was well over ignition temperature of the ethanol vapor. In spite of that, no internal ignition was observed and no explosion occurred contrary to the testing and comments by Dr. "G" and "P". The latter two plaintiff experts did not, in any way, experimentally demonstrate their incorrect opinions.

Testing the Integrity of the Diamond Natural Fire Starter Gel Plastic Container
The undersigned tested the integrity of a half filled container of the Diamond Natural Fire Starter Gel. The fire was started using the Diamond Natural Fire Starter Gel as the accelerant. Instead of a tablespoon of the material, approximately 1/3 of the container was sprayed on the wood nuggets and the vapors were ignited with a Scripto flame lighter. A continuous videotape was taken and the experiment was also documented with a series of photographs. A half filled container of the Diamond Natural Fire Starter Gel was placed approximately 14 inches away from the fire. The temperature of the heat emanating from the fireplace, at the point where the container was placed, was approximately 200 degrees Fahrenheit. No human being could withstand the temperature.

A half filled container of the Diamond Natural Fire Starter Gel was placed in front of the fireplace for over 30 minutes. After approximately 20 minutes, the area below the headspace that was facing the heat emanating from the fireplace swelled after approximately 25 minutes. There was no failure whatsoever and the bottom seam maintained its integrity contrary to the claims made by the plaintiff's experts. It is the opinion of the undersigned, with a reasonable degree of safety engineering and scientific certainty, that the plastic container is more than suitable for the formulation for the Diamond Natural Fire Starter Gel. In addition, unlike many formulations containing alcohol does not deform while in storage.

An additional important note was that the temperature emanating through the top hole in the cover of the garbage can where the undersigned squirted streams of the gel

into the fire below, was well over *300* degrees Fahrenheit. It was so hot, that the undersigned could not handle the closure for the top of the bottle after squirting the gel through the hole. Contrary to the claims made by the plaintiff's experts, there was no flame front that went through the opening in the spout that entered the internal headspace of the container. The reason was that the concentration of the ethanol vapor was above the upper explosion limit *and that the headspace was at the bottom of the container and not the top. When the container was placed in the upright position, and the spout left open until it cooled down, there was no ignition of the ethanol vapor escaping from the opening in the spout.*

The headspace above the half filled container contained mostly ethanol vapor. When exposed to the excessive heat in front of the exemplar fireplace, the gas expanded and the container responded by expanding as well. Contrary to the *defendants* allegations regarding the seam on the container, it would be predicted to anyone with knowledge of plastic containers, liquids and vapors within the container, that the most vulnerable area of the container when exposed to elevated temperatures would be the headspace within the container and not the bottom seam. *Even though this was foreseeable, it was also proven during the experiment by the undersigned.*

Furthermore, based on the demonstration by the undersigned, the plaintiff's claim that a leaked developed at the bottom seam of the container when the container was exposed in some unexplained way to their fireplace, is an impossibility. If it was exposed to excessive heat from the fireplace, the container would demonstrate an area of expansion above the headspace. The evidence in possession of the plaintiff documented that there was no expansion whatsoever. Therefore, it would be impossible for the accident to occur as explained under oath by "A", members of his family and the reports and experiments released by the plaintiff's experts.

- **The retail package provided sufficient barrier to prevent migration of the product or volatiles from the product.**

- **The Material Safety Data Sheet relating to this product was prepared in a reasonable manner and within all applicable standards.**

- **The delivery of this product, as delivered by PSC to Jarden/Hearthmark, was new and merchantable, free from any defects in workmanship or material, and conformed to applicable specifications, state of technology and the standard of care in the industry. Specifically, it is my opinion to a reasonable degree of engineering and scientific certainty, that the product met the specifications set forth by Jarden Home Brands / Hearthmark in the Performance Sepecifications for the Diamond Fire Starter Gel.**

Tests were conducted by the undersigned to determine the force required to open the closure that snapped in place. All measurements were over 8-pounds of force. That

would make the closure child resistant and very difficult, if not impossible for young children to open it up.

Conclusions

It is the opinion of the undersigned, with a reasonable degree of safety engineering and scientific certainty, that contrary to the allegations set forth by the plaintiffs and the plaintiffs expert's, that the cause of all of "A"'s injuries were directly related to his mother's *willful and intentional* disregard of the safety and welfare of her own son and had nothing to do with the formulation of the Diamond Natural Fire Gel or the integrity of the container. *The fact, at this time she is attempting to expiate her guilt by blaming her negligence on anyone of the defendants is a sham.* The conflicting statements by each member of her family, the EMTs and the medical specialist in the hospital more than document this opinion.

The flammable testing performed by the undersigned refuted and contradicted all of the allegations made by the plaintiff and by all of the plaintiff's experts. All of the testing performed by the plaintiff's experts was not realistic and did not relate, in any way, to the real world and what actually occurred at the time "A" was burned.

Summarily, the undersigned further opines that the Diamond Natural Fire Starter gel product is reasonably safe for its intended uses and reasonably foreseeable misuses, that the actions of PSC, Inc. in communicating with Jarden/Hearthmark and blending/packaging product were within all applicable standards of care, and PSC, Inc. did not cause or contribute to "A"'s incident.

The experiments performed and opinions made by the plaintiff's experts did not represent the actual behavior of "A" prior to and up to the time of the actual incident. All of the experts' opinions were *"validated" by the plaintiff's handpicked group of alleged group of "experts" that compared* based on the results of inaccurate testing and questionable testimony of "A" and his family, *but* not versus the real world and real science. Even so, the Plaintiff's experts still could not replicate the version of events provided by the family.

7. **Opinion**
    a. **With over forty years of experience in the area of flammable fabrics, explosions, fires, children's burn injuries, warnings and instructions I cannot find liability in Diamond's Natural Fire Starter Gel.**
    b. **"A"'s parents have a non-delegable duty to safeguard the safety welfare of "A.". They blatantly ignored that responsibility and allowed him to play unsupervised with a dangerous instrumentality. They are responsible for the consequences of their irresponsible behavior.**
    c. **The undersigned has compassion for "A" and his injuries. However, the liability and the cause of his burn injuries are directly related to the disregard of "A"'s's safety and welfare by his parents.**

C. J. Abraham,, PhD, JD, PE, CPC, FRSC, DEE, FTI, BCFE, DCEE, CChE, CChem has a varied background that can be found at: www.scientificadvisory.com. He also has over 45 patents in a variety of arenas as well. Many of the inventions were commercialized. The latest invention can be found on www.forcefieldheadbands.com

AUTHOR INDEX

AUTHOR INDEX